Waste Management in the Chemical and Petroleum Industries

Waste Management in the Chemical and Petroleum Industries

Alireza Bahadori

*School of Environment, Science and Engineering,
Southern Cross University, Australia*

WILEY

This edition first published 2014
© 2014 John Wiley & Sons, Ltd

Registered office
John Wiley & Sons Ltd, The Atrium, Southern Gate, Chichester, West Sussex, PO19 8SQ, United Kingdom

For details of our global editorial offices, for customer services and for information about how to apply for permission to reuse the copyright material in this book please see our website at www.wiley.com.

The right of the author to be identified as the author of this work has been asserted in accordance with the Copyright, Designs and Patents Act 1988.

All rights reserved. No part of this publication may be reproduced, stored in a retrieval system, or transmitted, in any form or by any means, electronic, mechanical, photocopying, recording or otherwise, except as permitted by the UK Copyright, Designs and Patents Act 1988, without the prior permission of the publisher.

Wiley also publishes its books in a variety of electronic formats. Some content that appears in print may not be available in electronic books.

Designations used by companies to distinguish their products are often claimed as trademarks. All brand names and product names used in this book are trade names, service marks, trademarks or registered trademarks of their respective owners. The publisher is not associated with any product or vendor mentioned in this book. This publication is designed to provide accurate and authoritative information in regard to the subject matter covered. It is sold on the understanding that the publisher is not engaged in rendering professional services. If professional advice or other expert assistance is required, the services of a competent professional should be sought.

The publisher and the author make no representations or warranties with respect to the accuracy or completeness of the contents of this work and specifically disclaim all warranties, including without limitation any implied warranties of fitness for a particular purpose. This work is sold with the understanding that the publisher is not engaged in rendering professional services. The advice and strategies contained herein may not be suitable for every situation. In view of ongoing research, equipment modifications, changes in governmental regulations, and the constant flow of information relating to the use of experimental reagents, equipment, and devices, the reader is urged to review and evaluate the information provided in the package insert or instructions for each chemical, piece of equipment, reagent, or device for, among other things, any changes in the instructions or indication of usage and for added warnings and precautions. The fact that an organization or Website is referred to in this work as a citation and/or a potential source of further information does not mean that the author or the publisher endorses the information the organization or Website may provide or recommendations it may make. Further, readers should be aware that Internet Websites listed in this work may have changed or disappeared between when this work was written and when it is read. No warranty may be created or extended by any promotional statements for this work. Neither the publisher nor the author shall be liable for any damages arising herefrom.

Library of Congress Cataloging-in-Publication Data

Bahadori, Alireza.
　Waste management in the chemical and petroleum industries / Alireza Bahadori.
　　1 online resource.
　Includes index.
　Description based on print version record and CIP data provided by publisher; resource not viewed.
　ISBN 978-1-118-73171-0 (MobiPocket) – ISBN 978-1-118-73172-7 (Adobe PDF) – ISBN 978-1-118-73173-4 (ePub) – ISBN 978-1-118-73175-8 (cloth) (print)　1. Petroleum industry and trade–Waste disposal.　2. Chemical industry–Waste disposal.　3. Petroleum refineries–Waste disposal.　I. Title.
　TD899.P4
　628.5′1–dc23

2013021840

A catalogue record for this book is available from the British Library.

ISBN: 9781118731758 (13 digits)

Set in 10/12pt Times by Aptara Inc., New Delhi, India.
Printed and bound in Malaysia by Vivar Printing Sdn Bhd

1　2013

Dedicated to the loving memory of my Parents, grandparents, and to all who contributed so much to my work over the years.

Contents

Preface		xv
Acknowledgments		xix
Biography		xxi

1 Wastewater Treatment — 1
  1.1 Characteristics of Wastewaters — 1
    1.1.1 Suspended Solids — 2
    1.1.2 Heavy Metals — 2
    1.1.3 Dissolved Inorganic Solids — 4
    1.1.4 Toxic Organic Compounds — 4
    1.1.5 Surfactants — 6
    1.1.6 Priority Pollutants — 6
    1.1.7 Volatile Organic Compounds — 6
  1.2 Treatment Stages — 7
    1.2.1 Sources of Wastewater — 8
    1.2.2 Discharge Options and Quality Requirements — 8
    1.2.3 Preliminary Wastewater Treatment — 9
    1.2.4 Primary Wastewater Treatment — 9
    1.2.5 Conventional Secondary Wastewater Treatment — 9
    1.2.6 Nutrient Removal or Control — 9
    1.2.7 Advanced Wastewater Treatment/Wastewater Reclamation — 10
    1.2.8 Toxic Waste Treatment/Specific Contaminant Removal — 10
    1.2.9 Sludge Processing — 10
  1.3 Treatment Processes — 11
    1.3.1 Selection of Treatment Processes — 14
  1.4 Chemical Oxygen Demand (COD) in Wastewater Systems — 22
    1.4.1 Determination of the COD — 23
    1.4.2 Calculation of Theoretical Oxygen Demand — 23

2 Physical Unit Operations — 25
  2.1 Flow Measurement — 25
  2.2 Screening — 25
  2.3 Comminution — 28
  2.4 Grit Removal — 28
  2.5 Gravity Separation — 29
    2.5.1 General — 29

	2.5.2 Application	30
	2.5.3 Oil–Water Separator General Design Considerations	30
	2.5.4 Conventional Rectangular Channel (API) Separators	31
	2.5.5 Parallel-Plate Separators	43
	2.5.6 Oil Traps	45
	2.5.7 Oil Holding Basins	46
2.6	Flow Equalization	46
	2.6.1 Application and Location	46
	2.6.2 Volume Requirements	48
2.7	Mixing	48
	2.7.1 Description and Type	48
	2.7.2 Application	48
2.8	Sedimentation	49
	2.8.1 Sedimentation Theory	49
	2.8.2 Application and Type	51
	2.8.3 Design Considerations	53
	2.8.4 Number of Basins	53
	2.8.5 Inlet Arrangements	54
	2.8.6 Short-Circuiting	54
	2.8.7 Outlet Arrangements	54
	2.8.8 Detention Time	54
	2.8.9 Surface Loading Rate	54
	2.8.10 Factors Affecting Sedimentation	55
2.9	Dissolved Air Flotation (DAF)	56
	2.9.1 General	56
	2.9.2 System Configuration	57
	2.9.3 Variables Affecting DAF Efficiency	58
	2.9.4 Treatability Testing	59
	2.9.5 Design Considerations	59
	2.9.6 Instruments and Control	64
	2.9.7 Piping	65
	2.9.8 Chemical Facilities	65
	2.9.9 Material	65
	2.9.10 Estimation of Air Concentration in Dissolved Air Flotation (DAF) Systems	66
2.10	Granular-Media Filters	71
	2.10.1 General	71
	2.10.2 Filter Types and Applications	72
	2.10.3 System Design Parameters	74
	2.10.4 Cycle Time	76
	2.10.5 Vessels and Appurtenances	77
	2.10.6 Instrumentation and Controls	78

3 Chemical Treatment — 81
3.1 Introduction — 81
    3.1.1 Chemical Precipitation — 81
    3.1.2 Chemical Coagulation — 81

	3.1.3	Chemical Oxidation and Advanced Oxidation	82
	3.1.4	Ion Exchange	82
	3.1.5	Chemical Stabilization	84
3.2	Definition and Application		84
	3.2.1	Activated Carbon Adsorption	85
3.3	Chemical Precipitation		87
3.4	Chemical Flocculation		87
	3.4.1	Definition and Applications	88
	3.4.2	Design Considerations	90
	3.4.3	Clarifier	90
	3.4.4	Chemical Addition Systems	93
3.5	Disinfection		94
	3.5.1	Chemical Agents	95
	3.5.2	Mechanical Means	95
3.6	Chlorination		95
	3.6.1	Application	96
	3.6.2	Chlorine Dosages	96
	3.6.3	Design Considerations	96

4 Biological Treatment — 99

4.1	Theory		99
	4.1.1	Biological Activated Carbon Process	101
	4.1.2	Biokinetic Theoretical Model	102
4.2	Biological Treatment Processes		104
	4.2.1	Major Differences in Aerobic and Anaerobic Treatment	106
	4.2.2	Aerobic Processes	107
	4.2.3	Anaerobic Waste Treatment	112
	4.2.4	Aerobic, Anaerobic (Facultative) Waste Treatment	112
4.3	Activated-Sludge Units		112
	4.3.1	Applications	113
	4.3.2	Effects of Activated-Sludge	114
	4.3.3	Feed Composition	115
	4.3.4	Process Design	120
	4.3.5	Design Considerations	120
4.4	Trickling-Filters		123
	4.4.1	Trickling-Filter Process Design	124
4.5	Rotating Biological Contactor System		126
4.6	Sewage Oxidation Ponds		126

5 Wastewater Treatment in Unconventional Oil and Gas Industries — 129

5.1	Background		129
	5.1.1	Dissolved and Dispersed Hydrocarbon Components	131
	5.1.2	Dissolved Mineral	131
	5.1.3	Production Chemicals	131
	5.1.4	Produced Solids	132
	5.1.5	Dissolved Gases	132

5.2		Toxicity Limitations of Coal Bed Water	132
5.3		Shale Gas and Coal Seam Gas Produced Water, Treatment and Disposal	135
	5.3.1	Evaporation Pond	136
	5.3.2	Surface Stream Disposal	136
	5.3.3	Ion Exchange	138
	5.3.4	Membrane Filtration Technology	138
	5.3.5	Freeze–Thaw Evaporation	140
	5.3.6	Adsorption	140
	5.3.7	Chemical Oxidation	140
	5.3.8	Filtration	141
	5.3.9	Constructed Wetlands	141
	5.3.10	Electrodialysis/Electrodialysis Reversal	141
	5.3.11	Deep Well Injection at Dedicated Onshore Sites	141
	5.3.12	Biological Aerated Filters	142
	5.3.13	Macro-Porous Polymer Extraction Technology	143
	5.3.14	Thermal Technologies	143
5.4		Re-Thinking Technologies for Safer Facing	147
5.5		Water Treatment for Oil Sands Mining	153
	5.5.1	Recycling and Water Treatment Options	153
	5.5.2	Oily Water Treatment in Oil Sands Mining	155

6 Wastewater Sewer Systems — 161

6.1		Stormwater Sewer System	162
6.2		Oily Water Sewer System	162
6.3		Non-Oily Water Sewer System	163
6.4		Chemical Sewer System(s)	164
	6.4.1	Disposal of Chemical Sewers	164
	6.4.2	Neutralization Systems	164
	6.4.3	Type of Chemical Wastes	164
6.5		Sanitary Sewer System	165
6.6		Special Sewer Systems	165
6.7		Effluent Sources and Disposals	165
6.8		Particular Effluents in Refinery and Petrochemical Plants	167
	6.8.1	Caustic Scrubs (Heavy Oils)	167
	6.8.2	Desalter Wastewater	168
	6.8.3	Foul or Sour Waters	168
	6.8.4	Spent Caustic Solutions	168
	6.8.5	MTBE or Leaded Contaminated Streams	170
	6.8.6	Benzene Contaminated Streams	171
	6.8.7	Spent Sulfuric Acid Products	171
	6.8.8	Nitrogen Base Components	172
	6.8.9	Cyanides	172
	6.8.10	Aluminum Chloride	173
	6.8.11	Polyelectrolyte	173
	6.8.12	Ferric Chloride	173
	6.8.13	Phosphoric Acid	173

	6.8.14 Hydrofluoric Acid	173
	6.8.15 Other Spent Catalysts	173
	6.8.16 Chemical Cleaning Wastes	174
	6.8.17 Sulfur Solidification and Crushing Facilities and Loading Systems Drainage	174
	6.8.18 Water Containing Solids, Emulsifying Agents, etc.	174
	6.8.19 Heavy Viscous Oils Drainage	174
	6.8.20 Toxic Metal Contaminated Streams	174
	6.8.21 Solvent Processes Drainage	174
	6.8.22 Treating Processes Drainage	175
6.9	Petrochemical Plants' Special Effluents	175
	6.9.1 Summary of Disposal/Treatment Methods	175
6.10	NGL, LNG, and LPG Area Effluents	178
	6.10.1 Liquefied Gas Spill	178
6.11	Gas Treatment Facilities' Effluents	178
6.12	Effluents from Terminals, Depots, and Product Handling Areas	178
6.13	General Considerations and Conditions for Release of Wastes	178
	6.13.1 Characteristics and Composition of Waste	179
	6.13.2 Characteristics of the Discharge Site and Receiving Environment	179
	6.13.3 Availability of Waste Technologies	180
6.14	Effluent Wastewater Characteristics	180
	6.14.1 Flow	180
	6.14.2 Temperature	181
	6.14.3 pH	181
	6.14.4 Oxygen Demand	181
	6.14.5 Phenol Content	182
	6.14.6 Sulfide Content	182
	6.14.7 Oil Content	182
	6.14.8 Light Hydrocarbon Solubility in Water	182
	6.14.9 Predicting Water–Hydrocarbon System Mutual Solubility	185
6.15	Wastewater Emissions	189
	6.15.1 Point Source Discharge	189
	6.15.2 Effluent Permissible Concentrations	193
7	**Sewage Treatment**	**195**
7.1	Sewage Effluents	196
	7.1.1 Receiving Water	196
	7.1.2 Final Effluents of Domestic Wastewater Plants	197
7.2	Methods of Sewage Treatment: General	197
	7.2.1 Conventional Methods	197
7.3	Choice of System: General	197
7.4	Design of Sewage Treatment Plants: General Guidances	198
7.5	Design of Small Sewage Treatment Plants	198
	7.5.1 Collection of Information	198
7.6	Preliminary Treatment	200

	7.7	Primary and Secondary Settlement Tanks	200
		7.7.1 Capacities of Primary Settlement Tanks	201
	7.8	Sludge Digesters	202
	7.9	Drying Beds	202
		7.9.1 Secondary Settlement Tanks	203
	7.10	Biological Filters	204
		7.10.1 Distribution	205
		7.10.2 Volume of Filter	205
		7.10.3 Mineral Filter Media	206
	7.11	Activated-Sludge Units	207
	7.12	Tertiary Treatment (Polishing) Processes	207
	7.13	Disposal of Final Effluent	207
	7.14	Advanced Wastewater Treatment	208
		7.14.1 Effects of Chemical Constituents in Wastewater	208
		7.14.2 Advanced Wastewater Treatment Operations and Processes	209
	7.15	Effluent Disposal and Reuse	212
		7.15.1 Direct and Indirect Reuse of Wastewater	212
8	**Solid Waste Treatment and Disposal**		**215**
	8.1	Basic Considerations	215
		8.1.1 Classification	215
		8.1.2 Methodology	215
		8.1.3 Sources	216
		8.1.4 Characteristics	219
		8.1.5 Quantities	223
	8.2	Sludge Handling, Treatment, and Reuse	223
		8.2.1 General	223
		8.2.2 Sludge and Scum Pumping	223
		8.2.3 Sludge Piping	226
		8.2.4 Preliminary Operation Facilities	229
		8.2.5 Thickening (Concentration)	230
	8.3	Stabilization	233
		8.3.1 Design Considerations	233
		8.3.2 Lime Stabilization	233
		8.3.3 Heat Treatment	234
		8.3.4 Anaerobic Sludge Digestion	235
		8.3.5 Composting	236
	8.4	Conditioning	237
	8.5	Disinfection	237
	8.6	Dewatering	237
		8.6.1 Sludge Dewatering Methods	238
		8.6.2 Vacuum Filtration	238
		8.6.3 Centrifugation	238
		8.6.4 Belt Filter Press	238
		8.6.5 Sludge Drying Beds	241
		8.6.6 Lagoons	241

8.7	Heat Drying		242
8.8	Thermal Reduction		242
	8.8.1	Thermal Reduction Process Applications	243
8.9	Land Application of Sludge		243
8.10	Chemical Fixation		245
8.11	Final Sludge and Solids Conveyance, Storage, and Disposal		245
	8.11.1	Conveyance Methods	245
	8.11.2	Environmental Considerations in Sludge Transportation	246
	8.11.3	Sludge Storage	246
	8.11.4	Final Disposal	246
	8.11.5	Incineration	247
	8.11.6	Ash Handling and Disposal	248
8.12	Disposal of Solid Waste		248
	8.12.1	Types of Waste	248
	8.12.2	Siting of Hazardous Waste Facilities	250
	8.12.3	Non-Hazardous Waste	251
	8.12.4	Sources, Segregation, Quantities, and Characteristics of Solid Waste in Refineries	251
	8.12.5	Source Reduction Methods	253
	8.12.6	Resource Recovery and Waste Minimization	255
	8.12.7	Hazardous Waste Reduction	258
	8.12.8	Treatment Prior to Ultimate Disposal	260
	8.12.9	Disposal of Waste Generated in Drilling Wells	266
	8.12.10	General Sampling Considerations	267
	8.12.11	Air Monitoring of Waste for Employee Protection	272
	8.12.12	Procedures	273
	8.12.13	Hazards	275
	8.12.14	Quality Assurance Consideration	275

Definitions and Terminology   277
Bibliography and Further Reading   307
Index   323

Preface

Oil and gas are major sources of energy and revenue for many countries today – their production has been described as one of the most important industrial activities in the twenty-first century – and obviously waste treatment and disposal assume a greater degree of importance in the petroleum, chemical processing, and unconventional oil and gas industries.

Wastewater quality and the quantity produced determine the means of disposal and the costs of disposal. Suspended solids, total dissolved solids, and oxygen demand of produced waters have the most impact on wastewater treatment.

Wastewater is a complex mixture of organic and inorganic compounds and the largest byproduct by volume generated during chemical processing and both conventional and unconventional oil and gas recovery operations. The potential of oilfield-produced water to be a source of freshwater for water-stressed oil-producing countries and increasing environmental concerns, in addition to stringent legislations on produced water discharge into the environment, have made produced water management a significant part of the oil and gas business.

In marginally economic coal bed projects, the water disposal costs and attendant environmental accounting are critical factors in the investment decision; water disposal costs can economically make or break a marginal project.

Before investing in a coal bed methane (CBM) process, multiple questions need to be answered concerning the water to be produced – questions concerning quantity, flow-rates, chemical content, disposal means, monitoring, and environmental regulations. Perhaps no other factor affects the economics and feasibility of CBM projects as much as water removal and disposal.

In heavy oil production, between 2 to 4.5 volume units of water are used to produce each volume unit of synthetic crude oil in an ex situ mining operation. Despite recycling, almost all of it ends up in tailings ponds. However, in Steam Assisted Gravity Drainage (SAGD) operations, 90–95% of the water is recycled and about 0.2 volume units of water is used per volume unit of bitumen produced.

A major hindrance to the monitoring of oil sands-produced waters has been the lack of identification of individual compounds present. By better understanding the nature of the highly complex mixture of compounds, including naphthenic acids, it may be possible to monitor rivers for leachate and also to remove toxic components. Such identification of individual acids has for many years proved impossible, but a recent breakthrough in analysis has begun to reveal what is in the oil sands-produced waters.

The extraction and use of shale gas can affect the environment through the leaking of extraction chemicals and waste into water supplies, the leaking of greenhouse gasses during extraction, and the pollution caused by the improper processing of natural gas.

A challenge to preventing pollution is that shale gas extractions vary widely in this regard, even between different wells in the same project; the processes that reduce pollution sufficiently in one extraction may not be enough in another.

Chemicals are added to the water to facilitate the underground fracturing process that releases natural gas. Fracturing fluid is primarily water and approximately 0.5% chemical additives (friction reducer, agents countering rust, agents to kill microorganisms). Since (depending on the size of the area) millions of liters of water are used, this means that hundreds of thousands of liters of chemicals are often injected into the soil.

Only about 50 to 70% of the resulting volume of contaminated water is recovered and stored in above-ground ponds to await removal by tanker. The remaining "produced water" is left in the earth where it can lead to contamination of groundwater aquifers, though the industry deems this "highly unlikely." However, the wastewater from such operations often leads to foul-smelling odors and heavy metals contaminating the local water supply above-ground.

This book unravels the essential requirements for the process design and engineering of the equipment and facilities pertaining to the wastewater treatment units, solid waste disposal, and wastewater sewer systems of oil and gas refineries, chemical plants, oil terminals, petrochemical plants, unconventional oil and gas industries (coal seam gas or coal bed methane, shale gas and oil sands production), and other facilities as required. Included within the scope are:

- Liquid and solid disposal systems.
- Primary oil/solids removal facilities.
- Further oil and suspended solids removal (secondary oil/solids removal), such as dissolved air flotation units.
- Granular media filters and chemical flocculation units.
- Chemical addition systems.
- Biological treatments.
- Filtration and/or other final polishing.
- Sewage systems handling domestic and medical sanitary appliances of buildings.
- Drainage systems carrying surface and rainwater.
- Wastewater gathering systems.
- Clean water drainage, e.g., from buildings and paved areas.
- Evaporation ponds and disposal by natural percolation into the subsoil in permeable ground.
- Sanitary sewage treatment.
- Sludge handling and treatment.

It is obvious that the aim of any drainage/effluent disposal system should be to segregate uncontaminated water from contaminated water or effluents and to segregate different types of effluents in order to reduce the size, complexity, and costs of any treatment units that may be required for handling the contaminated water and effluents before they are discharged from a unit.

All wastewater effluents from industry that are discharged to public and/or natural water sources or directed for recycling purposes inside the industry, and that may contain a wide variety of matters in solution or suspension, should be controlled according to the requirements imposed by the final destination. However, in any case, elimination of the waste or the hazard potential of the waste should be the ultimate goal in the management.

Under no circumstances should effluent water cause oil traces on the surface or embankments of the receiving water, or affect the natural self-purification capacity of the receiving water to such an extent that it would cause hindrance to other users.

Under no conditions should polluted streams be combined with unpolluted streams if the resultant stream will then require purification. In general, the main sewer systems in the industry will be segregated according to the following categories:

- Stormwater sewer systems.
- Oily water sewer systems.
- Non-oily water sewer systems.
- Chemical sewer systems.
- Sanitary sewer systems.
- Special sewer systems.

In all areas, including process, offsite, and utility units, provisions should be made to anticipate any of the above mentioned sewer systems as required.

The treatment of wastewaters involves a sequence of treatment steps. Each wastewater treatment process involves the separation of solids from water in at least some part of the operation and the removal of biochemical oxygen demand (BOD) to some extent.

The end of pipe treatment sequence can be divided into the following elements: primary or pre-treatment, intermediate treatment, secondary treatment and tertiary treatment plus ancillary, sludge dewatering, and disposal operations.

The key to optimizing the treatment sequence for provision of maximum water treatment at minimum cost is to identify the rule of each unit operation and optimize that operation. Optimizing the performance of specific unit operations, such as API separators, dissolved air flotation, biological treatment, etc., can best be achieved if:

1. The properties of influent streams are considered.
2. The chemical principles that are used in solids pre-treatment are understood.
3. The variety of chemicals available for solids treatment is recognized.
4. The properties of effluent water are established based on the local environmental regulations and final disposal.
5. The protocols for quantifying results are identified.

In general, most industries require water for processing or other purposes; much of this water after use is discharged either to public and/or natural water sources or directed for recycling purposes inside the industry.

Such discharge, which may contain a wide variety of matter in solution or suspension, should be controlled according to the requirements imposed by the final destination and/or environmental regulations.

Moreover, according to the type of plant and the method of plant operation, the sources of solids in a wastewater treatment plant can be uncovered. Solids may also be formed by interaction of waste streams in the sewer.

Wastewaters contain metal ions, such as iron, aluminium, copper, magnesium, and so on, from corrosion of the process equipment, chemicals used in treating cooling water, salts in the water intake, and chemicals used in processing.

Insoluble metal hydroxide floc may be formed when alkaline wastes are discharged and raise the pH of wastewater above neutral. The wastes, containing considerable concentrations of phenols, sulfides, emulsifying agents, and alkalines, should be segregated. In general, discharging any material to the oily sewer system or other drainage system should be investigated in line with the final waste treatment and disposal targets.

In view of the above, this book will unravel the fundamental engineering for waste recovery, treatment, and disposal systems in the petroleum, chemical, and unconventional oil and gas processing industries. These new fundamental discoveries will enable the development of practical solutions to these pressing environmental issues.

Dr. Alireza Bahadori
School of Environment, Science & Engineering, Southern Cross University, Lismore, NSW, Australia

25 July 2013

Acknowledgments

I would like to thank the Wiley editorial and production team Rebecca Stubbs, Emma Strickland and Sarah Tilley of John Wiley & Sons for their editorial assistance.

Biography

Alireza Bahadori, PhD is a research staff member in the School of Environment, Science & Engineering at Southern Cross University, Lismore, NSW, Australia. He received his PhD from Curtin University, Western Australia. For the better part of 20 years, Dr Bahadori had held various process engineering positions and involved in many large-scale projects at NIOC, Petroleum Development Oman (PDO), and Clough AMEC PTY LTD.

He is the author of over 200 articles and 6 books. His books have been accepted/published by prestigious publishers such as John Wiley & Sons, Springer, Taylor & Francis and Elsevier. Dr Bahadori is the recipient of highly competitive and prestigious Australian Government's Endeavour International Postgraduate Research award as part of his research in oil and gas area. He also received top-up award from State Government of Western Australia through Western Australia Energy research Alliance (WA:ERA) in 2009. Dr Bahadori Serves as a member of editorial board for a number of journals such as *Journal of Sustainable Energy Engineering* which is published by Wiley-Scrivener.

1
Wastewater Treatment

Wastewater treatment refers to the treatment of sewage and water used by residences, business, and industry to a sufficient level that it can be safely returned to the environment. It is important to treat wastewater to remove bacteria, pathogens, organic matter, and chemical pollutants that can harm human health, deplete natural oxygen levels in receiving waters, and pose risks to animals and wildlife.

1.1 Characteristics of Wastewaters

A number of chemical and physical characteristics are used to describe wastewater. The most common are:

- Biochemical Oxygen Demand (BOD). This is a measure of the amount of unstable organic matter in the water. It measures how much oxygen is required by the available microorganisms to break down the readily available organic matter into simpler forms, such as carbon dioxide, ammonia, and water.
- Total Nitrogen (TN) and Total Phosphorus (TP). These are the sum of all forms of nitrogen and phosphorus in the water, respectively.
- Fecal microbes (which include viruses, bacteria, and protozoans). These are found in wastewater and may cause disease.
- Suspended solids, biodegradable organics, nutrients, refractory organics, heavy metals, dissolved inorganic solids, and pathogens are important contaminants that may be found in the oil, gas, and chemical processing industry's wastewaters. Table 1.1 presents a list of important wastewater contaminants and reasons for their importance.

Suspended solids can be removed by physical treatment to some extent. Removal of biodegradable organics, suspended solids, and pathogens is achieved through the secondary treatment operation units.

Table 1.2 shows typical waste compounds classified as priority pollutants. The more stringent rules deal with the removal of nutrients and priority pollutants. When wastewater

Table 1.1 Contaminant importance in wastewater treatment.

Contaminants	Reason for Importance
Physical suspended solids	Suspended solids are important for esthetical reasons and because they can lead to the development of sludge deposits and anaerobic conditions
Chemical biodegradable organics	Composed principally of proteins, carbohydrates, and fats, biodegradable organics are measured most commonly in terms of BOD (Biochemical Oxygen Demand) and COD (Chemical Oxygen Demand). If discharged untreated to the environment, the biological stabilization of these materials can lead to the depletion of natural oxygen resources and to the development of septic conditions.
Nutrients	Carbon, nitrogen, and phosphorus are essential nutrients for growth. When discharged to the aquatic environment, these nutrients can lead to the growth of undesirable aquatic life. When discharged in excessive amounts on land, they can also lead to the pollution of groundwater.
Refractory organics	These organics tend to resist conventional biological methods of wastewater treatment. Typical examples include surfactants, phenols, and agricultural pesticides
Heavy metals	Due to their toxic nature, certain heavy metals can negatively impact upon biological waste treatment processes and stream life.
Dissolved inorganic solids	Inorganic constituents, such as calcium, sodium, and sulfate, are added to the original domestic water supply as a result of water use and may have to be removed if the wastewater is to be reused.
Biological pathogens	Communicable diseases can be transmitted by the pathogenic organisms in wastewater.

is to be reused, rules normally include requirements for the removal of refractory organics, heavy metals, and in some case dissolved inorganic solids.

1.1.1 Suspended Solids

Typically, suspended solids carry a significant portion of organic material, thus significantly contributing to the organic load of the wastewater (solids can contribute up to 60% of the BOD of a wastewater). Hence, effective solids removal can significantly contribute to wastewater treatment. A widely-accepted means of testing a wastewater for suspended solids is to filter the wastewater through a 0.45 μm porosity filter. Anything left on the filter after drying at about 103 °C is considered a portion of the suspended solids. Table 1.3 provides another classification system for the solids found in wastewater.

1.1.2 Heavy Metals

Any cation having an atomic mass (weight) greater than 23 (atomic mass of sodium) is considered a heavy metal. Motivations for controlling heavy metal concentrations in gas streams are diverse. Some of them are dangerous to health or to the environment (e.g., mercury, cadmium, lead, chromium), some can cause corrosion (e.g., zinc, lead), some are

Table 1.2 Typical waste compounds classified as priority pollutants.

Name (Formula)	Concern
Non-metals	
Arsenic (As)	Carcinogen and mutagen. Long term: sometimes cause fatigue and loss of energy; dermatitis.
Selenium (Se)	Long term: red staining of fingers, teeth, and hair; general weakness; depression; irritation of nose and mouth.
Metals	
Barium (Ba)	Flammable at room temperature in powder form. Long term: increased blood pressure and nerve block.
Cadmium (Cd)	Flammable in powder form. Toxic by inhalation of dust or fume. A carcinogen. Soluble compounds of cadmium are highly toxic. Long term: concentrates in the liver, kidneys, pancreas, and thyroid; hypertension suspected effect.
Chromium (Cr)	Hexavalent chromium compounds are carcinogenic and corrosive on tissue. Long term: skin sensitization and kidney damage
Lead (Pb)	Toxic by ingestion or inhalation of dust or fumes. Long term: brain and kidney damage; birth defects.
Mercury (Hg)	Highly toxic by skin absorption and inhalation of fume or vapor. Long term: toxic to central nervous system; may cause birth defects.
Silver (Ag)	Toxic metal. Long term: permanent gray discoloration of skin, eyes, and mucus membranes.
Organic compounds	
Benzene (C_6H_6)	Carcinogen. Highly toxic. Flammable, dangerous fire risk.
Ethylbenzene ($C_6H_5C_2H_5$)	Toxic by ingestion, inhalation, and skin absorption; irritant to skin and eyes. Flammable, dangerous fire risk.
Toluene ($C_6H_5CH_3$)	Flammable, dangerous fire risk, Toxic by ingestion, inhalation, and skin absorption.
Halogenated compounds	
Chlorobenzene (C_6H_5Cl)	Moderate fire risk. Avoid inhalation and skin contact.
Chloroethene (CH_2CHCl)	An extremely toxic and hazardous material by all avenues of exposure. A carcinogen.
Dichloromethane (CH_2Cl_2)	Toxic. A carcinogen, narcotic.
Tetrachloroethene (CCl_2CCl_2)	Irritant to eyes and skin.
Pesticides, herbicides, insecticides (Pesticides, herbicides, and insecticides are listed by trade name. The compounds listed are also halogenated organic compounds.)	
Endrin ($C_{12}H_8OCl_6$)	Toxic by inhalation and skin absorption, carcinogen.
Lindane ($C_6H_6Cl_6$)	Toxic by inhalation, ingestion, skin absorption.
Methoxychlor [$Cl_3CCH(C_6H_4OCH_3)2$]	Toxic material.
Toxaphene ($C_{10}H_{10}Cl_8$)	Toxic by ingestion, inhalation, skin absorption.
Silvex [$Cl_3C_6H_2OCH(CH_3)COOH$]	Toxic material; use has been restricted.

Table 1.3 General classification of wastewater solids.

Particle Classification	Particle Size, mm
Dissolved	Less than 10^{-6}
Colloidal	10^{-6} to 10^{-3}
Suspended	Greater than 10^{-3}
Settleable	Greater than 10^{-1}
Supracolloidal	10^{-3} to 10^{-1}

harmful in other ways (e.g., arsenic may pollute catalysts). Unlike organic pollutants, heavy metals do not decay and thus pose a different kind of challenge for remediation.

Currently, plants or microorganisms are tentatively used to remove some heavy metals such as mercury. Plants that exhibit hyper accumulation can be used to remove heavy metals from soils by concentrating them in their bio-matter. Some treatment of mining tailings has occurred where the vegetation is then incinerated to recover the heavy metals.

1.1.3 Dissolved Inorganic Solids

Total dissolved inorganic solids (TDIS) are a calculated value to assess the actual inorganic salt content of a water or process water.

The following procedure can be used to determine the inorganic dissolved solids in wastewaters. A sample of wastewater is filtered through a 0.45 μm filter, filtrate is collected, the water is vaporized first (at 103 °C) and then the organic fraction (at 550 °C) from the filtrate. The amount of material left in the vessel after incineration at 550 °C is referred to as the fixed or inorganic dissolved solids level.

1.1.4 Toxic Organic Compounds

Wastewater systems are known to contain toxic metals, organic micro pollutants, and pathogens that may add constraints to their beneficial uses. Environmental risks related to toxic inorganics, dioxins, furans, and pathogens can be controlled by:

1. Selecting a wastewater system with a low content of regulated contaminants that respects the local legislation for land application.
2. Application of a decontamination process to remove toxic metals.
3. The necessary step of sterilization for monocultures that eliminates pathogens.

These toxic organic compounds eventually reach sewage treatment plants and can be concentrated in wastewater systems. Disposal of wastewater systems is one way that these pollutants can be introduced into the environment. The presence of these toxic organic compounds can add constraints to the ultimate disposal of these sludges and/or reduce the possibilities for their beneficial use.

Tables 1.4 and 1.5 provide some organic compounds that are considered toxic and/or carcinogenic.

Table 1.4 Toxic organic compounds; occupational exposure to carcinogenic substances.

Compound	Site	Comment
Organic substances for which there is wide agreement on carcinogenicity		
4-Aminodiphenyl	Bladder	A contaminant in diphenylamine
Benzidine	Bladder	Ingredient of aniline dyes, plastics, and rubber
Beta-naphthylamine (2-NA)	Bladder	Dye and pesticide ingredient; synonym, 2-naphthylamine exposed workers have 30 to 60 times more cases of bladder cancer
Bis (chloromethyl) ether	Lung	Used in making exchange resins; exposed workers have 7 times more cases of lung cancer; synonym, BCME
Vinyl chloride	Liver	Angiosarcoma cases among PVC workers
Additional organic substances on USDA-OSHA cancer-causing substances list		
Alpha-naphthylamine (1-NA)	Bladder	Human case implicated; used in making dyes, herbicides, (1-NA) food colors, color film; an antioxidant
Ethyleneamine	Unknown	Carcinogenic in animals; used in paper and textile processing and manufacture of herbicides, resins, rocket and jet fuels
3,3-Dichlorobenzidine	Unknown	Carcinogenic in animal species; exposure accompanies benzidine and betanaphthylamine
Methyl chloromethyl methyl ether	Unknown	Carcinogenic in animals; synonym, CMME; BCME contaminants CMME; used in resin-making, textile, and drug production.
4,4-Methylene bis (2-chloroaniline)	Unknown	Synonym MOCA. Tumorigenic in rats and mice. Skin absorption may be the hazard. Curing agent for iso-cyanate polymers.

Table 1.5 Industrial substances suspected of carcinogenic potential for humans.

Industrial Substance	Industrial Substance
Antimony trioxide production	Epichlorhydrin
Benzene (skin)	Hexamethyl phosphoramide (skin)
Benzo(a) pyrene	Hydrazine
Beryllium	4,4-Methylene bis (2-chloroaniline) (skin)
Cadmium oxide production	4,4-Methylene dianiline
Chloroform	Monomethyl hydrazine
Chromates of lead and zinc	Nitrosamines
3,3-Dichlorobenzidine	Propane sulfone
1,1-Dimethyl hydrazine	Beta-propiolactone
Dimethyl sulfate	Vinyl cyclohexene dioxide
Dimethylcarbamyl chloride	

1.1.5 Surfactants

Surfactants, or surface-active agents, are large organic molecules that are slightly soluble in water and cause foaming in wastewater treatment plants and in the surface waters into which the waste effluent is discharged.

The surfactants present in detergent products remain chemically unchanged during the washing process and are discharged down the drain with the dirty wash water. In the vast majority of cases, the drain is connected to a sewer and ultimately to a wastewater treatment plant, where the surfactants present in the sewage can be removed by biological and physical-chemical processes.

During aeration of wastewater, these compounds are collected on the surface of the air bubbles and thus create a very stable foam. The determination of surfactants is accomplished by measuring the color change in a standard solution of methylene blue active substance (MBAS).

1.1.6 Priority Pollutants

Priority pollutants (both inorganic and organic) are selected on the basis of their known or suspected carcinogenicity, mutagenicity, teratogenicity, or high acute toxicity. Many of the organic priority pollutants are also classified as volatile organic compounds (VOCs).

Representative examples of the priority pollutants are shown in Table 1.2. Within a wastewater collection and treatment system, organic priority pollutants may be removed, transformed, generated, or simply transported through the system unchanged. Five primary mechanisms are involved: (1) volatilization (also gas stripping); (2) degradation; (3) adsorption to particles and sludge; (4) transport through the entire system; (5) generation as a result of chlorination or as byproducts of the degradation of precursor compounds.

1.1.7 Volatile Organic Compounds

Wastewaters are collected and treated in a variety of ways, some of which result in the emission of volatile organic compounds (VOCs) from the wastewater to the air. Water may come into direct contact with organic compounds during a variety of different chemical processing steps, thus generating wastewater streams that must be discharged for treatment or disposal. Direct contact wastewater includes:

- Water used to wash impurities from organic compound products or reactants.
- Water used to cool or quench organic compound vapour streams.
- Condensed steam from jet eductor systems pulling vacuum on vessels containing organic compounds.
- Water from raw material and product storage tanks.
- Water used as a carrier for catalysts and neutralizing agents (e.g., caustic solutions).
- Water formed as a byproduct during reaction steps.

Direct contact wastewater is also generated when water is used in equipment washes and spill clean-ups. This wastewater is normally more variable in flow-rate and concentration than the streams listed above and may be collected in a way that is different from process wastewater. Wastewater streams generated by unintentional contact with organic compounds through equipment leaks are defined as "indirect contact" wastewater. Indirect contact wastewater may become contaminated as a result of leaks from heat exchangers, condensers, and pumps.

Organic compounds that have a boiling point ≤ 100 °C and/or a vapor pressure > 1 mm Hg (or 133.3 Pa) at 25 °C are generally considered to be volatile organic compounds (VOCs), e.g., vinyl chloride. The release of these compounds in sewers and treatment plants, especially at the head works, is of particular concern with respect to the health of collection system and treatment plant workers.

1.2 Treatment Stages

Generally, the terms "preliminary" and/or "primary" refer to physical unit operations; "secondary" refers to chemical and biological unit processes; and "advanced" or "tertiary" refer to combinations of all three.

The application and definition of the various stages of treatment and methods to perform specific functions are described in the following sections. Figure 1.1 shows a schematic of wastewater treatment stages.

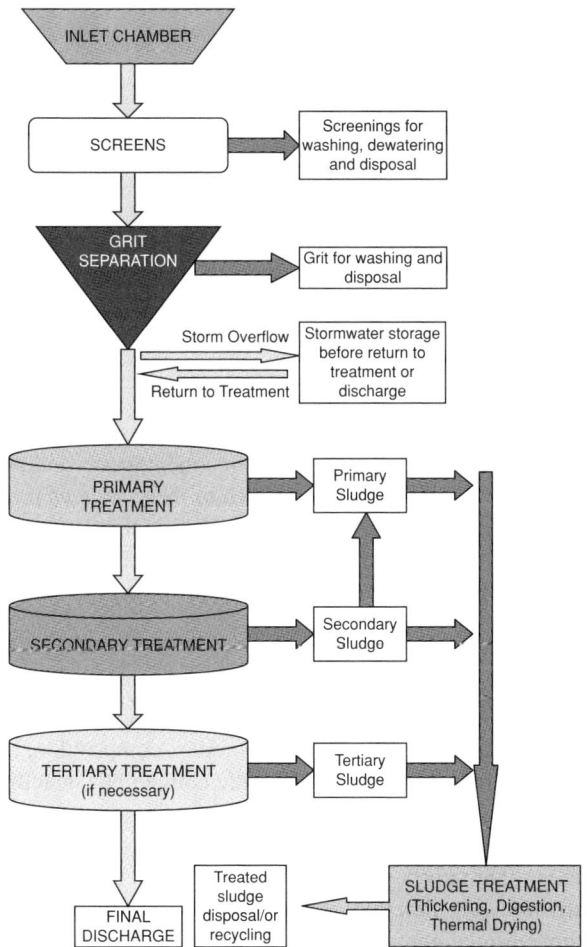

Figure 1.1 A simplified schematic of wastewater treatment stages.

Table 1.6 Typical wastewater qualities.

Parameter	Unit	Oily Wastewater	Stripped Sour Water	Combined High TDS Waters (Ion Exchange Waste, Boiler Blowdown, RO Reject)	Cooling Tower Blow-down
Temperature	°C	30–60	30–35	30–40	–
pH	–	7–8	7–8	7–8	8
TDS	mg/L	150–5000	50–150	500–2500	5000–6000
TSS	mg/L	300–800	10–20	50–100	16 000–19 000
Cl2 Residual	–	–	–	–	0.3–0.5
BOD	mg/L	300–500	100–300	5–150	–
COD	mg/L	300–1200	200–500	100–500	–
TOC	mg/L	–	–	<100	–
Hardness	mg/L as $CaCO_3$	–	–	–	1200–1400
Total Alkalinity	mg/L as $CaCO_3$	–	–	–	100–125
Ca^{2+}	mg/L	–	–	–	1000
Cl^-	mg/L	50–2000	–	–	1000–1500
NH_3	mg/L	20–50	40–80	–	<5
Cyanides	mg/L	1–3	–	–	–
Phenols	mg/L	5–20	20–80	–	–
H_2S	mg/L	5–10	10–40	–	–

1.2.1 Sources of Wastewater

Sources of wastewater in the oil, gas, and chemical processing industries include oily wastewater, sour water, stripped sour water, water treatment waste, and blow-down streams (cooling tower, boiler, and gasifier) and so on. Each of these sources produces wastewater with slightly different characteristics and treatment requirements.

Table 1.6 provides typical wastewater qualities for some of the wastewater streams in the oil, gas, and chemical processing industries.

1.2.2 Discharge Options and Quality Requirements

Produced water in the oil and gas industry has a complex composition, but its constituents can be broadly classified into organic and inorganic compounds including: dissolved and dispersed oils, grease, heavy metals, radionuclides, treating chemicals, formation solids, salts, dissolved gases, scale products, waxes, microorganisms, and dissolved oxygen.

The four discharge alternatives listed below are all technically feasible. Selection of the preferred alternative is a function of the selected process, recycling opportunities, economics, regulatory limitations, and social requirements. Process effects, which relate primarily to dissolved solid concentrations and financial implications, will be examined here.

- Physical and biological treatment followed by discharge to a river.
- Physical, biological, and chemical treatment followed by discharge to a river.
- Physical and biological treatment and recycling with deep well injection, thus no surface discharge.
- Physical and biological treatment, evaporation and crystallization, thus no discharge.

1.2.3 Preliminary Wastewater Treatment

Preliminary wastewater treatment is defined as the removal of wastewater constituents that may cause maintenance or operational problems within the treatment operations, processes, and ancillary systems.

Screening and comminution for the removal of debris and rags, grit removal for the elimination of coarse suspended matter that may cause wear or clogging of equipment, and flotation for the removal of large quantities of oil and grease are examples of preliminary operations.

1.2.4 Primary Wastewater Treatment

In primary treatment a portion of the suspended solids and organic matter is removed from the wastewater. This removal is usually accomplished with physical operations such as screening and sedimentation.

The effluent from primary treatment will ordinarily contain considerable organic matter and will have a relatively high BOD. The principal function of primary treatment continues to be as a precursor to secondary treatment.

Following primary treatment, the treated water is suitable for use as cooling water and utility water but will require further treatment to be used as boiler feed water.

1.2.5 Conventional Secondary Wastewater Treatment

Secondary treatment is directed principally toward the removal of biodegradable organics and suspended solids. Disinfection is frequently included in the definition of conventional secondary treatment.

Conventional secondary treatment is defined as the combination of processes customarily used for the removal of these constituents, and includes biological treatment by activated sludge, fixed film reactors, or lagoon systems and sedimentation.

1.2.6 Nutrient Removal or Control

Nutrient removal or control is generally required for:

1. Discharges to confined bodies of water where eutrophication might be caused or accelerated.
2. Discharges to flowing streams where nitrification could tax oxygen resources or where rooted aquatic plants could flourish.
3. Recharge of groundwaters that might be used indirectly for public water supplies.

The nutrients of principal concern are nitrogen and phosphors, and they may be removed by biological, chemical, or a combination of these processes. In many cases, the nutrient removal processes are coupled with secondary treatment; for example, biological denitrification may follow an activated-sludge process that produces a nitrified effluent.

1.2.7 Advanced Wastewater Treatment/Wastewater Reclamation

Advanced wastewater treatment or tertiary treatment is normally defined as the level of treatment required beyond conventional secondary treatment to remove constituents of concern, including nutrients, toxic compounds, and increased amounts of organic material and suspended solids.

In addition to the nutrient removal processes, unit operations or processes frequently employed in advanced wastewater treatment are chemical coagulation, flocculation, and sedimentation followed by filtration and activated carbon. Less used processes include ion exchange and reverse osmosis for specific ion removal or for reduction of dissolved solids.

Advanced wastewater treatment is also used in a variety of reuse applications where quality water is required, such as for industrial cooling water and groundwater recharge.

1.2.8 Toxic Waste Treatment/Specific Contaminant Removal

The removal of toxic substances and specific contaminants is a complex subject and concentrations of toxic pollutants are usually controlled by pre-treatment prior to discharge to the final treatment system. Many toxic substances, such as heavy metals, are reduced by some form of chemical-physical treatment such as chemical coagulation, flocculation, sedimentation, and filtration.

Some degree of removal is also accomplished by conventional secondary treatment. Wastewaters containing volatile organic constituents may be treated by air stripping or by carbon adsorption. Small concentrations of specific contaminants may be removed by ion exchange. Table 1.7 presents a list of typical pollutants that have an inhibitory effect on the activated-sludge process.

1.2.9 Sludge Processing

When a liquid sludge is produced, further treatment may be required to make it suitable for final disposal. Typically, sludges are thickened (dewatered) to reduce the volumes transported off-site for disposal.

Processes for reducing water content include lagooning in drying beds to produce a cake that can be applied to land or incinerated; pressing, where sludge is mechanically filtered, often through cloth screens, to produce a firm cake; and centrifugation where the sludge is thickened by centrifugally separating the solid and liquid. Sludges can be disposed of by liquid injection to land or by disposal in landfill. For the most part, the methods and systems reported in Table 1.8 are used to process the sludge removed from the liquid portion of the wastewater.

Table 1.7 Threshold concentrations of pollutants inhibitory to the activated-sludge process.

Pollutant	Concentration, mg/L	
	Carbonaceous Removal	Nitrification
Aluminum	15–26	–
Ammonia	480	–
Arsenic	0.10	–
Borate (Boron)	0.05–100	–
Cadmium	10–100	–
Calcium	2500	–
Chromium (hexavalent)	1–10	0.25
Chromium (trivalent)	50	–
Copper	1	0.005–0.50
Cyanide	0.1–5	0.34
Iron	1000	–
Manganese	10	–
Magnesium	–	50
Mercury	0.1–5	–
Nickel	1–2.5	0.25
Silver	5	–
Sulfate	–	500
Zinc	0.8–1	
Phenols:	200	4–10
Phenol	–	10–16
Cresol	–	150
2-4 Dinitrophenol		

1.3 Treatment Processes

Industrial wastewater treatment processes cover the mechanisms and processes used to treat waters that have been contaminated in some way by anthropogenic industrial or commercial activities, prior to its release into the environment or its reuse.

The oil, gas, and chemical industries produce some contaminants in wastewater that should be removed by physical, chemical and/or biological means. Figure 1.2 shows a schematic of wastewater treatment in chemical industries.

Unit operations and processes that are commonly used in wastewater treatment are listed in Table 1.9. The following instructions should be taken into consideration for the selection of treatment technologies:

- Technologies should be categorized into those that work, those that have the potential to work, and those that have no place in the particular application.
- Technologies should be evaluated based on their effectiveness (ability to reliably attain treatment goals), implementability (availability of materials and services), and costs (capital and operation and maintenance).
- Viable technologies should be identified for each of the individual wastewater streams. Streams that use the same technologies should be combined to create composite waste

Table 1.8 Sludge processing and disposal methods.

Processing or Disposal Function	Unit Operation, Unit Process, or Treatment Method
Preliminary operation	Sludge pumping Sludge grinding Sludge blending and storage Sludge degritting
Thickening	Gravity thickening Flotation thickening Centrifugation Gravity belt thickening Rotary drum thickening
Conditioning	Chemical conditioning Heat treatment
Disinfection	Pasteurization Long-term storage
Dewatering	Vacuum filter Centrifuge Belt press filter Filter press Sludge drying beds Lagoons
Heat drying	Dryer variations Multiple-effect evaporator
Thermal reduction	Multiple-hearth incineration Fluidized bed incineration Co-incineration with solid wastes Wet air oxidation Vertical deep well reactor
Ultimate disposal	Land application Distribution and marketing Landfill Lagooning Chemical fixation

treatment trains. These should be compared to current manufacturing and waste treatment practices to identify possible candidates for waste segregation and independent treatment.
- The level of wastewater treatment and method of effluent discharge should be established to protect the receiving body of water or the water table and its usages.

The level of treatment of the facility to be designed should be determined by the ability of the receiving waters to accept residual wastes and the allocation set up by effluent standards. The degree of treatment can be determined by comparing the influent wastewater characteristics to the required effluent wastewater characteristics.

In the case of wastewater reuse applications, the quality of water used as make-up will govern the wastewater treatment needed and the degree of reliability required for the

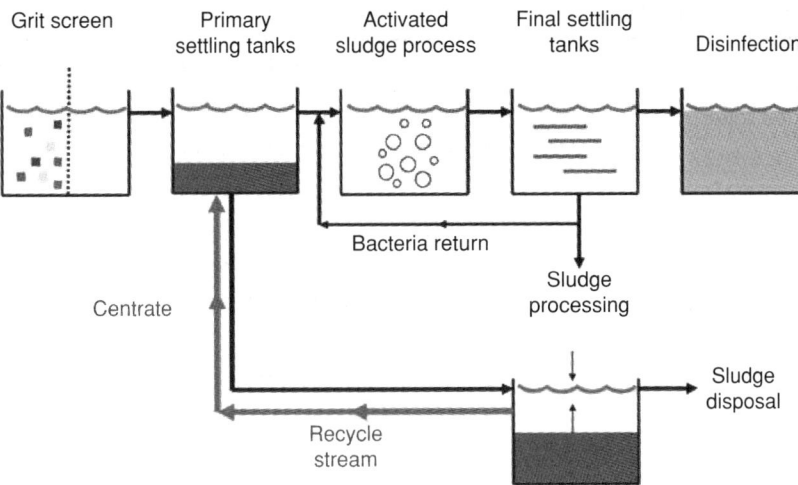

Figure 1.2 *A schematic of wastewater treatment (Reproduced with permission from [10] © Elsevier, 2011).*

treatment processes and operations. The reliability of the proposed treatment processes and operations must be evaluated to provide a continuous supply of water with consistent water quality.

All toxic and highly chemically active materials should be treated at source and not discharged in any active state into the sewers loading to the waste treatment plant. This may include removal of soluble and insoluble forms of metals such as lead, zinc, copper, or their derivatives, and other similarly dangerous classified metals and their byproducts.

It should be required that highly active metals, inclusive of finely divided magnesium and aluminium alloys, are not discharged in the sewers but are treated and removed by special methods and equipment at source.

It should be required that all highly toxic inorganic chemicals, inclusive of cyanides, fluorides, and related objectionable anions, must be treated and removed from the water at or near the source to the degree specified in the code regulations. This includes the chromates and other special complex anion derivatives.

Another group excluded from discharge of waste in the sewer should be acting oxidizing agents, particularly peroxides of organic and inorganic structures. This group should also include other powerful oxidizing agents, inclusive of chlorates, perchlorates, nitric acid, and other similar products.

The discharge of volatile organic materials into the waste should also be restricted and these materials should be isolated and treated at source. This restriction is a must, because disastrous explosions can occur in sewer systems where the volatilization of the organic matter creates an explosive mixture or some other conditions set off chemical reactions.

In general, all toxic materials, particularly of the organic family that are known to be dangerous to plant, animal, or human life, should be treated at source.

All solutions containing radioactive products must also be kept isolated and treated at source.

Table 1.9 Unit operations, unit processes, and systems for wastewater treatment.

Contaminant	Unit Operation, Unit Process, or Treatment System
Suspended solids	• Sedimentation • Screening and comminution • Filtration • Flotation • Chemical-polymer addition • Coagulation/sedimentation
Biodegradable organics	• Activated sludge variations • Fixed film: trickling-filters • Fixed film: rotating biological contractors • Lagoon and oxidation pond variations • Intermittent sand filtration • Land treatment systems • Physical-chemical systems
Pathogens	• Chlorination • Hypochlorination • Ozonation • Land treatment systems
Nutrients: a) Nitrogen	• Suspended-growth nitrification and denitrification variations. • Fixed-film nitrification and denitrification variations • Ammonia stripping • Ion exchange • Breakpoint chlorination • Land treatment systems
b) Phosphorus	• Metal-salt addition • Lime coagulation/sedimentation • Biological-chemical phosphorus removal • Land treatment systems
c) Nitrogen and phosphorus	• Biological nutrient removal • Carbon adsorption
Refractory organics	• Tertiary ozonation • Land treatment systems
Heavy metals	• Chemical precipitation • Ion exchange • Land treatment systems
Dissolved inorganic solids	• Ion exchange • Reverse osmosis • Electrodialysis

1.3.1 Selection of Treatment Processes

1. The removal of all wastewater contaminants will be achieved only through the various treatment operation units. Selection of the most probable appropriate treatment sequences will provide more desirable treated wastewater.
2. The two general categories of approach to develop the treated wastewater are physical/chemical treatment and biological treatment. The essential difference between the

Table 1.10 Selective list of unit processes used for particular waste constituents.

Unit Process	Removal Mechanism	Waste Constituent
Sedimentation/ flotation	Gravity	Solid phase organics/inorganics
Coagulation/ sedimentation	Particle aggregation/gravity	Solid phase organics/inorganics
	Chemical bonding	Colloidal phase organics/inorganics
		Colloidal phase inorganics
Biological treatment	Particle aggregation/ biological	Solid phase organics/inorganics
		Colloidal phase organics/inorganics
	Metabolism/gravity	Soluble phase biodegradable organics
Filtration	Entrapment	Solid phase organics/inorganics
	Particle aggregation/ adsorption	Colloidal phase organics/inorganics
Carbon adsorption	Adsorption	Soluble phase adsorbable organics/inorganics
	Entrapment	Solid phase organics/inorganics
	Particle aggregation/ adsorption	Colloidal phase organics/inorganics

capabilities of a physical/chemical process and a biological process is the ability of each to remove certain types of organic materials.

The physical/chemical process is subject to apparent inefficiencies caused by a certain amount of non-adsorbable organics in the wastewater. The biological process is subject to apparent inefficiencies as a result of non-biodegradable organics in the wastewater. A selective listing of unit processes and the waste constituents for which they are generally applied and/or are effective is shown in Table 1.10.

Assembly of the most applicable process train is based on a full knowledge of the wastewater's condition and constituents.

In general, chemical/physical treatment is a suitable alternative:
- for a waste having a high particulate organic concentration, provided the soluble organic concentration, following chemical coagulation, sedimentation, and filtration, is less than 50 mg/L BOD5;
- for wastewater treatment systems where no influent flows will be received for substantial periods of time, for example batch treatment or systems experiencing significant flow variations;
- if land space is limited or toxic substances present in the raw wastewater.

Care should be exercised in the application of chemical-physical treatment systems to medium to high strength wastes (BOD_5 greater than 200 mg/L). For this situation, on-site pilot studies are desirable to determine the effluent quality that can be obtained and to ascertain if the biological activity anticipated in the carbon column will be more of a detriment (odor, plugging) than an asset (higher organic removal).

3. Land application of wastewater is viewed as an alternative to other secondary treatment schemes or as a final add-on step for liquid disposal and convenient water use. Alternative land disposal methods include various modes of surface and subsurface percolation and deep well injection.

Table 1.11 *Treatment processes used for the removal of toxic compounds.*

Process	Removal Application
Activated-carbon adsorption	Natural and synthetic organic compounds including VOCs; pesticides; PCBs; heavy metals
Activated-sludge powdered activated carbon	Heavy metals; ammonia; selected refractory priority pollutants
Air stripping	Volatile organic compounds (VOCs) and ammonia
Chemical coagulation, sedimentation, and filtration	Heavy metals and polychloro-biphenyls (PCBs)
Chemical oxidation	Ammonia; refractory and toxic halogenated aliphatic and aromatic compounds
Conventional biological treatment (activated-sludge, trickling-filter)	Phenols; PCBs; selected hydrogenated hydrocarbons

    Combination land disposal and wastewater reclamation methods include infiltration-percolation, overland flow, irrigation, and groundwater recharge.

4. Many treatment methods can be used for toxic compounds. Because of the complex nature of toxicity, the treatment method must consider the specific characteristics of the wastewater and the nature of the toxic compounds.

    Treatment processes used to remove some of these specific compounds or groups of compounds are summarized in Table 1.11.

Various combinations of unit operations and processes and their interaction for the treatment of refinery wastewater are identified in Table 1.12. A summary of treatment methods for petrochemical wastes are also presented in Tables 1.13, 1.14, and 1.15. The selection of a process train or alternative process trains should be made on the basis of the ability of the individual unit processes to remove specific waste constituents.

1.3.1.1 *Important Factors in Process Selection*

The most important factors that must be considered when evaluating and selecting unit operations and processes are identified below:

1. **Process applicability**

   The applicability of a process should be evaluated on the basis of past experience, published data, data from full scale plants and from pilot plant studies. If new or unusual conditions are encountered, pilot plant studies are essential.

2. **Applicable flow range and flow variation**

   The process should be matched to the expected range of flow-rates. For example, stabilization ponds are not suitable for extremely large flow-rates. Most unit operations and processes have to be designed to operate over a wide range of flow-rates.

   Most processes work best at a relatively constant flow-rate. If the flow variation is too great, flow equalization is necessary. Table 1.16 identifies critical design and sizing factors for secondary treatment plant facilities and describes the potential performance impacts of flow-rate and constituent mass-loading variations.

Table 1.12 Refinery treatment sequence options.

Pre-treatment	Primary Treatment	Secondary Treatment	Tertiary Treatment	
Removal of phenolics, S-, NH_3, RSH, F-, acid sludge, oil, etc., water reuse; waste equalization	Removal of free oil suspended solids	Removal of emulsified oil, suspended and colloidal solids	Removal of dissolved organics (variable) produced biological sludge	
		Processes		
Unit separators Steam stripping Fuel gas stripping Air oxidation Neutralization Surge ponds	API separators CPI, PPI separators (Note: CPI = corrugated plate interceptor: PPI = parallel plate interceptor.)	Chemical coagulation and air flotation Chemical coagulation and filtration pH control Reduction of intermediate oxygen demand Equalization of wastes	Trickling-filter Activated oxidation pond Aerated lagoon Rotating biological contactors	Chemical coagulation and air flotation Chemical coagulation and filtration Carbon adsorption Carbon adsorption
Inhibitory contaminant surge control	↓Sludges	↓Sludges	↓Sludges	↓Sludges

Design provisions for flow-rate variations, in addition to flow equalization, may include flow splitting and unit process bypassing under certain peak flow-rate conditions. Minimum treatment requirements, if permitted by regulatory authorities, may include primary treatment and disinfection of the entire flow and secondary treatment of a portion of the flow. The advantages of a unit process flow-splitting and bypassing strategy are that:

o the biomass of the secondary treatment process can be preserved during peak storm conditions and not lost due to washout;
o the quality of the treatment plant effluent can be restored quickly after the storm event; and
o the entire treatment facilities need not be oversized to handle unusual events.

A disadvantage of flow-splitting and bypassing is that the effluent quality may violate the discharge permit for short periods of time.

However, any treatment sequence designed for flow-splitting and unit bypassing should be investigated in advance to ensure it meets with environmental regulation requirements.

Table 1.13 Summary of physical treatment methods for petrochemical wastes classified by plant product.

Plant Product	Physical Treatment						
	Sedimentation	Filtration	Separators (API)	Stripping	Absorption and Extraction	Evaporation	Submerged Comousion
General chemicals	✓			✓	✓		
Nylon							
Nylon chemical intermediates	✓	✓		✓		✓	✓
Organic chemicals			✓			✓	
Photochemicals		✓					
Resins			✓		✓		
Rocket fuels			✓				
Synthetic rubber							

Table 1.14 Summary of chemical and biological treatments methods for petrochemical wastes classified by plant product.

Plant Product	Chemical Treatment		Biological Treatment		
	pH Adjustment	Chemical Oxidation	Biological Filters	Activated Sludge	Lagoons
General chemicals	✓	✓			✓
Nylon			✓		
Nylon chemical intermediates		✓	✓		✓
Organic chemicals	✓				
Oxygenated hydrocarbons				✓	
Photochemicals			✓		
Powders			✓		
Resins			✓		✓
Rocket fuels	✓		✓		
Rubber, textiles, and plastics				✓	

3. **Influent wastewater characteristics**
   The characteristics of an influent wastewater affect the types of processes to be used (e.g., chemical or biological) and the requirements for their proper operation.
4. **Inhibiting and unaffected constituents**
   Identification should be made of:
   - The constituents that are present.
   - The constituents that may be inhibitory to the treatment processes.
   - The constituents that are not affected during treatment.

Table 1.15 Summary of ultimate disposal methods for petrochemical wastes classified by plant product.

Plant Product	Ultimate Disposal						
	Controlled Dilution to Streams and Bays	At Sea	On Land Surfaces	Dumping or Burial	Deep Wells	Incineration	Salvage
Nylon			✓		✓	✓	✓
Nylon chemical intermediates	✓					✓	
Synthetic rubber	✓					✓	

Table 1.16 Effect of flow-rates and constituent mass loadings on the selection and sizing of secondary treatment plant facilities.

Unit Operation or Process	Critical Design Factor(s)	Sizing Criteria	Effects of Design Criteria on Plant Performance
Wastewater pumping and piping	Maximum hour flow-rate	Flow-rate	Wet well may flood, collection system may surcharge, or treatment units may overflow if peak rate is exceeded.
Screening	Maximum hour flow-rate	Flow-rate	Head losses through bar rack and screens increase at high flow-rates.
	Minimum hour flow-rate	Channel approach velocity	Solids may deposit in the approach channel at low flow-rates.
Grit removal	Maximum hour flow-rate	Overflow rate	At high flow-rates, grit removal efficiency decreases in Flowthrough-type grit chambers causing grit problems in other processes.
Primary sedimentation	Maximum hour flow-rate	Overflow rate	Solids removal efficiency decreases at high overflow rates; increases loading on secondary treatment system.
	Minimum hour flow-rate	Detention time	At low flow-rates, long detention times may cause the wastewater to become septic.
Activated sludge	Maximum hour flow-rate	Hydraulic residence time	Solid washout at high flow-rates; may need effluent recycling at low flow-rates.
	Maximum daily organic load	Food/microorganism ratio	High oxygen demand may exceed aeration capacity and cause poor treatment performance.
Tickling-filters	Maximum hour flow-rate	Hydraulic loading	Solids washout at high flow-rates may cause loss of process efficiency.
	Minimum hour flow-rate	Hydraulic and organic loading	Increased recycling at low flow-rates may be required to sustain process.
	Maximum daily organic load	Mass loading/media Volume	Inadequate oxygen during peak load may result in loss of process efficiency and cause odors.
Secondary sedimentation	Maximum hour flow-rate	Overflow rate or detention time	Reduced solids removal efficiency at high overflow rates or short detention times.
	Minimum hour flow-rate	Detention time	Possible rising sludge at long detention times.
	Maximum daily organic load	Solids loading rate	Solids loading to sedimentation tanks may be limiting.
Chlorine-contact tank	Maximum hour flow-rate	Detention time	Reduced bacteria extermination at reduced detention time.

5. **Climatic constraints**
   Temperature affects the rate of reaction of most chemical and biological processes. Temperature also affects the physical operation of facilities. Warm temperatures may accelerate odor generation and also limit atmospheric dispersal.
6. **Reaction kinetics and reactor selection**
   Reactor sizing should be based on the governing reaction kinetics. Data for kinetic expressions are usually derived from experience, published literature, and the results of pilot plant studies.
7. **Performance**
   Performance is usually in terms of effluent quality, which must be consistent with the effluent discharge requirements.
8. **Treatment residuals**
   The types and amounts of solid, liquid, and gaseous residual produced should be known or estimated.
9. **Sludge processing**
   If there are any constraints that would make sludge processing and disposal infeasible or expensive, these should be identified. The extent of recycle loads from sludge processing that affect the liquid unit operations or processes should also be clarified.
10. **Environmental constraints and regulations**
    Environmental factors, such as prevailing winds, wind direction, and proximity to residential areas, may restrict or affect the use of certain processes, especially where odors may be produced.
    Noise and traffic may affect selection of a plant site. The receiving waters may have special limitations, requiring the removal of specific constituents. The characteristics of the treated water imposed by the final destination and/or environmental regulations will dictate special unit operations and processes for treatment of wastewater.
11. **Chemical requirements**
    The resources and the amounts that must be committed for a long period of time for successful operation of the unit operation or process need to be clarified. The effects that the addition of chemicals might have on the characteristics of the treatment residuals and the cost of treatment should also be determined.
12. **Energy requirements**
    The energy requirements, as well as probable future energy costs, must be known if cost-effective treatment systems are to be designed.
13. **Operating and maintenance requirements**
    Any special operating or maintenance requirements needed, along with the necessary spare parts and their availability and cost, should be determined.
14. **Reliability**
    The question of what the long-term reliability of the unit operation or process under consideration is must be answered. Is the operation or process easily upset? Can it withstand periodic shock loadings? If so, how do such occurrences affect the quality of the effluent? Because of variations in effluent quality performance, a treatment plant should be designed to produce an average effluent concentration below the permit requirements.
15. **Compatibility**
    The unit operation or process should be able to be used successfully within the existing facilities. Plant expansion should be able to be accomplished easily.

16. **Land availability**
    Sufficient space should be allocated to accommodate either the facilities currently under consideration or possible future expansion.
17. **Equipment availability**
    Most of the equipment used in wastewater treatment is custom manufactured, except for items such as small pumps, motors, and valves. Some items of equipment may require special manufacturing techniques or are proprietary and only available from limited sources. Therefore, the equipment components that make up the process or system should be considered carefully to determine their potential effects upon the design, construction, and operation and maintenance of the facilities.
18. **Personnel requirements**
    The selection of a treatment process should consider not only the number of operating and maintenance personnel needed but also the skills required.

    The extent and complexity of the control systems and the staffing levels required must be evaluated carefully.

1.4   Chemical Oxygen Demand (COD) in Wastewater Systems

This section first discusses the background of the COD method, then the method to calculate the theoretical oxygen demand. Chemical oxygen demand (COD) is the equivalent amount of oxygen consumed under specified conditions in the chemical oxidation of the organic and oxidizable inorganic matter contained in wastewater, corrected for the influence of chlorides. In American practice, unless otherwise specified, the chemical oxidizing agent is hot acid dichromate.

Microorganisms in natural water bodies consume oxygen when they degrade organic matter to biomass. Biochemical oxygen demand (BOD) is a parameter that describes the oxygen consumption when microorganisms "eat" organic mass. The more organic load enters the water body, the more oxygen the microorganisms will use for conversion. Large amounts of pollutants may cause oxygen loss in a water body and lead to harmful effects in nature so it is useful to know the oxygen consumption of wastewater before discharge.

BOD requires days for results and COD is used because it is a lot faster. So called standard methods for COD determination are open and closed reflux systems, where a strong oxidant is added to the water solution and is boiled in open or closed reflux for a few hours and then the amount of used oxidant is measured by titration methods. COD determination with standard methods usually takes a few hours, but colorimetric determination of COD requires less than an hour. COD measures the organic content of a solution.

Due to its rapidity, COD is used to measure the total amount of pollutants in water media. It is often used in water and wastewater quality determination.

The difference between COD and BOD is that, in COD, the amount of oxidant used for oxidation of organic components is measured, while BOD expresses the oxygen consumed by microorganisms when the sample is kept for five days at 20 °C. The dissolved oxygen is measured at the beginning and end of the test and the oxygen consumption is the difference between the amount of dissolved oxygen at the beginning and end. There is a correlation factor between BOD and COD so, when COD is known, the biochemical oxygen consumption in nature can also be determined.

1.4.1 Determination of the COD

Usually a strong oxidant, like potassium dichromate, is used in this measurement. The standard method for determination of the COD is titration with open or closed reflux, where the sample is heated either in a closed or an open vessel in the laboratory. The sample is boiled for two hours in the presence of the oxidant.

The amount of dichromate consumed can be determined when the difference in oxidant concentration at the beginning and end is determined. Potassium dichromate is considered to be the best oxidant because it has a strong oxidizing capability, it is applicable to many kinds of organic and inorganic matter, and it is easy to manipulate.

Oxidation of inorganic components interferes with COD determination because they also consume dichromate. In the oxidation process, the dichromate ion (Cr^{6+}) is oxidized to chromate (Cr^{3+}).

The study of Baker et al. (1999) declared that the COD for the chemicals, present in group one and group two well-correlated data groups, directly equalled the ThOD.

For chemicals in group three and group four (potentially well-correlated data) the COD could be estimated from the ThOD; and no correlation for chemicals presented in group five and group six could be determined.

1.4.2 Calculation of Theoretical Oxygen Demand

The Theoretical Oxygen Demand (ThOD) can be obtained for different chemicals from different studies. Because theoretical oxygen demand is not usually determined for all chemicals found in the wastewaters, the ThOD could be determined for every chemical with the method presented in the study of the Baker et al. (1999).

According to Baker et al. (1999) the amount of oxygen consumed by single component i can be determined with equations 1.1 and 1.2. Letters $n, m, k, j, i, h,$ and e can be determined with equation 1.1. Equation 1.1 assumes that all compounds are oxidized completely to end products. Letters tell us how many molecules there are of element i in component i.

$$C_n H_m O_e X_k N_j S_i P_h + bO_2 \rightarrow nCO_2 + \left(\frac{m-k-3j-2i-3h}{2}\right) H_2O + kH_x \\ + jNH_3 + iH_2SO_4 + hH_3PO_4 \quad (1.1)$$

$$b = n + \frac{m-k-3j-2i-3h}{4} - \frac{e}{2} + 2i + 2h \quad (1.2)$$

To determine the $ThOD_i$, the oxygen demand per one gram component i has to be determined. Oxygen demand per one gram component i is represented by $ThOD_{O,i}$ and determined with equation 1.3, where b_i is the amount of oxygen consumed in moles per one mole component i, M_i is the molar mass of a single component i, M_{O2} is the molar mass of an oxygen molecule, and c_i is the concentration of component "i" in water.

$$ThOD_{O,i} = b_i \frac{M_{O2}}{M_i} \quad (1.3)$$

$$ThOD_i = ThOD_{o,i} \times c_i \quad (1.4)$$

$$\sum ThOD = \sum_i^n ThOD_i = \sum_i^n \left(ThOD_{o,i} \times c_i\right) \quad (1.5)$$

Table 1.17 provides calculated oxygen demand per one mole component i, molar mass, and theoretical oxygen demand for various components.

Table 1.17 Oxygen demand per one mole component i, molar mass, and theoretical oxygen demand.

Component	Composition	Oxygen Demand Per One Mole Component i b	Molecular Weight M_i (g/mol)	Theoretical Oxygen Demand $ThOD_i$ (gO_2/g_i)
Methanol	CH_4O	1.5	32.03	1.5
Phenol	C_6H_6O	7	94.1	2.38
Acetone	C_3H_6O	4	58.07	2.2
2-Aminoethanol	C_2H_7NO	2.5	61.08	1.31
Benzene	C_6H_6	7.5	78.11	3.07
Cumene	C_9H_{12}	12	120.2	3.19
Propanols	C_3H_8O	5	44.1	3.63
Toluene	C_7H_8	9	92.14	3.12
Total Nitrogen	NH_3	0	17.03	0
Xylenes	C_8H_{10}	10.5	106.17	3.16

For example, Phenol's chemical formula is C_6H_6O. If the concentration of phenol is 360 mg/L in a wastewater sample, the theoretical oxygen demand determination for this phenol concentration is shown in the calculations below.

According to equation 1.1: $n = 6, m = 6, e = 1, k = 0, j = 0, i = 0, h = 0$

When n, m, k, j, i, h, and e are used in equation 1.2, the amount of oxygen molecules needed to oxidize phenol into end products can be determined.

$$b_{C_6H_6} = 6 + \frac{6 - 0 - 3 - 2 - 3}{4} - \frac{1}{2} + 2 + 2 = 7 \text{ mol } O_2/(mol \, C_6H_6) \quad (1.6)$$

$$ThOD_{O,i} = 7 \frac{mol \, O_2}{mol \, C_6H_6} \left(\frac{31.98 \frac{g C_6H_6}{mol \, C_6H_6}}{94.1} \right) = 2.38 \frac{g \, O_2}{g \, C_6H_6} \quad (1.7)$$

The total theoretical oxygen demand is determined by summing the single theoretical oxygen demands together.

$$ThOD_{C_6H_6} = ThOD_{o,C_6H_6} \times C_{C_6H_6} = 2.38 \frac{mg \, O_2}{mg \, C_6H_6} \times 360 \frac{mg \, C_6H_6}{Liter \, H_2O} = 856.8 \frac{mg \, O_2}{Liter \, H_2O}$$

$$(1.8)$$

2

Physical Unit Operations

Physical unit operations are those used for treatment of wastewater in which change is brought about by means of or through the application of physical forces.

The unit operations most commonly used in wastewater treatment and their applications are shown in Table 2.1.

2.1 Flow Measurement

Important criteria that must be considered in the selection of flow metering devices include the type of application, proper sizing, fluid composition, accuracy, headloss, installation requirements, operating environment, and ease of maintenance.

Because of rapid advances made in metering device electronics and converters, up to date information should be obtained from meter manufacturers.

Among the important criteria that must be considered in the selection of flow metering devices are accuracy and repeatability, which are critical especially when the readings from the metering devices are to be used for process control.

2.2 Screening

Suspended particles greater than 2 to 6 mm should be removed by screening. A screen should have openings of a uniform size of any shape. The screening element should consist of parallel bars, rods or wires, wire mesh, grating, or a perforated plate. Figure 2.1 shows a typical screening process.

An adequate number of units should be provided to facilitate continuous screening with permission of maintenance type, size classification, and application of screening devices as follows in Table 2.2.

Figure 2.2 is a typical bar screen that consists of a series of parallel bars or a perforated screen placed in a channel. The flow passes through the screen and the large solids are

Table 2.1 Applications of physical unit operations in wastewater treatment.

Operation	Application
Flow metering	Process control, process monitoring, and discharge reports
Screening	Removal of coarse and settleable solids by interception (surface straining)
Comminution	Grinding of coarse solids to a more or less uniform size
Gravity separation	Settling of oil droplets or solid particles by relative density (specific gravity) differences
Flow equalization	Equalization of flow and mass loadings of BOD and suspended solids
Mixing	Mixing chemicals and gases with wastewater, and maintaining solids in suspension
Flocculation	Promotes the aggregation of small particles into larger particles to enhance their removal by gravity sedimentation
Sedimentation	Removal of settleable solids and thickening of sludges
Flotation	Removal of finely divided suspended solids and particles with densities close to that of water. Also thickens biological sludges
Filtration	Removal of fine residual suspended solids remaining after biological or chemical treatment
Micro-screening	Same as filtration. Also removal of algae from stabilization-pond effluent
Gas transfer	Addition and removal of gases
Volatilization and gas stripping	Emission of volatile and semi-volatile organic compounds from wastewaters

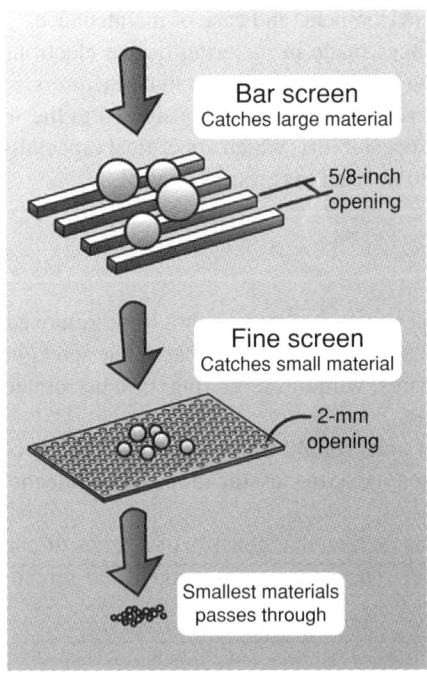

Figure 2.1 Illustration of screening process in water and wastewater treatment.

Table 2.2 Screening devices.

Type of Screening Device	Size Classification	Size Range (mm)	Application
Bar rack Screens:	Coarse	15–35	Pre-treatment
Inclined (Fixed)	Medium	0.25–2.5	Primary treatment
Inclined (Rotary)	Coarse	0.76 × 2.29 × 50	Pre-treatment
Drum (Rotary)	Coarse	2.5–5	Pre-treatment
	Medium	0.25–2.5	Primary treatment
	Fine	6–35 µm	Removal of residual secondary suspended solids
Rotary disk	Medium	0.25–10	Primary treatment
	Fine	0.025–0.5	Primary treatment
Centrifugal	Fine	0.05–0.5	Primary treatment, secondary treatment with settling tank, and the removal of residual secondary suspended solids

trapped in the bars for removal. The bar screen may be coarse (2–4 inch openings) or fine (2–5 cm openings). It may be manually cleaned or mechanically cleaned. Manual or mechanical cleaning is performed frequently enough to prevent solids buildup that would reduce flow into the plant. The water velocity through the screen is very important and should be around 1.5 ft/s (0.45 m/s). If the velocity decreases below 1 ft/s (0.3 m/s) or slower, grit will drop out of the flow and into the screening channel.

Figure 2.2 A typical bar screen that consists of a series of parallel bars or a perforated screen placed in a channel (Courtesy of City of Phoenix (Arizona) Water Services Department).

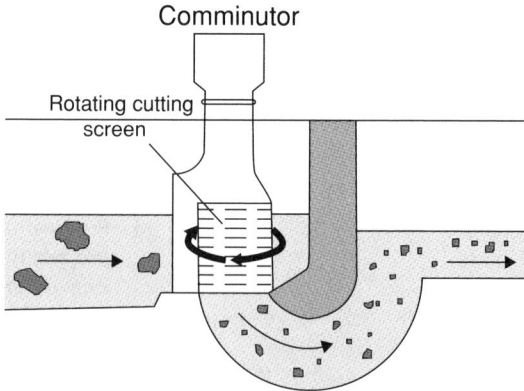

Figure 2.3 A schematic of a comminutor.

2.3 Comminution

As an alternative to racks or coarse screens, comminutors can be used to grind up coarse solids without removing them from the flow.

In this device, all of the wastewater flow passes through the grinder assembly. The grinder consists of a screen or slotted basket, a rotating or oscillating cutter, and a stationary cutter. Solids pass through the screen and are chopped or shredded between the two cutters. The comminutor will not remove solids that are too large to fit through the slots, and it will not remove floating objects. These materials must be removed manually. Figure 2.3 shows a schematic of a comminutor.

Comminutors should be constructed with a bypass arrangement so that a manual bar screen is used in case flow-rates exceed the capacity of the comminutor or in case there is a power or mechanical failure. Stop gates and provisions for draining should also be included to facilitate maintenance.

Manufacturer's data and rating tables for these units should be consulted for recommended channel dimensions, capacity ranges, upstream and downstream submergence, and power requirements. Because the manufacturer's capacity ratings are usually based on clean water, the ratings should be decreased by approximately 20 to 25% to account for partial clogging of the screen.

2.4 Grit Removal

Grit removal may be accomplished in grit chambers or by the centrifugal separation of sludge. The removal of grit is essential ahead of centrifuges, heat exchangers, and high pressure diaphragm pumps.

The purpose of grit removal is to remove the heavy inorganic solids, which could cause excessive mechanical wear. Grit includes sand, gravel, clay, egg shells, coffee grounds, metal filings, seeds, and other similar materials. There are several devices or processes used for grit removal. All of them are based on the fact that grit is heavier than the organic solids

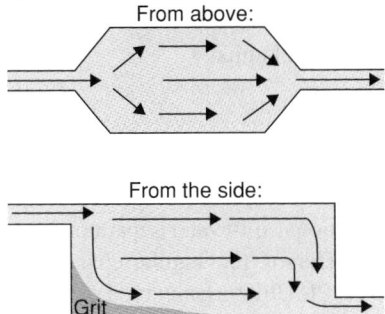

Figure 2.4 A schematic of a grit chamber in both above and side view.

that should be kept in suspension for subsequent treatment. Grit removal processes use gravity/velocity, aeration, or centrifugal force to separate the solids from the wastewater. Figure 2.4 shows a schematic of a grit chamber for grit removal.

2.5 Gravity Separation

2.5.1 General

The API separator is normally the first, and is arguably the most important, wastewater treatment step in most petroleum refineries. For years, refineries have attempted to use other technologies or treatment scenarios as an alternative to the API separator. But most refineries ultimately select, or return to, the API separator as the technology of choice for their wastewater treatment primary oil/solids separation stage.

Gravity differential type separators are used to remove most of the oil from the plant wastewater before discharging the water to the further treatment units. The oil–water separator is basically a holdup basin that reduces wastewater velocity and provides holdup time to allow the oil to rise to the surface, where it is removed by skimming.

The primary function of a properly designed API separator is to remove gross quantities of oil and suspended solids from refinery wastewater prior to subsequent downstream wastewater treatment processes, normally a second oil–water separator polishing step and some form of advanced treatment for the removal of dissolved organic compounds (typically biological treatment, though other treatment technologies have been used).

In some special cases, oil traps, oil holding basins, and/or water retention basins may be provided in order to lower the final oil–water separator load.

The effectiveness of the treatment facility depends on the design flow-rate, the water temperature, the density and size of the oil globules, and the amount of suspended matter. It also relies on the operating techniques, proper supervision, and maintenance.

The gravity-type separators will not separate or retain substances in solution and will not remove soluble BOD, nor will they break stable emulsions and therefore they should not be used for such applications. Oil–water separators will only remove free oil; emulsified and dissolved oils require additional treatment.

Oil–water separators are as follows:

1. Conventional, rectangular-channel units.
2. Parallel-plate separators.
3. Oil traps.
4. Oil holding basins.
5. Other types.

In this book, the term "conventional oil–water separator" is sometimes used in place of the term "API separator" and refers to rectangular channel units designed in accordance with the criteria published by API. Utilizing any type of oil–water separator in the oil and gas industries should be congruent with the local environmental regulations and have the approval of the relevant company. To design oil–water separators, the API-421 latest edition should be used in addition to the instructions stipulated in this book.

2.5.2 Application

Oil–water separators are designed to remove free oil only. If emulsified or dissolved oil is present, an oil–water separator will not be able to remove it, and additional downstream treatment will be required. The principal function of the oil–water separator is to remove gross quantities of free oil before future treatment.

In this capacity, the oil–water separator protects more sensitive downstream treatment processes from excessive amounts of oil.

Parallel-plate separators require less space than do conventional oil–water separators and are theoretically capable of achieving lower concentrations of effluent oil.

In some applications, the oil–water separator is provided as a protective device for the containment of spills and leaks (e.g., on once-through cooling water).

It should be stressed that, whenever an oil–water separator is considered for an application where it must stand alone, the amount of emulsified and dissolved oils in the wastewater stream must be properly quantified, because these oils will not be removed by the separators. Figure 2.5 shows a highland rectangular API oil–water separator.

2.5.3 Oil–Water Separator General Design Considerations

The following requirements should be taken into consideration at the design stage.

1. **Location:**
    - safety distance;
    - access from the road;
    - prevailing wind (when applicable);
    - possible future extensions.
2. **Function:**
    - two way access /escape for operator;
    - hand railing and grating;
    - easy access to skimmers, pumps, and filters;
    - proper ventilation for depressed locations;
    - lamp posts;
    - surface drainage;

Figure 2.5 A highland rectangular API Oil water separator (Reproduced with permission from © Can-Am Instruments Ltd).

- kerbs around pump foundations;
- water supply for flushing and/or pump seal cooling;
- maintenance access for vacuum truck/mobile crane;
- signboard with unit number and instructions;
- high level alarm in oil collecting sumps in the control room;
- heating coils in cold climate conditions.

3. **Safety distances:**
   - distance to the edge of public roads: 30 m
   - distance to the edge of main roads: 15 m
   - distance to a fixed source of ignition: 15 m.

4. **Air pollution control:**
   Effective means of control should be instituted to minimize losses of hydrocarbon and other contaminants, such as sulfur compounds, to the atmosphere from the large exposed surface area of the separators if required by the environmental regulations.

   Control of hydrocarbon or contaminant emissions from oil–water separators may be achieved by the covering of the forebays or primary separator sections by either fixed roofs or floating roofs. The roof should be as vapor tight as possible.

5. **Screening:**
   Screens should be installed at the oily sewer outlet (oil–water separator influent) in order to stop and manually remove rags, stones, and other debris that would interfere with the operation. The screens should be equipped with a hot dip galvanized steel frame, lateral rails, a movable box for solids removal, and a stainless steel removable screen. A hoist for screen removal for the cleaning operation should be provided.

2.5.4 Conventional Rectangular Channel (API) Separators

2.5.4.1 Basic Design Considerations

Conventional oil–water separators (Figures 2.5 and 2.6) are simple rectangular channel tanks that provide a nominal retention time to remove larger oil molecules. This type of

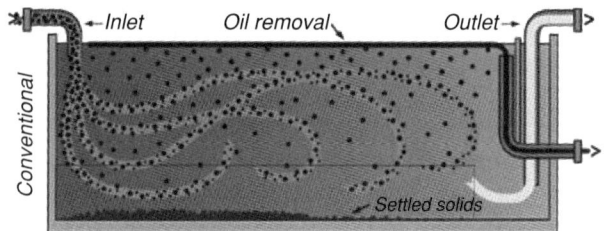

Figure 2.6 A conventional API Oil water separator (Image provided courtesy of Brentwood Industries Inc. All rights reserved).

equipment has been found to only remove free oil droplets with a diameter greater than or equal to 150 microns. These separators typically do not achieve effluent oil concentrations of less than 100 ppm because of the abundance of smaller oil droplets that are harder to remove. Ideal separators are those that have no turbulence, short circuiting, or eddies.

If particles fall through the viscous fluid by their own weight due to gravity, then a terminal velocity, also known as the settling velocity, is reached when this frictional force, combined with the buoyant force, exactly balances the gravitational force. The resulting settling velocity (or terminal velocity) is given by:

$$v_s = \frac{2\left(\rho_P - \rho_f\right)gR^2}{9\mu} \qquad (2.1)$$

Where:

- v_s is the particles' settling velocity (m/s) (vertically downwards if $\rho_p > \rho_f$, upwards if $\rho_p < \rho_f$)
- g is the gravitational acceleration (m/s^2)
- ρ_p is the mass density of the particles (kg/m^3)
- ρ_f is the mass density of the fluid (kg/m^3)
- R is the radius of the spherical object (in m) and
- μ is the dynamic viscosity (N s/m^2)

Stokes' Law makes the following assumptions for the behavior of a particle in a fluid:

- laminar flow
- spherical particles
- homogeneous (uniform in composition) material
- smooth surfaces
- particles do not interfere with each other.

Free oil can be removed by gravity separation given proper quiescent flow conditions. Gravity oil–water separators exploit the differences in specific gravity between the fluids by providing an adequate retention time for the less dense oil globules to rise to the surface of the water. The theory of this type of separation is based on the rise rate of oil droplets in the water and the surface-loading rate of the separator. The surface-loading rate is the ratio of the flow-rate to the separator and the surface area of the separator. If the oil globule rises towards the surface of the separator faster than the surface loading rate, the oil will reach the surface and can be skimmed off mechanically.

The required rise rate can be found using Stokes' Law.

The design of oil–water separators should carefully consider these characteristics of the wastewater and the oil itself. Additionally, the separator design should take into account the amount of settleable solids in the wastewater and other contaminants, such as surfactants, which might hinder the treatment of the system. Often, many of these characteristics must be assumed based on the type of facility generating the wastewater.

The minimum requirements for the design of API separators should be as per the procedure outlined in API publication 421 (latest edition) and the following design notes:

1. A minimum of two parallel channels should be included, so that operation can be continued when one channel is removed from service for repair or cleaning. The separator shall be designed for the design flow with all the channels in service.
2. Provision should be made for separator expansion to meet possible future requirements in design of separators for new installations.
3. The unit should be particularly amenable to expansion, such that additional channels can be added without interruption of service or major construction changes to the existing facility.
4. Design of the separators should be made based on the worst case ambient conditions (e.g., minimum ambient temperature, affects of wind velocity, etc.).
5. The main separators should be designed to guarantee an outlet oil that is "Suspectible To Separation (STS)" in the range of 50–70 mg/L in worst case ambient conditions. The maximum outlet oil content should be reduced to 50 mg/L for areas with a minimum ambient temperature of about zero degrees Celsius or higher. Effluent free oil concentration should be 50 mg/L maximum for crude desalter oil–water separators for all areas.
6. The total oil and grease content (only free oil) of the streams concerned should be determined by United States Environmental Protection Agency's (EPA's) Methods 413.1 "Gravimetric Separation"; and 413.2, "Infrared Spectrophotometry." ASTM Method of D-3921, "Test Method for Oil and Grease for Petroleum Hydrocarbons in Water" can also be used in absence of the EPA's Methods. The EPA 413.1 method is applicable for an oil concentration range of (50–1000) mg/L and non-volatile hydrocarbons, and method EPA 413.2 is applicable for an oil concentration range of (0.2–1000) mg/L and volatile hydrocarbons.
7. The necessity for separator covers and the type of coverage (fixed or floating) will be advised in accordance with the environmental pollution regulations. However, special attention should be made for provision of the covers for oil–water separators that receive more than 800 L per day of volatile organic compounds.
8. The separators shall be installed such that oily water from the industry oily water sewer can flow via gravity to the separators. Influents to the separators shall be below grade.
9. In order to minimize the emulsification of oil and the remixing of the separated oil in the wastewater flow, pumping of separator influent should be avoided.
10. A holdup basin should be provided at upstream of the main separator(s) for the dumping of sudden fluctuations in the receiving oily waters. The basin should be designed for

a minimum capacity of 3 hours holdup of influent oily water to the separator(s) at maximum flow-rate conditions.
11. The basin should be designed with a suitable cover for environmental pollution control purposes.
12. Each separator should consist of either a pre-separation or an entry flume and two parallel channels. The pre-separation flume should be common to the both channels.
13. Provision of coalescers in the separators is not acceptable.
14. The effective length of the channels should not be less than 40 m for oil and gas processing industry main separator(s). However, the design should be based on separating oil droplets of sizes down to 100 micrometres (μm) in diameter for all oil–water separators.
15. The following characteristics of wastewater should be evaluated for the design of oil–water separators as a minimum requirement:
    - free oil;
    - solids content;
    - relative densities (specific gravities) of oil and water phases;
    - absolute viscosity of wastewater.
16. The relative densities (specific gravities) and viscosity should be evaluated at the minimum design temperature.
17. The separators should be provided with floors (other than earth). Floors should allow the use of sludge scrapers, provide leakage of groundwater, and avoid contaminating the groundwater.
18. In materials selection, consideration should be given to corrosion, leakage, structural strength, buoyancy (height with local water table vs. mass of separator), etc. All parts in contact with fluid should be manufactured in dip galvanized steel and/or red wood.
19. Antispark material should be used for metal to metal contacts in moving parts. Figure 2.6 shows a schematic of a conventional API oil–water separator.

2.5.4.2 An API Separator's Components

Figure 2.7 shows an oil–water separator. API and oil–water separators should contain, but not be limited to, the following components:

2.5.4.2.1 Pre-Separator Flume.
The pre-separator flume is the transition between the end of the inlet sewer and the separator forebay and will be common to both channels. It serves two functions: reduction of flow velocity and collection of floating oil.

The transition and separator sections of the pre-separator flume must be covered to reduce evaporation loss. To achieve this, the vapor space should be enclosed by a barrier or wall at the downstream end of the cover.

The transition between the sewer outlet and the pre-separator section should be designed to accomplish velocity reduction with a minimum of turbulence. The pre-separator section should be designed to reduce the horizontal flow velocity to about 3 to 6 m per minute (m/min).

The pre-separator section contains two kinds of equipment as a minimum: a trash or bar screen and a floating oil skimmer.

Figure 2.7 An oil–water separator (Reproduced with permission from © Skim oil™, Inc).

1. **Trash rack**

   Trash racks or bar screens, which are provided to remove sticks, rags, stones, and other debris, should consist of a series of bars or rods and should be located at the entrance to each separator flume. The bars should be 1 mm by 5 mm with clear openings of 1.9 mm to 2.5 mm. The bars should be spaced at 25–50 mm center to center, at an angle of 45 to 60° from the horizontal, depending upon the depth of the flume and space availability. A pan or trough should be provided at the top of the trash rack to receive the refuse when the trash rack is cleaned. The refuse pan should be perforated to allow liquid to drain back into the flume.

2. **Oil skimmer**

   An oil skimmer should be installed downstream of the trash rack for removal of oil from the inlet section. A rotatable slotted-pipe skimmer (1.2.2 below) and a floating type oil skimmer (2.1 below) should both be employed in the inlet section.

   2.1. **Floating oil skimmers**

   Floating oil skimmers are applicable for installation in the pre-separator section particularly when the liquid level is expected to vary significantly. The specifications stipulated below are typical and should be followed for all other types of floating oil skimmers as applicable.

   2.1.1. **Floating vortex oil skimmer**

   The floating vortex oil skimmer installed in the pre-separation flume should be equipped with a vortex oil device and have the following characteristics:
   - It should remove at least 85% of oil entering each basin.
   - It should be of the floating type and each one should consist of a vortex generator and drive, generator support structure, float assembly, control station, pump and pump drive.
   - The vortex generator should be equipped with a propeller set on a vertical axis driven by an electric motor to create a vortex to collect floating oil, which is then recovered by pumping.

- Each unit should be equipped for automatic, unattended operation by addition of a simple reliable oil level indicator electrode control system, which serves as an automatic switch to operate the pump when the vortex pocket contains oil. The vortex oil skimmers should be capable of continuous operation.
- The vortex generator should consist of a cylindrical skirt surrounding a vortex generating impeller. The impeller should utilize a conical inlet section and a lower zero pitched propeller. The bottom of the generator should consist of a flat plate separated from the skirt section, both forming an annular opening to flow through the generator.
- The generator should be mounted on a float structure with three equally spaced support rods. The impeller, skirt, and base plate should be of carbon steel construction with two coat epoxy paint for corrosion resistance. A hollow steel pipe should be provided to connect the impeller to its drive motor. The pipe should be seated at the base of the impeller in a Teflon bearing mounted on the base plate. The pipe should be perforated at its upper end and fitted with a sleeve to control the flow level into the pipe. The flow into the pipe should be directed through the base plate and piped laterally to the pump mounted on the generator skirt.
- The pump should be a direct coupled centrifugal pump. The pump casing and bearing cover should be made of bronze. The impeller should be dynamically balanced and keyed to the pump shaft. The pump should have sufficient head to send the oil to the slops tanks.
- The float assembly should consist of three fiberglass float structures attached to the vertical legs of the support structure. The fiberglass floats should be adjustable in the vertical plane, to allow for proper skimming depth below the surface of the water. Changes in waste stream composition, which cause liquid density variations, should be taken into consideration in the design of the float assembly.
- All electrical equipment and wiring should be fabricated and installed for Class I, Group D Division 1 Area hazardous locations in accordance with API-RP-500A (latest edition). All motors should be furnished with drip covers and lifting lugs. Motor drain plugs should be located at the lowest point so that accumulated moisture may be drained. Each motor should be have a stainless steel nameplate.
- Control panels should be explosion-proof and suitable for installation within non-ventilated, 220 volts, 50 Hz motor control circuitry.
- Space heaters or other means of heating should be provided on both motors and the control panel. The space heating elements should be independently controlled by integral thermostats set at predetermined temperatures to prevent excessive condensation within the control panels during periods of cold weather when the unit is stored and not in use.
- All bolts, nuts, screws, and other necessary connecting devices should be stainless steel.
- The pump discharge should be provided with 6 m of DN 50 (2 inch) PVC hose with quick disconnect couplings. A rigid single point lifting

fixture should be attached to the assembly for permanent positioning and handling.
- The following requirements should be guaranteed by the vendor when the equipment is operated in accordance with the written operating instructions:
  - The vortex oil skimmers should remove not less than 85% of oil entering each basin.
  - The equipment should be free from fault in design, workmanship, and material to fulfil satisfactorily the specified operating conditions.

2.1.2. Rotary drum oil skimmer

A rotary drum oil skimmer (Figure 2.8) can provide large-scale solutions for pit and sump oil removal for metal foundries and oil refineries, as well as industrial facilities and consumer product manufacturers.

The rotary drum oil skimmer consists of a drum or unslotted pipe, mounted in a horizontal position, partially submerged below the surface of the settled oil layer. The drum is rotated by an external motor; as the drum rotates it picks up a film of oil that adheres to the drum surface. The oil film is removed by a scalpel and directed to a trough.

A rotary drum oil skimmer may be used in the following applications:
- for continuous and automatic operation;
- for minimizing water skimmed off with the oil.

Figure 2.8 *A rotary drum oil skimmer (Reproduced with permission from © Skim oil™, Inc).*

The disadvantages of the rotary drum oil skimmer are as follows:
- limited oil removed; cannot handle massive spills;
- has problems with heavy greases and heavy objects, which slide off;
- floating debris can interfere with oil pick-up;
- requires maintenance for rotating machinery;
- does not skim floatable solids.

The rotary drum skimmer should not be used if a large amount of oil is expected. When rotary drum skimmers are selected, a variable speed drive and a Teflon wiper blade shall be provided.

2.1.3. Horseshoe-type floating skimmer

The horseshoe-type floating oil skimmer consists of a floating collecting pan buoyed by hollow chambers on three sides (see API Publ.-421). The fourth side is open, contains an oil-skimming weir, and faces in the upstream flow direction to skim on-coming oil. This skimmed oil flows out through a pipe or a hose to a sump or other reservoir.

The advantages and disadvantages of a horseshoe-type floating skimmer are as detailed Table 2.3.

The slotted-pipe skimmer is normally the preferred type of skimmer for separator channel sections. It has the ability to remove the large amounts of oil that could be encountered in the event of a massive spill. The recovered oil should be drained to a slop oil pit at one side of the unit.

2.5.4.2.2 Separator Forebay. The pre-separator flume discharges into the separator forebay, which distributes the influent to the separator channels. If an upstream grit collector is lacking, sludge is likely to be deposited in the forebay. In the absence of an upstream grit collector, means should be provided for removing or transferring the sludge to the separator for subsequent removal, particularly if a reaction-jet diffusion device is used. Water jets can be used to flush solids from the forebay into the separator zone. This practice can affect the quality of the effluent unless the separator channel is blocked off.

Alternatively, the reaction-jet diffusion device can be put at the floor level of the forebay to allow solids to be scoured out of the forebay to the separator channels for collection. A slotted-pipe skimmer should be provided in the forebay, depending on whether or not oil is trapped by the flow distribution devices.

Table 2.3 Horseshoe-type floating skimmer advantages and disadvantages.

Advantages	Disadvantages
Simple	Requires manual operation
Economical	Removes relatively large amounts of water with the oil
Low maintenance	Usually not continuous
No utilities required	Operated with only a limited variation in liquid level
High capacity	
Can remove floating solids in addition to oil	

1. **Oil retention baffle**
   An oil retention baffle should be located immediately downstream of the slotted-pipe oil skimmer.
   Spacing between the oil skimmer and baffle should not exceed 150 mm. The oil retention baffle should be high enough to prevent oil from flowing or splashing over it. Submergence of this baffle should not exceed 450 mm. The baffle should be extended to the top of the channel.
2. **Separation section**
   2.1. **Gateways**
   Each channel should be provided with one or more sluice gates at its inlet to allow the flow to the channel to be shut off when desired. A hoist should be provided for gate removal. The gate frame and slots should be of suitable corrosion resistant and erosion resistant materials.
3. **Velocity head diffusion devices**
   A diffusion device should be immediately downstream of the inlet gateway, to distribute the flow equally over the cross-sectional area of the channel and to reuse flow turbulence. Two types of diffusion devices are available: vertical slot baffles and reaction jets. Only the reaction jet type should be used.
   3.1. **Vertical slot baffle**
   A vertical slot baffle is formed from vertical posts.
   This type of distributor suffers from severe fouling and should not be used.
   3.2. **Reaction jet inlet**
   The reaction jet inlet, which is the preferred diffusion device, introduces and distributes the influent appropriately.
   The reaction jet consists of a tube or orifice and a dished baffle; the concave surface of the baffle faces the tube or orifice. Water flows through the tube (orifice), is reversed by the baffle, and impinges on the inlet wall of the separator, dissipating the velocity head and distributing the flow.
   Orifices are generally used if the separator forebay is large and the normal direction of flow is toward the inlet wall. Tubes are used when approach velocities come from abnormal directions toward the inlet wall and exceed 9 m/min.
   Reaction jets have a number of advantages:
   - They are less subject to clogging than vertical slot baffles.
   - They provide a good distribution over a wide range of flow-rates.
   - They are cheaper than vertical slot baffles.
   - They can be shut off easily, since they are relatively small orifices in solid barriers, thus obviating gates or dams for interrupting the flow for maintenance.
   - They may result in less oil in the channel effluent than vertical slot baffles.

   Reaction jet inlets should be designed on the following basis in addition to the requirements outlined in API-421.
   - Reaction jets should be made up of a # 10 gauge thick stainless steel plate formed to proper radius, and should be completely mounted in front of the inlet pipes.
   - At least four reaction jets should be provided for each API channel and two for each oil/water channel.
   - Reaction jet wall sleeves and inlet guide vanes should be provided to mount the reaction jets to the influent wall.
   - The maximum spacing between the reaction jets should be 1.5 m.

2.5.4.2.3 Oil and Sludge Moving Devices. The separator channels contain a mechanical device to move the separated oil and sludge to the collecting area. The floating oil is moved to the downstream end of the separator and the settled sludge is moved to the upstream end. Two oil and sludge moving devices are available: the traveling bridge (span) type and the flight scraper or chain type.

1. **Traveling bridge or span type**

   This type of moving device consists of one or two blades extending across the width of the channel and hung from a beam or truss spanning the channel.

   The span rests on wheels in a carriage arrangement; the wheels rise on rails at the sides of the channel and travel the length of the basin. The wheels are chain driven and the rails are located on top of the channel walls.

   The blades are adjusted to sequentially skim oil on the downstream movement and scrape sludge on the upstream flight. The one-blade arrangement accomplishes this by adjusting the height of the blade to either the oil level or channel bottom using a hoist or cam mechanism. The blade pivots on the underside of the bridge as the height is changed.

   The two blade arrangement achieves the sequential movement of material by one of two means. During oil movement, the sludge blade can be moved out of the channel or it can be feathered parallel to the channel bottom. At the end of the downstream run, the oil moving blade is raised out of the water and the sludge-moving blade is properly positioned to move sludge upstream on the return trip.

   Surface travel is typically on the order of 0.6 m per minute and bottom travel is on the order of 0.3 m per minute.

   This device can span and operate in one or several channels. Operation can be manual or continuous; but is preferably automatic and actuated intermittently by a cycle time.

   The traveling bridge type oil/sludge moving device offers the following advantages:
   - the parts requiring lubrication are located above the water;
   - it allows different travel speeds on the forward and reverse runs.

   The disadvantages are as follows:
   - it is more expensive than other types;
   - it requires a movable power cable; usually either a cable reel or a festoon system using a cable looped from a supporting wire;
   - it complicates covering the channels;
   - the design is not common and has not been used on units equipped with covers.

   A design specification for a traveling bridge type device should include the following features:
   - provide full width skimming;
   - provide totally enclosed, explosion-proof, weather-proof drivers;
   - provide overload protection;
   - provide facilities for leveling the blades;
   - provide a means to raise blades out of the water for maintenance.

2. **Flight scraper or chain type**

   The flight scraper type of oil and sludge-moving device consists of two parallel endless chains, one at each side of the channel, with flights connected to the chains across the channel width. The assembly is moved at a flight speed on the order of 0.3 to 0.6 m per minute by motor driven sprockets.

Flights can be spaced uniformly along the entire length of the chains, but flights on just half the chain length are usually sufficient and are preferred because the smaller the number of flights, the less the turbulence. The flights should span the entire width of the channel. Typically, flights are 200 mm high and spaced on 3 m centers.

Only one oil/sludge moving device per channel should be installed. The motor and speed reducer assembly may be mounted directly on the concrete. Each oil/sludge-moving device in a parallel channel installation should have its own drive unit operating independently of the others. Operation of these devices can be either automatic or manual.

Advantages of the flight scraper type oil/sludge-mover are:
- lower initial cost;
- can be used on channels equipped with covers.

The disadvantages of this device are:
- it requires underwater bearings;
- sludge can accumulate on the chain and sprockets;
- moves at the same speed on the top and the bottom of the separator;
- chain sag can redistribute oil beneath the water surface.

A design specification of flight scraper type oil/sludge-moving devices should include the following features:
- provide full width skimming;
- provide totally enclosed, explosion-proof, weather-proof drivers;
- provide overload protection of the driving sprocket;
- provide shaft aligning facilities;
- provide flight leveling facilities;
- provide chain guards for personnel protection on chains above grade or near stairs, ladders, platforms, access-ways, etc.
- provide chain tighteners;
- flight cleaner chains should have an average ultimate strength of 18 200 kg with the plain and attachment links assembled with DN 20 ($^3/_4$ inch) heat treated high carbon steel pins and rivets.
- provide angle tracks for supporting the surface run of sludge collectors;
- flights to be spaced at 3 m intervals and to be 200 mm high;
- provide squeegees of spark-proof construction on two flights for positive cleaning of the tank wall at the water surface and at the tank bottom;
- the flight cleaner should consist of 75 mm × 203 mm (3 inch × 8 inch) scrapers on two strands of chain;
- the chains should run over four sets of sprocket wheels so as to clean sludge from the basin bottoms and move floating oil and material to the downstream end of the separator channel;
- the flight cleaner speed should not exceed 36.6 m/h (10.15 mm/s).
- all flights should be accurately drilled and notched at the factory and be carefully grouped and banded together for safe shipment and storage;
- provide rails flush with the tank bottom for wear surfaces;
- use corrosion resistant anchor bolts;
- the sprockets for drive and collector chains should have a hardness of not less than 360 Brinell at the tooth bearing surfaces. All sprockets should be stress relieved before

42 *Waste Management in the Chemical and Petroleum Industries*

machining. The sprocket should be keyed firmly to the head shaft. All flight cleaners should be of the double life type;

- all shafting should be of solid, cold finish steel, straight and true, should extend across the full width of the basin, and should be held in alignment with a set of screwed set collars. The shafting should contain key ways with fitted keys where necessary and should be of sufficient size to transmit the power required;
- all underwater bearings should be babbitted, of water lubricated, ball and socket, self-aligning type, especially designed to prevent accumulation of settled solids in their surfaces. These bearings should be bolted directly to the concrete walls in a manner that will permit their easy adjustment;
- each motor should have ample power for starting and operating the cleaner mechanism without overloading. Each drive unit speed reducer should be of the worm gear type, fully enclosed, running on oil and of an approved make with antifriction throughout. The reducer unit and the electric motor should be mounted as a common unit directly on the concrete.

2.5.4.2.4 Sludge Collection and Removal. A sludge hopper should be provided independently for each channel at the downstream base of the inlet baffle. A screw conveyor for the positive removal of the sludge should be provided for each channel. The hoppers should consist of inverted pyramids with sides sloped at least 45°. With an underdraw system, each hopper contains an exit pipe at the apex. The exits discharge into a sludge withdrawal pipe.

Screw conveyors should include a wet well-type drive complete with a helical gear reducer and a chain and sprocket drive arrangement. The screw should be at least 9.5 mm thick and run in water lubricated bearings and should be supported by a suitable diameter pipe shaft. The sludge should be directed to the sludge pit.

The sludge pit will be common to both separator channels. Accumulated sludge should be removed by sludge pumps. The sludge pumps should be started automatically when the level in the sump reaches a high liquid level and switched off when the level comes down to a low liquid level. Two vertical sludge pumps (one operating and one spare) should be provided.

High and low level alarms in the control room should also be provided for the sludge sump.

2.5.4.2.5 Skimmed Oil Collection. The skimmed oil from the main separator channels as well as from the forebay section should be routed by gravity to a common sump. Both separator channels have a common oil sump. The sump provides a reservoir for the pump-out pump, and in addition can be used to allow water to be separated from the oil.

The sump should be equipped with two vertical centrifugal pumps (one operating and one spare). All the oil should be pumped to the slops tank. Level controls of the float, electric probe, or air differential pressure type should actuate the pumps at high levels and switch them off at low levels. High and low level alarms should also be provided in the control room to guard against sump overflow and pump cavitation, respectively.

The sump and the pumps should be large enough to avoid continuous operation of the pumps, and should also be large enough to handle a big spill. The size of the sump pumps can be estimated on the following basis if data are unavailable:

- assume incoming oil can reach up to 2000 ml/m^3 (2000 ppm by volume) on dry weather separator influent;
- water withdrawn with the skimmed oil can be assumed to be about seven times the skimmed oil volume when a slotted-pipe oil skimmer is used; and can be assumed to be the same as the skimmed oil volume when a rotary drum oil skimmer is used;
- provide about 4 hours holdup time in the sump before pump-out;
- provide for sump pump-out of oil in 20 to 30 minutes and sump pump-out of water in 30 to 90 minutes (when the oil and water pumps are not the same);
- provide for sump pump-out of oil and water in 1 hour when the same pump serves to pump mixture of oil and water to the slops tank.

2.5.4.2.6 Oil Skimming Device. A slotted-pipe oil skimmer should be provided at the end of each separator channel. The requirements of the skimming device should be as outlined above. The rotatable oil skimmer-pipes for each channel should be connected end to end in a line that drains to a slop oil pit at one side of the unit.

2.5.4.2.7 Oil Retention Baffle. An oil retention baffle shall be provided just downstream of the oil skimming device, spaced not more than 300 mm downstream of the skimming device. The baffle is installed with a maximum submergence of 55% of water depth. It should extend to the top of the channel.

2.5.4.2.8 Effluent Weir. The effluent weir wall is located downstream of the oil retention baffle, no more than 600 mm downstream of the oil retention baffle.

The weir wall extends from the channel floor to a height equal to the normal water depth minus the depth of normal flow over the crest (top edge) of the weir. The crest corresponds to the head on the weir. A sharp-crested or notched weir plate is attached to the downstream face of the weir wall along its top. Bolt holes in the plate should be elongated vertically so that the weir plate can be made absolutely level when installed and minor adjustments in elevations can be made.

Provision should be made to prevent leakage between the weir plate and the weir wall.

2.5.5 Parallel-Plate Separators

2.5.5.1 General

Parallel-plate oil–water separators offer improved performance by increasing the horizontal surface area of the separator and creating less turbulent flow. Conventional channel units can often be retrofitted with parallel plates to either improve the wastewater effluent or allow for higher flow-rates of wastewater. Separators with parallel-plate coalescing media can allow flow-rates up to three times that of conventional units.

They can also remove free oil droplets with smaller diameters than conventional oil–water separators. Effluent from parallel-plate separators is reported to have up to 60% less oil than that from conventional separators and the oil collected from these units contains less water.

In conventional rectangular oil–water separators, more separation (higher flow-rate or better effluent) is accomplished only by adding volume to the tank. Adding volume in an open-channel rectangular tank is the only way to create the necessary surface area for improved treatment.

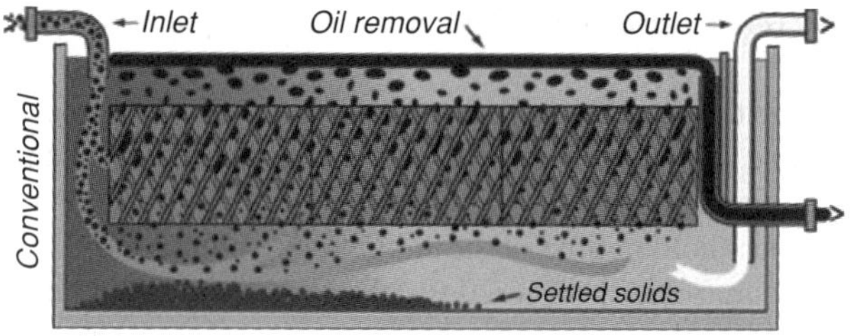

Figure 2.9 A parallel plate API oil–water separator (Image provided courtesy of Brentwood Industries Inc. All rights reserved).

Since these tanks are traditionally manufactured from welded steel, increased volume translates into increased steel at a tremendous material cost. Adding PVC parallel-plate coalescing media is an inexpensive and easy way to increase treatment without adding volume to existing steel separators. Figure 2.9 shows a parallel plate API oil–water separator.

In cases where the available space for a separator is limited, the extra surface area provided by a more compact parallel-plate unit makes the parallel-plate separator an attractive alternative to the conventional separator. The separator's surface area can be increased by the installation of parallel plates in the separator chamber.

The resulting parallel-plate separator will have a surface area increased by the sum of the horizontal projections of the plates added. Flow through a parallel-plate unit can be two to three times that of an equivalent conventional separator.

In addition to increasing the separator surface area, the presence of parallel plates may decrease tendencies toward short-circuiting and reduce turbulence in the separator, thus improving efficiency. Oil content in the effluent water can be up to 60% lower for parallel-plate systems with a higher proportion of small oil droplets recovered with respect to the conventional separators. This type of separator is not suitable for removing emulsified or dissolved oils.

2.5.5.2 Design Considerations

1. An oil-globule diameter of 60 microns (60 μm) should be assumed for the design of separator.
2. The relative density (specific gravity) and viscosity should be calculated based on the worst case ambient conditions for either wastewater or oil in the design of separator.
3. Process problems, such as oil and solids removal and clogging, should be diagnosed and taken into account in equipment selection and separator design. The plate inclination and plate-to-plate spacing should be selected such as to avoid any clogging problem. Solids removal devices should be provided preceding the separator to avoid clogging.
4. Mechanical sludge-removal equipment should be provided to avoid raising the plate pack from the separator for sludge removal. The system should also include oil draw-off equipment.
5. The plates should be of the corrugated type. However, necessary coordination should be made with well-known vendors for optimizing the plate spacing and configuration.

6. In oil and gas industries with mainly oily water effluent, a minimum of two parallel separators with one inlet in combination with a flow diverter (sluice gate) and independent outlets should be provided, so that operation can be continued when one separator is removed from service for repair or cleaning. The separators should be designed to handle design flow-rate when all separators are in service.
7. Separators should be designed to guarantee outlet oil which is "Susceptible to Separation (STS)" to be 40 mg/L in worst case ambient conditions.
8. The separators should be installed such that oily water from the industry oily water sewer can flow by gravity to the separators. Pumping of separator influents should be avoided.
9. A pre-separator/pre-sedimentation basin is often required. Pre-separators are normally constructed in two bays and consist of an overflow weir, a retention baffle, a sedimentation zone, and an overflow into the inlet channels of the parallel-plate separators.
10. Floating oil skimmers should be installed in front of the retention baffle, one per bay. The recommended overflow rate under maximum flow conditions is 10 mm/s. The retention time for the maximum flow-rate should be about 15 minutes. It should be noted that the composition of the effluent may vary by location. Samples should be taken and tested to verify the above flow and retention time values. The bottom slab of the pre-separator should have a slope of 1:50 to simplify cleaning by vacuum truck.

2.5.6 Oil Traps

2.5.6.1 General

An oil trap is a facility designed to retain floating oil only and should not be used for segregation of dispersed oil. The necessity of installing an oil trap should be instructed based on regulations. Oil traps can be used as:

- an inlet to a stormwater pond to keep the surface clean;
- an outlet of e.g., a lagoon, where separated oil is prevented from entering public waters.

2.5.6.2 Design

An oil trap consists of:

- an approach channel with a minimum length of 60 m in which the velocity should be limited to 0.25 m/s under normal conditions and 0.45 m/s during maximum rainfall;
- a transition part;
- a final weir section in which the velocity should be limited to 0.20 m/s under normal conditions and 0.35 m/s during maximum rainfall. However, the velocity underneath the baffle should not exceed 0.08 m/s and 0.15 m/s respectively.

For calculation purposes using the above velocities, a sediment layer with a depth of 0.30 m on the bottom of the complete oil trap should be taken into consideration.

2.5.6.3 Construction

The weir section should be constructed in a V-shape, each leg forming one compartment. However, depending on the throughput, a construction of one leg only may be considered.

In the transition part, the cross-section of the approach channel should change gradually into the cross-section(s) of the weir section. The underside of the baffle wall should be at least 0.60 m below the water level in the oil trap.

The oil trap approach channel and transition part should preferably be built in two bays in order to facilitate cleaning and repair when required.

2.5.7 Oil Holding Basins

A holding basin should be designed to hold the amount of oil spilled by a large accidental spillage. This amount of oil should be assumed to be at least 100 cubic meters for one spillage. The holding basin should be constructed at the end of an accident drainage system.

2.5.7.1 Design

The maximum velocity of effluent through the holding compartment should not exceed 0.05 m/s under normal conditions and 0.075 m/s during maximum rainfall. The velocity underneath the baffle should in no circumstances exceed 0.15 m/s. When calculating velocities, a sediment layer of 0.30 m on the bottom of the entire holding basin should be taken into consideration.

The overflow rate, (i.e., the discharge quantity of the oil holding basin divided by its horizontal surface) should be 0.005 m/s, unless otherwise specified. It is recommended to assume a theoretical design water depth of 1.20 m. The underside of the baffle should be at least 0.60 m below the water level in the holding basin.

2.5.7.2 Construction

The holding compartment should be preceded by an oil-retention baffle, oil skimmer, and oil sump in the transition part in order to keep the surface of the holding basin free from oil under normal conditions. The underside of the oil-retention baffle should be 0.20 m below the water level. The holding compartment should preferably be built in two or more bays in order to facilitate its cleaning and repair, when required.

2.6 Flow Equalization

The equalization basin (Figure 2.10) is used to steady the flow throughout the day. The flow coming into a wastewater plant is very inconsistent. As the flow comes into the plant during the day at a high rate, the equalization basin will fill up. When the flow drops off at night, the equalization basin will start to drain.

It is used to overcome the operational problems caused by flow-rate variations, to improve the performance of the downstream processes, and to reduce the size and cost of downstream treatment facilities.

2.6.1 Application and Location

Flow equalization should be used for damping flow-rate variations so that a constant or nearly constant flow rate is achieved. The design of equalization basins must take into

Figure 2.10 *An equalization basin (Reproduced with permission from © Hickory wastewater treatment).*

account both dry and wet weather flows, as well as variations caused by operational problem.

Sufficient mixing and aeration should be provided in the equalization basin(s) to equalize various wastewater streams and prevent solids deposition, septicity, and odor problems.

In order to achieve an equalized flow and a considerable amount of constituent concentration and flow-rate damping, the equalization basin(s) should be provided in an in-line arrangement so that all of the flow passes through the basin(s).

Depending on the type of treatment and the characteristics of the collection system and the wastewater, detailed studies should be performed to locate the equalization basin(s) in the optimum position. In refineries and petrochemical plants the appropriate location is downstream of the oil–water separators and upstream of the flotation units. However, locating the basin(s) in any place or in the off-line arrangement should take into account all economical and operational points of views and should be based upon the Company's approval.

Application of flow equalization will principally result in the following benefits:

- biological treatment is enhanced, because shock loadings are eliminated or can be minimized, inhibiting substances can be diluted, and pH can be stabilized;
- the effluent quality and thickening performance of secondary sedimentation tanks following biological treatment is improved through constant solids loading;
- effluent filtration surface-area requirements are reduced, filter performance is improved, and more uniform filter-backwash cycles are possible;
- in chemical treatment, damping of mass loadings improves chemical feed control and process reliability.

2.6.2 Volume Requirements

The following factors should be taken into account in determining the required volume for the equalization basin(s):

- sufficient depth of liquid in the basin(s) should be provided to allow continuous operation of aeration and mixing equipment;
- sufficient volume must be provided to accommodate the concentrated plant recycle streams that are expected;
- additional volume should be provided for unforeseen changes in diurnal flow and additional damping of the BOD mass-loading rate;
- the volume shall be sufficient to accommodate a minimum of 16 hours retention time for all incoming wastewater flow-rates to the plant based on daily average flows.

2.7 Mixing

2.7.1 Description and Type

Mixing is an important unit operation in many phases of wastewater treatment, including:

- the mixing of one substance completely with another;
- the mixing of liquid suspensions;
- the blending of miscible liquids;
- flocculation;
- heat transfer.

Most mixing operations in wastewater can be classified as:

1. Continuous rapid (30 seconds or less) mixing, which is used most often where one substance is to be mixed with another.
2. Continuous mixing, which is used to maintain the contents of a reactor or holding tank or basin in a completely mixed state.

Typical mixers used in wastewater treatment plants are:

- propeller mixer;
- turbine mixer;
- static in-line mixer;
- in-line turbine mixer;
- pneumatic mixer.

2.7.2 Application

Mixers should be selected on the basis of laboratory or pilot plant tests or similar data provided by manufacturers. Mixers with small impellers operating at high speeds are best for dispersing gases or small amounts of chemicals in wastewater. Mixers with slow-moving impellers are best for blending two fluid streams for flocculation.

Paddle mixers should be used as flocculation devices when coagulants, such as aluminium or ferric sulfate, and coagulant aids, such as polyelectrolytes and lime, are added to wastewater or sludges. Mechanically, flocculation is promoted by gentle stirring with slow-moving paddles.

A paddle-tip speed of approximately 0.6 to 0.9 m/s achieves sufficient turbulence without breaking up the floc. In-line static mixers are commonly used for the mixing of chemicals, whereas over and under baffled channels are used for flocculation. In dissolved air flotation units, flocculation is achieved by introducing air bubbles in the bottom of the tank (pneumatic mixing).

2.8 Sedimentation

Sedimentation, or clarification, is the process of letting suspended material settle by gravity. Suspended material may be particles, such as clay or silts, originally present in the source water.

More commonly, suspended material or floc is created from material in the water and the chemical used in coagulation or in other treatment processes, such as lime softening.

Sedimentation is accomplished by decreasing the velocity of the water being treated to a point below which the particles will no longer remain in suspension. When the velocity no longer supports the transport of the particles, gravity will remove them from the flow.

2.8.1 Sedimentation Theory

As per Figure 2.11 for a discrete spherical particle, settling velocity/terminal velocity is obtained when Gravitational force (F) is equal to Frictional Drag force (F_D):

So we will have:

$$(\rho_S - \rho_W)gV = C_D A_C \rho_W v_S^2/2 \tag{2.2}$$

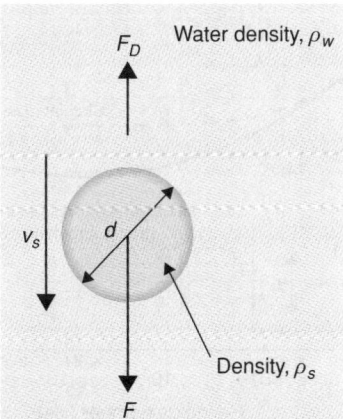

Figure 2.11 Force balance on a discrete spherical particle.

Settling velocity will be:

$$v_S = \sqrt{\frac{2gV(\rho_S - \rho_W)}{C_D \rho_W A_C}} \qquad (2.3)$$

V = volume of particle = $\dfrac{\pi d^3}{6}$

A_c = X-sectional area of particle = $\dfrac{\pi d^2}{4}$

V_S = settling velocity

C_D = drag coefficient

After revising equation 2.3 we will have:

$$v_S = \sqrt{\frac{4g(\rho_S - \rho_W)d}{3 C_D \rho_W}} \qquad (2.4)$$

2.8.1.1   Determination of Drag Coefficient, CD

Drag coefficient can be estimated using the below chart in Figure 2.12:

Typical Reynolds numbers for particles in water treatment are:

$$C_D = 24/\text{Re for Re} < 1 \qquad (2.5)$$
$$C_D = 24/\text{Re} + 3/\text{Re} + 0.34 \text{ for } 1 < \text{Re} < 10\,000 \qquad (2.6)$$
$$C_D = 0.4 \text{ for } 10\,000 < \text{Re} < 100\,000 \qquad (2.7)$$

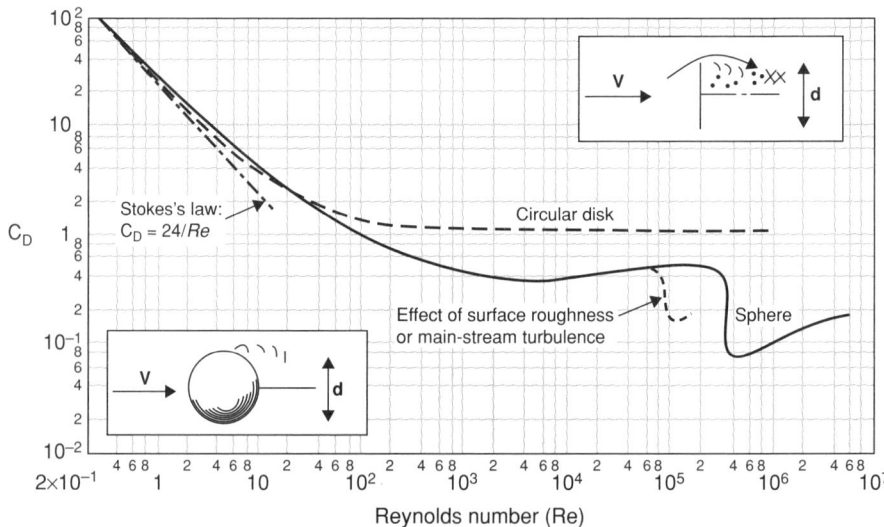

Figure 2.12   Chart to estimate drag coefficient.

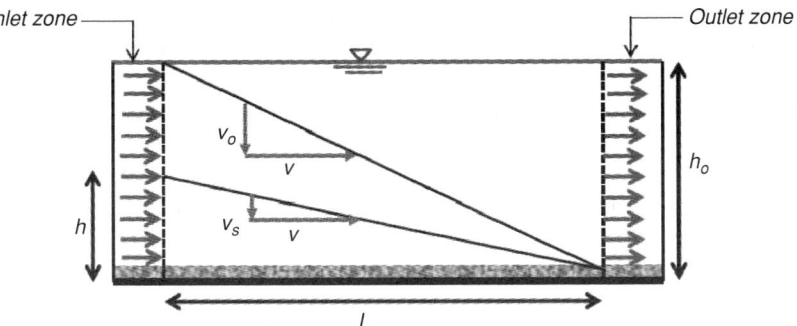

Figure 2.13 Ideal sedimentation basin.

Stokes' Law makes the following assumptions about the behavior of a particle in a fluid:

- laminar flow
- spherical particles
- homogeneous (uniform in composition) material
- smooth surfaces
- particles do not interfere with each other.

$$v_S = \frac{g(\rho_S - \rho_W)d^2}{18\mu} \qquad (2.8)$$

- v_s is the particles' settling velocity (m/s) (vertically downwards if $\rho_p > \rho_f$, upwards if $\rho_p < \rho_f$),
- g is the gravitational acceleration (m/s^2),
- ρ_p is the mass density of the particles (kg/m^3), and
- ρ_w is the mass density of the fluid (kg/m^3),
- d is the diameter of the spherical object (in m), and
- μ is the dynamic viscosity (N s/m^2),

2.8.1.2 Ideal Sedimentation Basin

Figure 2.13 shows an ideal sedimentation basin.

In an ideal sedimentation basin, an overflow rate identifies the smallest settling velocity attributable to the class of particles which experience complete removal.

$$v_O = h_O/t_O = h_O Q/V = Q/A \qquad (2.9)$$

For $v_S < v_O$, the removal percentage will be $100(v_S/v_O)$ because:

$$h/h_O = v_S/v_O \qquad (2.10)$$

2.8.2 Application and Type

Sedimentation is the separation from water, by gravitational settling, of suspended particles that are heavier than water. Sedimentation may be used for grit removal, particulate-matter removal in the primary settling basin, biological-floc removal in the activated-sludge settling basin, and chemical-floc removal when the chemical coagulation process is used.

It is also used for solids concentration in sludge thickeners. However, in most cases, the primary purpose is to produce a clarified effluent. In designing sedimentation basins, consideration must be given to production of both a clarified effluent and a concentrated sludge.

On the basis of the concentration and the tendency of particles to interact, four types of settling can occur: discrete particle, flocculent, hindered (also called zone), and compression (see Table 2.4 for a description of these settling phenomena).

Table 2.4 Types of settling phenomena involved in wastewater treatment.

Types of Settling Phenomenon	Description	Application/Occurrence
Discrete particle (type 1)	Refers to the sedimentation of particles in a suspension of low solids concentration. Particles settle as individual entities, and there is no significant interaction with neighbouring particles.	Removes grit and sand particles from wastewater.
Flocculant (type 2)	Refers to a rather dilute suspension of particles that coalesce, or flocculate, during the sedimentation operation. By coalescing, the particles increase in mass and settle at a faster rate.	Removes a portion of the suspended solids in untreated wastewater in primary settling facilities, and in upper portions of secondary settling facilities. Also removes chemical floc in settling tanks.
Hindered, also called zone (type 3)	Refers to suspensions of intermediate concentration, in which inter-particle forces are sufficient to hinder the settling of neighbouring particles. The particles tend to remain in fixed positions with respect to each other, and the mass of particles settles as a unit. A solids–liquid interface develops at the top of the settling mass.	Occurs in secondary settling facilities used in conjunction with biological treatment facilities.
Compression (type 4)	Refers to settling in which the particles are of such concentration that a structure is formed, and further settling can occur only by compression of the structure. Compression takes place from the mass of the particles, which are constantly being added to the structure by sedimentation from the supernatant liquid.	Usually occurs in the lower layer of a deep sludge mass, such as in the bottom of deep secondary settling facilities and in sludge-thickening facilities.

2.8.3 Design Considerations

1. Pre-sedimentation basins may be constructed in excavated ground or out of steel or concrete. However, the basins should be equipped with a continuous mechanical sludge removal apparatus. The minimum detention time for pre-sedimentation should be 4 hours. Chemical feed equipment should be provided ahead of pre-sedimentation to provide pre-chlorination or partial coagulation.
2. Settling basins are usually provided for chemical coagulation or softening with a wide variety of shapes and flow mechanisms. The selection of the particular form or shape for a given plant depends upon area available and conformity with adjacent structures. The basins should preferably be circular in shape and of reinforced concrete construction. However, rectangular or square basins can also be used upon space availability and the Company's approval.
3. The basins should be provided with sloping bottoms to facilitate the removal of deposited sludge. The bottom slopes should be 1% in rectangular tanks and 8% in circular or square tanks. The water depth is recommended to be in the range of 3.5 to 5.3 m. The length to width ratio for the rectangular basins should be in the range of 3:1 to 5:1.
4. To minimize the effects of short-circuiting and turbulent flow, care should be taken in the effective hydraulic design of inlet and outlet structures in all tanks. Inlet structures should be designed to:
    - uniformly distribute flow over the cross-section of the settling zone;
    - initiate parallel or radial flow;
    - minimize large-scale turbulence;
    - preclude excessive velocities near the sludge zone.
5. Flow through a sedimentation basin normally enters at the top of the basin, but in circular basins, the flow may enter a central flocculating chamber of the basin. Effluent flows vertically out over perimeter weirs. An efficiently designed vertical tank is more stable than a horizontal one.
6. The rating of a vertical tank as related to the bulk settling velocity is usually 1 to 3 m/h. The volumetric capacity can be increased with the introduction of inclined tubes into the basins.
7. Sedimentation processes should be preceded by coagulation and followed by filtration.
8. Each clarifier should consist of a scraped clarifier mechanism with a bridge, with a drive moving along the length/circumference (depending on the type of the basin) to scrape the accumulated scum on the surface of the water. The rotating blades just above the clarifier floor should be provided to collect the separated sludges from the mixed liquor. Each blade should be provided with hard rubber squeegees that scrape the basin bottom once per revolution, collecting the sludge towards the discharge hopper.

2.8.4 Number of Basins

The following considerations should be taken into account in the selection of the number of sedimentation basins:

1. The effect upon the production of treated wastewater if one basin is removed from service for cleaning, repairs, or any other reason.
2. The largest size that can be expected to produce satisfactory result.

For any supply that requires coagulation and filtration for the production of safe water, a minimum of two basins should be provided.

2.8.5 Inlet Arrangements

The inlets should be arranged such as to distribute the coagulated water uniformly among the basins and uniformly over the cross-section of each sedimentation basin and to avoid-short-circuiting through the basin. The permissible velocity range for any water and floc can be determined by tests, but the recommended velocities are between 0.20 and 0.55 m/s.

Where inlet pipelines or flumes are required, the headloss at each opening should be large compared to the maximum difference of energy head available at the inlets. However, velocities must be maintained sufficiently low to prevent break up of floc.

2.8.6 Short-Circuiting

When arranging baffles in sedimentation basins for reduction of short-circuiting and improving settling efficiency, special attention should be paid to avoiding the formation of dead spaces, the production of eddy currents, and causing disturbance of the deposited solids.

2.8.7 Outlet Arrangements

Water leaving the sedimentation basin should be collected uniformly across the width/circumference of the basin to prevent high velocities of approach and the consequent lifting of the settled sludge over the weir. Weirs may be constructed across the basin, or slots, or provided effluent ports.

Combinations of effluent orifices ahead of the submerged weir provide efficient outlet arrangements and reduce short-circuiting. Special attention should be made to provision of sufficient weir length in the construction of the basins, to avoid excessive overflow rates and consequent high velocities of approach to the weir.

2.8.8 Detention Time

The time theoretically required for a unit volume of water to flow through a mixing or sedimentation basin is called the detention time. It is equal to the time required to fill the basin at a given rate of flow and can be computed by dividing the volume of the basin by the rate of flow through it.

Detention time must be distinguished from retention time, which is the minimum time required for a particle of water to pass through the basin. Depending on the purpose of the basin, a detention time of 2 to 4 hours should be applied. The detention time should be higher when a high degree of suspended solids removal is desired.

2.8.9 Surface Loading Rate

The surface overflow rate for any given sedimentation tank can be determined by jar test studies, wherein the best coagulant, optimum dosage, and the best flocculation are used. The maximum surface loading rate for suspended solids removal should be 25 m^3/ day/m^2 when proper coagulant and flocculation are used.

2.8.10 Factors Affecting Sedimentation

Several factors affect the separation of settleable solids from water. Some of the more common factors to consider are as follows.

2.8.10.1 Particle Size

The size and type of particles to be removed have a significant effect on the operation of the sedimentation tank. Because of their density, sand or silt can be removed very easily. The velocity of the water-flow channel can be slowed to less than one foot per second, and most of the gravel and grit will be removed by simple gravitational forces.

In contrast, colloidal material, small particles that stay in suspension and make the water seem cloudy, will not settle until the material is coagulated and flocculated by the addition of a chemical, such as an iron salt or aluminium sulfate.

The shape of the particle also affects its settling characteristics. A round particle, for example, will settle much more readily than a particle that has ragged or irregular edges.

All particles tend to have a slight electrical charge. Particles with the same charge tend to repel each other. This repelling action keeps the particles from congregating into flocs and settling.

2.8.10.2 Water Temperature

Another factor to consider in the operation of a sedimentation basin is the temperature of the water being treated. When the temperature decreases, the rate of settling becomes slower. The result is that as the water cools, the detention time in the sedimentation tanks must increase.

As the temperature decreases, the operator must make changes to the coagulant dosage to compensate for the decreased settling rate. In most cases temperature does not have a significant effect on treatment.

A water treatment plant has the highest flow demand in summer, when the temperatures are highest and settling rates the best. When the water is colder, the flow in the plant is at its lowest and, in most cases, the detention time in the plant is increased so the floc has time to settle out in the sedimentation basins.

2.8.10.3 Sedimentation Basin Zones

Most sedimentation tanks are divided into these separate zones (Figure 2.14):

1. **Inlet zone**
   The inlet or influent zone should provide a smooth transition from the flocculation zone and should distribute the flow uniformly across the inlet to the tank. The normal design includes baffles that gently spread the flow across the total inlet of the tank and prevent short-circuiting in the tank. (Short-circuiting is the term used for a situation in which part of the influent water exits the tank too quickly, sometimes by flowing across the top or along the bottom of the tank.) The baffle could include a wall across the inlet, perforated with holes across the width of the tank.

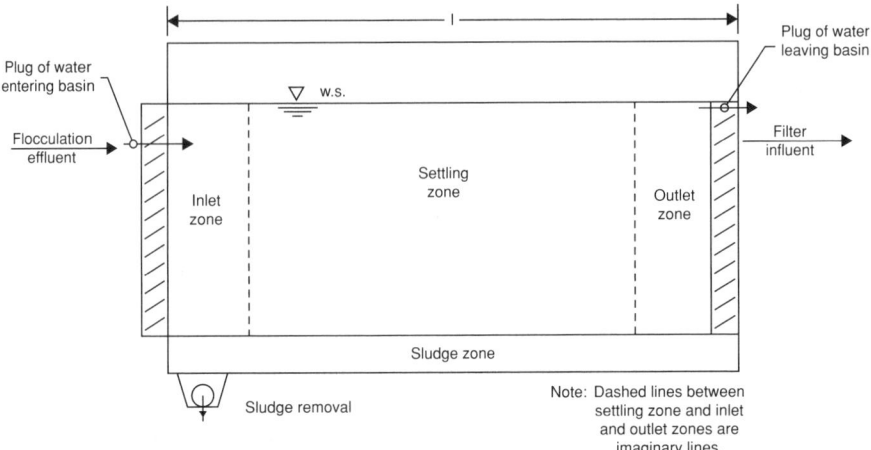

Figure 2.14 *Sedimentation basin zones.*

2. **Settling zone**

   The settling zone is the largest portion of the sedimentation basin. This zone provides the calm area necessary for the suspended particles to settle.

3. **Sludge zone**

   The sludge zone, located at the bottom of the tank, provides a storage area for the sludge before it is removed for additional treatment or disposal.

2.9 Dissolved Air Flotation (DAF)

A key application of DAF units is the removal of free and emulsified hydrocarbons from petrochemical and similar wastewaters upstream of biological processes. This is done in an effort to prevent toxic or inhibitory materials from hindering the biological processes downstream of the DAF unit.

Other industrial applications of DAF systems include the treatment of concentrated fish farming wastewaters, the pre-treatment of food and meat processing effluents, and the treatment of effluents generated by the pulp and paper industries. In certain circumstances, DAF can also be a substitute for gravity settlement of solids generated by a biological treatment process.

Advantages linked with DAF systems include the fact that they are high-rate processes when compared with more traditional gravity-based settlement systems. This means that a reduction in space requirements can be achieved, and in terms of sludge thickening, a thicker sludge can be produced. Additionally, DAF systems offer the operator some degree of flexibility, subject to design, with regard to the system's operating parameters.

2.9.1 General

Dissolved Air Flotation is a liquid–solid separation process in which microscopic air bubbles (10–100 μm) become attached to solid particles suspended in liquid, causing the

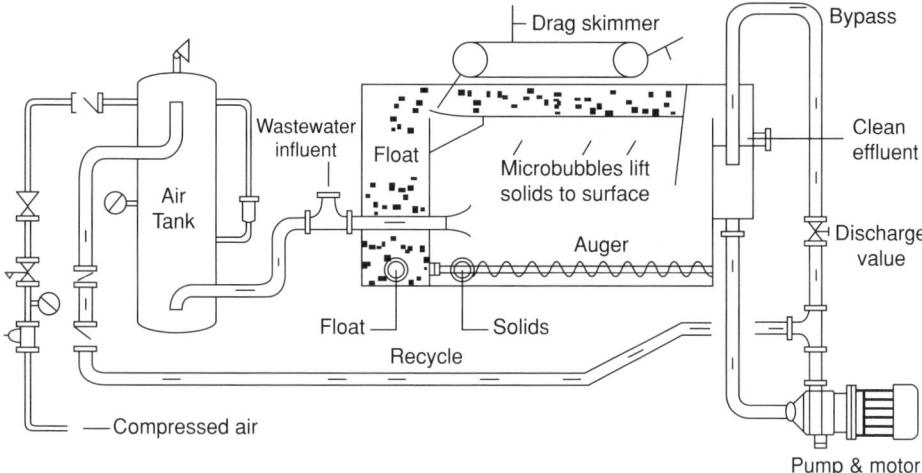

Figure 2.15 A schematic of a conventional dissolved air flotation system (Reproduced with permission from © Skim oil^{TM}, Inc).

solid particles to float. In a DAF system, air is dissolved into liquid under pressure. The dissolved air remains in solution until the pressure is released to atmospheric pressure, causing the air to come out of solution in the form of microscopic air bubbles.

The bubbles are mixed intimately with the wastewater and become attached to the solids in the waste stream, causing the air–solid agglomerate to float to the liquid surface where a solids (float) blanket is formed. Surface skimmers then remove the float blanket.

Figure 2.15 shows a conventional dissolved air flotation unit. It should be designed such as to produce an effluent containing less than 15 mg/L (max) of free oil with chemical aids. Reduction of the oil content below the solubility level of hydrocarbons in water should not be expected in DAF units.

Dissolved gas flotation (DGF), in which gas (such as natural gas) is employed instead of air, can be applied to maintain a reducing action, or otherwise to avoid oxidation of waste material.

The primary purpose of further oil removal by DAF units is to meet desired effluent quality and/or pre-treatment for downstream biological units. DAF units are normally located downstream of API separators and upstream of biological oxidation units.

2.9.2 System Configuration

DAF units should be constructed in concrete tanks located partially or totally above grade. A minimum of two parallel units, each designed for 50% of total design flow-rate as fresh feed, plus the pressurized recycle water flow-rate, should be provided. The water recycled to each unit should be designed to be in the range of 50 to 75% of each DAF unit design fresh feed flow-rate. Multiple DAF units should be supplied when the dimension of a rectangular unit would exceed 6 m width or 30 m length or when the diameter of a circular flotator would exceed 24 m.

Each unit should consist of, but not be limited to, the following facilities:

- pressurized flow control system and pressure retention tank;
- flocculation chamber;
- control valves and instrumentation for effective control of all flow conditions throughout the recycle system;
- telescopic valves;
- scum skimmer including drive and supports;
- scum screw conveyor and scum trough;
- scum retention baffles;
- bottom sludge collector and drive;
- perforated recycle pipe and baffle for distribution of the pressurized recycle across the flotation channel;
- deflection baffle made of stainless steel;
- scum pit agitator and drive;
- effluent weir and scum retention baffles;
- all booster and pressurizing pumps, including motors and drivers;
- air compressors;
- inlet sluice gates of 0.1 square meter free area;
- adjustable outlet weirs.

2.9.3 Variables Affecting DAF Efficiency

1. **Influent characteristics**

   Certain characteristics of the influent affect DAF efficiency, such as globule size, pH, surface activity, and suspended solids concentration. Coagulation agents can significantly affect globule size and the chemicals used may alter surface activity; floating all the suspended solids can be achieved by introducing sufficient air bubbles through the water.

   The pH should be adjusted to within the proper range prior to the addition of coagulating chemicals. The pH adjustment facilities should be provided upstream of DAF units.

2. **Design variables**

   The following major design variables should be taken into account:
   - overflow rate (rise rate);
   - recycle rate;
   - air pressure.

3. **Operating variables**

   The unit should be sized such as to provide sufficient operating flexibility to allow effluent quality to be maintained despite the expected variations in influent characteristics. The operator should be able to adjust the following major operating variables:
   - pH;
   - coagulant type and dose;
   - coagulant aid type and dose;
   - air pressure;
   - recycle rate.

2.9.4 Treatability Testing

Studies should be made on the wastewater to determine if it is amenable to cleaning up by DAF process. If it is established that DAF is a suitable technique for the particular waste, pilot plant tests or laboratory bench-scale tests can be used to establish the following design parameters (refer to API, Manual on Disposal of Refinery Wastes, Volume on Liquid Wastes, 1969, for bench-scale treatability test procedures for the DAF process):

1. Overflow rate or rate of particles of mass.
2. Optimum recycle rate.
3. Pressure level in the air–water mix drum.
4. Type of chemical conditioning required such as:
   - type and number of chemicals;
   - optimum dosages of chemicals;
   - best points for chemical application;
   - flash-mix and flocculation detention times.

If the wastewater cannot be tested before design, data from a similar refinery or plant should be used where possible.

2.9.5 Design Considerations

The hydraulic loading rate (HLR) is a practically important characteristic of unit operations in a waste treatment process as it is directly related to the tank size required for the utility to meet water demand. The HLR for a DAF clarifier is calculated as shown in Equation 2.11.

$$\text{HLR} = \frac{\text{Treated Flow}}{\text{Total Clarifier Footprint}} \qquad (2.11)$$

Where:

HLR = hydraulic loading rate (m/h)
Treated Flow = clarifier effluent flow, not including recycle flow (m^3/h)
Total Clarifier Footprint = footprint area including contact and separation zones of clarifier (m^2)

DAF is a high-rate clarification process with nominal HLRs ranging from 5 to 40 m/h, as opposed to sedimentation loading rates which are 0.5 to 5 m/h, both depending on factors such as water quality, temperature, and basin configuration.

Temperature affects water viscosity, such that at warm temperatures higher rise velocities can be achieved. Theoretical floc-bubble aggregate rise rate is derived from Stokes' Law (equation 2.12), and is shown below in equation 2.13. Ensuring floc-bubble aggregate rise velocities are sufficient to reach the top of the clarifier before being drawn into the subnatant effluent flow is crucial to DAF performance.

$$v_b = \frac{g(\rho_w - \rho_b)d_b^2}{18\mu_w} \qquad (2.12)$$

$$v_{fb} = \frac{4g(\rho_w - \rho_b)d_{fb}^2}{3K\mu_w} \qquad (2.13)$$

Where:

v_b = rise velocity of bubble
v_{fb} = rise velocity of floc-bubble aggregate
g = gravitational constan,
ρ_w = density of water
ρ_b = density of air
μ_w = water dynamic viscosity
d_b = diameter of bubble
d_{fb} = diameter of floc-bubble aggregate
K = coefficient: varying from 24 for flocs < 40 microns to 45 for 170 microns.

2.9.5.1 System Design

1. The system should be designed for continuous service and an uninterrupted operation of 2 years (minimum).
2. All equipment should be suitable for unsheltered outdoor installation for the specified climatic zone.
3. Enclosure for electrical equipment should be appropriate for the specific area classification and environment.
4. Effluent from the flocculation tank to the DAF unit, and the float/sludge from the DAF unit to the receiver tank, should be moved by gravity.
5. Design retention time should be a minimum as follows:
   - Neutralization tank: 3 minutes
   - Flocculation tank: 20 min (influent feed)
   - Flotation basin: 40 min (influent feed + recycle)
   - Air–water mix drum: 3 min at maximum recycle rate.
6. The design flow-rate should be 1.5 L /s/m² (max) based on the DAF flotation area.
7. A continuous supply of air to the pressure retention tank should be provided at a pressure not less than 500 kPa (ga) or more than 700 kPa (ga).

2.9.5.2 DAF Section

1. The Side Water Depth (SWD) should be a minimum of 2.5 m and a maximum of 3.0 m. In addition to the water depth, 500 mm of free board should also be provided.
2. A perforated recycle distribution pipe and baffle should be provided to uniformly distribute the flow across the depth and width of the flotation compartment. The perforations should be spaced and designed for proper flow distribution across the width of the chamber. The pipe should be flanged at one end for connection to the recirculation piping.
3. The unit should be designed to remove floatable free oil and/or suspended solids with a rise rate of 7.2 m/h or greater when the unit is working at the design flow-rate.
4. Fluid velocity in the flotation chamber should not exceed 36 m/h.
5. The scum skimmer should consist of two strands of chain running above the liquid surface over two sets of sprockets with pivoted skimming flights. The flights should enter the floated material blanket at the effluent end of the tank and skim towards the

influent end. The skimmer should be driven by a suitable variable speed drive. Skimmer chains should be supported where required.

The skimmer blade should include a depth adjustment feature. Skimmer chains should be provided with a self-lubricating device for rectangular units. For circular units, blades should be interconnected with transverse bracing to reduce the potential for excessive horizontal deflection and rebound.

6. A helical flight screw conveyor should be provided in the flotation basin to remove the floatable solids from the scum trough to a sump beyond the basin wall. The conveyor should be 225 mm in diameter with full pitch flights mounted on a standard mass (weight) steel pipe. The conveyor should be driven by a suitable electric drive. An agitator and drive should be provided at the scum sump to keep the slurry in suspension.
7. The bottom sludge collector should consist of a 50 mm × 152 mm nominal size redwood scraper, mounted on two strands of chain no more than 3000 mm apart. The collector chains should run over three sets of sprocket wheels and be driven by a suitable variable speed drive. The chains should be C720, heavy pintle type, with the plain and attachment links assembled with heat-treated high carbon steel pins and rivets. At least 25# tee rails (as per Table-3 of DIN 997) should be provided for the tank bottom for the collector. Tracks for the return run of the scraper, where required, should be made of 76 mm × 51 mm × 9.5 mm thick steel angles with 6.25 mm steel supporting brackets.
8. Skimmer/sludge scraper tip speed should be a maximum of 0.025 m/s for circular units and 0.015 m/s for rectangular units.
9. The use of belt transmissions for skimmers and scrapers should be approved by the Company. Limitations on the use of belt transmissions should be as follows:
   o maximum drive power rating: 110 kW;
   o drive service factor: 1.5;
   o belt type: jointed (multiple V-belt).
10. Belts should be heavy duty or premium quality with oil resistant, static conducting characteristics. Utilization of variable speed transmissions should be approved by the Company. Couplings should be of forged steel, flexible type.
11. The float skimmer and sludge scraper should have cycle timers with on/off switches. A remote torque indicator and manual torque lifting device should be provided for the circular scraper. If design load is reached, the drive motor should shut-off and visual and audible alarms should be activated. A shear pin should be provided on the skimmer/scraper drive shaft.
12. Provisions should be included to remove settled sludge from the flotation tank. Generally, a number of hoppers or a trough at the inlet end of the tank may be provided. Settled sludge is usually scraped into the hoppers or trough and periodically removed either by pumping or by allowing the hydrostatic head in the flotation tank to pressure the sludge out.
13. The float/sludge receiving tank should be sized to accommodate a minimum of 3 hours retention of a float production based on 0.5 volume percent of raw wastewater feed. The tank should include facilities for steam or air injection, or both. Float boxes should be sized to accommodate the maximum amount of float produced. A minimum of two float boxes should be provided. The Vendor's design basis, including the number of boxes, should be approved by the Company.

14. Float/sludge transfer pumps should have as a minimum the following characteristics:
    - A minimum of two screw pumps (one operating and one spare) should be provided. If single screw type pumps are used, spare stators should be provided for each pump.
    - Pump suction strainers should be provided to screen out any solid particles which the pump cannot handle. The screen material should be a copper alloy. Strainers should be provided with flushing connections.

2.9.5.3 Neutralization/Flocculation Section

2.9.5.3.1 Neutralization Section

1. Neutralization tank design should be as follows:
    - The tank should be baffled.
    - For circular tanks, the liquid height should be approximately equal to the diameter.
    - For non-circular tanks, the liquid volume should have approximate cubical dimensions.
2. The feed stream inlet piping layout should allow for a free fall of about 600 mm above liquid level to promote pre-mix condition.
3. Influent and effluent connections should be diametrically opposite each other. The effluent connection should be as per API 650 low type.
4. Acid and caustic injection into the neutralization tank should be adjacent to the raw wastewater influent.
5. The mixer should be the propeller or axial flow type. Liquid pumping or moving capacity should be 10 to 20 times the feed flow-rate with the flow directed toward the bottom of the tank.
6. Neutralization facilities should preferably be located downstream of the equalization basin.
7. The pH control system should incorporate the following:
    - at least proportional plus a reset feedback control mode should be provided;
    - the pH analyzer should have a self-cleaning feature. The pH control probe should be located on the neutralization tank effluent line as close as practicable to the tank;
    - a pH indicator and alarm high in the control room should be provided;
    - the pH control system should be automatic.

2.9.5.3.2 Flocculation Section. The flocculation equipment should be installed within an adequately sized compartment. A freeboard of 610 mm at the head of each unit should be provided. The equipment should consist of a paddle wheel mounted on a shaft and driven by a suitable variable speed drive. Each paddle assembly should consist of nominal size heart redwood blades supported by steel angle arms.

The flocculation tank should incorporate the following:

1. The circular tanks should be baffled.
2. The mixer should have a maximum tip speed of 0.6 m/s.
3. The tank should include an air sparger.

A manually operated rack and pinion scum collection pipe should be installed in the flocculator compartment for the removal of floating material trapped in the coalescing area.

The scum pipe should be capable of being rotated backward as well as forward to insure the removal of scum between the pipe and the basin wall.

A 60° slot should be cut symmetrically about the vertical axis of the pipe with the edges of the slot serving as a weir over which the scum flows into the pipe when the pipe is rotated. A suitable watertight seal should be provided for the open end of the pipe. The seal should be renewable without removing the pipe from the supporting brackets and should not bind or impede the smooth action of the revolving pipe.

2.9.5.4 Pressurized Flow Recycle System

1. Each pressurized flow recycle system should consist of a pressurized tank, a pressurizing pump, a flow meter, regulating valves for water and air, and necessary instrumentation such as indicators, alarms, on/off valves, etc.

    Individual pressurized flow systems comprising an individual pressurized tank, flow meter, regulating valve, connecting piping, instrumentation, etc., should be provided to serve each DAF unit. When two DAF systems are provided, three pressurizing pumps should be foreseen, such that each pump is capable of serving both DAF units. In this case, each pressurized flow system should also be designed to be operable for both flotation basins.

    Each pressurized recycle system, with components such as pressurized tanks, pumps, piping, etc., should be designed to handle at least 75% of the design's fresh feed to each DAF unit. The recycle pumps should be horizontal centrifugal types.

2. The pressurized tank should be designed to ensure that a minimum of 80% saturation at operating conditions is achieved. Air spargers on the air inlet line and water spray nozzles on the water inlet line should be provided. The equivalent liquid hold-up time should be 3 minutes. A minimum 1 m of outage should be provided to allow for frothing and for the air to disengage. A means to drain the drum should be provided.

3. Float switches and solenoid valves mounted on the pressurized tank should maintain the proper air–liquid ratio in the tank. The float switches should activate an alarm in the event of high or low water levels. The pressure tank should also have a pressure gauge, level gauge, low pressure alarm, air vent with needle valve, and pressure relief safety valve. The level and pressure control system proposed by Vendor should be submitted to the Company for approval. Any proposed level control should be of the external type.

4. Suitable packing should be provided in the pressurized tank to increase the water surface and improve air solubility

5. The air–water mix drum should be made of killed carbon steel with a 3 mm corrosion allowance and a suitable coating or lining as specified in Article 7.9.10 below. The mechanical design pressure of the drum should be sufficiently high to permit future operation at a process pressure higher than that initially anticipated.

6. The back pressure (let-down valve) should be located as close to the flotation tank and to the influent stream as possible. The valve must be suitable for the mechanical design pressure of the system, should not be easily plugged by solids, and should minimize turbulence.

The valve should not impart high shear to prevent breaking the floc. Downstream of the back pressure valve at the entrance to the flotation chamber, a proper inlet distributor with suitably sized nozzles should be provided.
7. Injection of air to the suction and/or interstage of the recycle pump is not acceptable. When the air is added to the discharge side of the recycle pump, it should be injected through a porous diffuser.
8. Take-offs for recycling and effluent should be oriented to satisfy recycle feed water requirements so that upon loss of raw feed, the DAF unit will go into total recycle.
9. The recycle line should ensure full flow so that air entrainment will not bind the pump. The recycle line should be fabricated of corrosion resistant material.
10. The air supply should provide at least 0.12 Nm^3/m^3 of raw wastewater feed. If plant air is unavailable, and a separate compressor is included with the system, a 100% installed spare should be furnished.
11. The final configuration of the air–water mix drum control and instrumentation system should be approved by the Company.

2.9.5.5   Effluent Chamber

Effluent water shall leave the effluent chamber by gravity. A common effluent chamber should be provided for the flotation units. The minimum residence time for the chamber should be 12 minutes. An adjustable weir plate at the end of each flotation chamber should be provided to permit the regulation of the water level in order to control the penetration of the skimmer blades into the scum blanket in the flotation area.

A scum retention baffle should be provided to prevent flotation solids from passing under and out of the flotation chamber. A wiper should be provided to clean the skimmer blades. Recycle flow should be designed to be delivered to the suction side of the pressurizing pump from the effluent chamber at a rate equal to 50 to 75% of the design fresh feed to the flotation basin.

2.9.6   Instruments and Control

1. All instrumentation for the equipment provided should be suitable for service and electrical area classification.
2. All instruments should be tagged with identifying tag numbers that will be assigned by the Company. A stainless steel nameplate should be provided that is stamped or engraved with the tag number and permanently mounted with stainless steel drive screws.
3. The following instruments should be provided in addition to the instruments specified in Section 2.9 of this book:
   - a flow-rate recorder for fresh influent;
   - ratio controllers to control chemical dosages;
   - a recycle flow-rate meter;
   - an air flow-rate meter (usually a rotameter);
   - an oil-in-water detector for effluent water;
   - a turbidity meter for effluent total suspended solids.

2.9.7 Piping

1. Gravity line(s) between the flocculation tank and DAF should be sized for a maximum velocity of 0.6 m/s.
2. Sample connections and relevant valves should be DN 20 for the wastewater lines, DN 50 for the sludge line, and DN 25 for the feed recycle sample line. Connections should be located for easy access.
3. The float and sludge line to the float/sludge receiver tank should include rodding and flushing connections.
4. Facilities should be provided for let-down system blow-down to the flotation tank via compressed air.
5. Drains should be included for all vessels and tanks.
6. Corrosion allowances should be specified.
7. All lines and accessories in contact with aerated water should be made of carbon steel with a corrosion resistant lining. The minimum acceptable lining for corrosion protection is coal tar epoxy.

2.9.8 Chemical Facilities

1. When feasible, the types and quantities of chemicals to be added to a DAF feed should be determined by tests before the chemical facilities are designed.
2. In addition to the chemicals (acid, caustic, etc.) used for the neutralization section, facilities should be provided for at least two chemicals: a coagulant and a coagulant aid. Necessary equipment for chemical storage, solution or dilution, addition, metering, flash mixing, and flocculation should be provided. If solid chemicals are handled, a suitable enclosure or shed to protect the solids and their addition facilities from the weather should be provided.
3. The floc must not be pumped, and lines carrying the floc should be sized to avoid both its break-up and settlement. Design for a velocity of about 0.6 m/s.
4. The flash-mixing drum should be designed for a holdup time of 4 to 5 minutes, and should be provided with rapid agitation. The length to diameter ratio should be about 1.0.
5. The flocculation tank should be provided with gentle agitation by a slow-speed mixer with a maximum tip speed of 0.6 m/s.
6. A suitable shelter or building should be provided to cover the chemical facilities.

2.9.9 Material

The material that the vessels, piping, and all other equipment is made of should be selected in accordance with the Company's relevant standards unless otherwise specified below:

1. Carbon steel tanks and vessels should have an at least 3 mm corrosion allowance and a protective coating system of two coats of coal tar polyamide epoxy. Surface preparation for the protective coating should be by White Metal Blast Cleaning. The nominal dried thickness for each coat should be 200 micrometer (8 mils).
2. The carbon steel float/sludge receiver tank(s) that include facilities for the live steam should be lined with gunite of 25 mm thickness. Coal tar polyamide epoxy is not acceptable.

66 Waste Management in the Chemical and Petroleum Industries

3. All internal surfaces of the vessels/tanks in contact with aerated water should be stainless steel.
4. Piping in contact with aerated water should be made of carbon steel with a minimum corrosion allowance of 3 mm.
5. Rubber and plastic parts should be resistant to attack by aromatic solvents.
6. The bolting for skimmer and weir adjustment should be type 316 stainless steel.
7. Antispark material should be used for metal to metal contacts in moving parts.

2.9.10 Estimation of Air Concentration in Dissolved Air Flotation (DAF) Systems

Dissolved air flotation (DAF) is a water treatment process that clarifies wastewaters (or other waters) through the removal of suspended matter. Removal is achieved by dissolving air in the water or wastewater under pressure and then releasing the air at atmospheric pressure in a flotation tank or basin. Figure 2.16 shows a schematic of a dissolved air flotation (DAF) system. Like other gravity separation processes, raw water is coagulated and flocculated prior to entering the DAF basin.

The water is introduced into the contact zone of the basin near the floor. A baffle wall separates the contact zone from the clarification zone and limits short-circuiting. In the DAF basin, a cloud of air bubbles, called white water, typically 10 to 100 μm in diameter, adheres to floc particles, causing them to float by reducing the net specific gravity of the floc–bubble aggregate below that of surrounding liquid.

Air bubbles are introduced near the bottom of the basin containing the water to be treated. As the bubbles move upward through the water, they become attached to particulate matter and floc particles, and the buoyant force of the combined particle and air bubbles will cause the particles to rise to the surface.

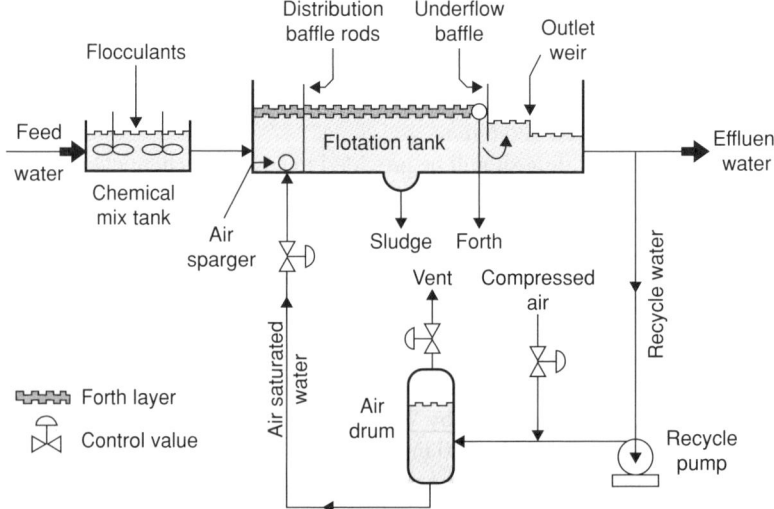

Figure 2.16 Schematic of dissolved air flotation system (DAF) (Reproduced with permission from [30], © Elsevier, 2013).

The released air forms tiny bubbles that adhere to the suspended matter causing it to float to the surface of the water where it may then be removed by a skimming device. Thus, particles that have a higher density than the liquid float. Particles that rise to the surface are removed for further processing as residuals, and the clarified liquid is filtered to remove any residual particulate matter.

The increased dissolved air concentration in water at elevated pressure is the fundamental principle that allows the formation of micro-bubbles. In this chapter, a simple predictive tool is developed to estimate air saturated concentration in DAF systems.

Effluent turbidity declines with increasing air loading until a breakpoint is reached where the application of additional air provides no corresponding increase in process performance. The mass concentration of air released can be calculated from the following expression

$$C_b = \frac{C_r - C_{fl}}{1 + r} r \tag{2.14}$$

The air concentration in the saturator can be expressed either as the bubble number concentration (N_b) or the bubble volume concentration (φ_b). These values can be determined from the expressions:

$$\varphi_b = \frac{C_b}{\rho_{air}} \tag{2.15}$$

$$N_b = \frac{10^{12} \times 6\varphi_b}{\pi d_b^3} \tag{2.16}$$

In the above equations, C_b is mass concentration of air released, C_{fl} is mass concentration of air in floc tank effluent, C_r is mass concentration of air in recycle flow, and d_b is mean bubble diameter.

It has been found that flotation performance increases as N_b increases because there are more collision and attachment opportunities between the bubbles and particles. Attached air bubbles provide lower floc particle density and larger volume, producing floc particle–bubble aggregates that have high upward velocities.

In view of the above, it is necessary to develop an accurate and simple correlation that is less complicated than existing approached and requires fewer computations to predict air saturated concentration.

Equation 2.17 represents the proposed governing equation in which four coefficients are used to correlate dissolved air concentration in DAF as a function of pressure and temperature, where the relevant coefficients have been reported in Table 2.5.

$$\ln(x) = a + \frac{b}{P} + \frac{c}{P^2} + \frac{d}{P^3} \tag{2.17}$$

Where:

$$a = A_1 + B_1 T + C_1 T^2 + D_1 T^3 \tag{2.18}$$

$$b = A_2 + B_2 T + C_2 T^2 + D_2 T^3 \tag{2.19}$$

$$c = A_3 + B_3 T + C_3 T^2 + D_3 T^3 \tag{2.20}$$

$$d = A_4 + B_4 T + C_4 T^2 + D_4 T^3 \tag{2.21}$$

Table 2.5 Tuned coefficients used in equations 2.18–2.21.

Coefficient	Value
A_1	$5.112376352927 \times 10^1$
B_1	$-3.767964895874 \times 10^{-1}$
C_1	$1.060069664864 \times 10^{-3}$
D_1	$-1.011565521006 \times 10^{-6}$
A_2	$-3.283012529565 \times 10^5$
B_2	$3.360561486772 \times 10^3$
C_2	$-1.150208224506 \times 10^1$
D_2	$1.311220065117 \times 10^{-2}$
A_3	$1.021345559148 \times 10^8$
B_3	$-1.048743022561 \times 10^6$
C_3	$3.594675738720 \times 10^3$
D_3	-4.104275097337
A_4	$-7.036559254429 \times 10^9$
B_4	$7.218196550836 \times 10^7$
C_4	$-2.473400471536 \times 10^5$
D_4	$2.824070339246 \times 10^2$

In equations 2.14–2.21 the list of symbols are:

A, B, C, and D Tuned parameter;
C_b: mass concentration of air released, mg/L;
C_{fl}: mass concentration of air in floc tank effluent, mg/L;
C_r: mass concentration of air in recycle flow, mg/L;
d_b: mean bubble diameter, μm;
N_b: bubble number concentration, no./Ml; P: pressure, kPa(abs);
r: recycle ratio, dimensionless;
T: Temperature, K;
x: dissolved air concentration mg/L;
φ_b: bubble volume concentration, L/L;
ρ_{air}: density of air saturated with water vapor, mg/L.

These optimally tuned coefficients help to cover temperatures up to 40 °C and pressures up to 700 kPa (abs). The optimally tuned coefficients given in Table 2.5 can be further retuned quickly according to the proposed approach if more data become available in the future.

Here, our efforts are directed at formulating a correlation that can be expected to assist engineers in rapid calculation of dissolved air concentration in DAF systems as a function of pressure and temperature using an exponential function. The proposed novel tool developed in this chapter is a simple and unique expression that cannot be found in the literature. Furthermore, the selected exponential function to develop the tool leads to well-behaved (i.e., smooth and non-oscillatory) equations, enabling reliable and more accurate predictions.

Figure 2.17 shows the proposed method results in comparison with data in the literature. Figures 2.18 and 2.19 show the results from the proposed method and its smooth performance in the prediction of dissolved air concentration in DAF systems as a function of pressure and temperature.

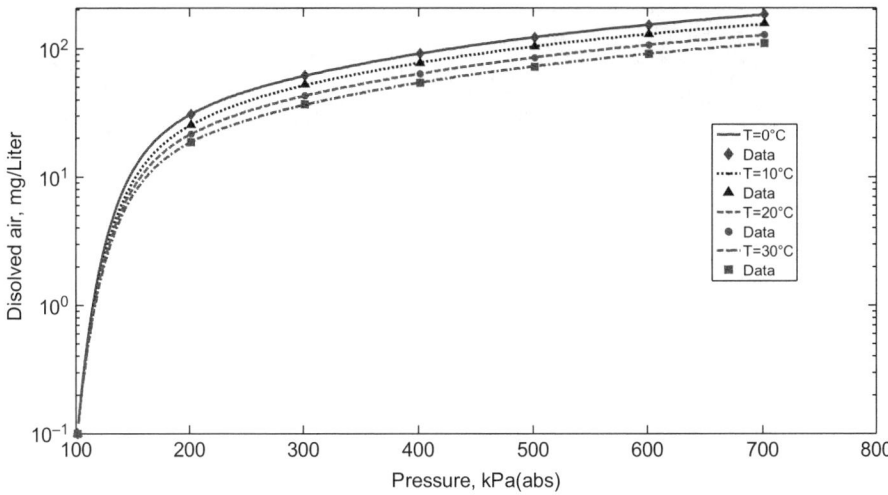

Figure 2.17 *The performance of the predictive tool in comparison with data for calculating dissolved air concentration in DAF systems (Reproduced with permission from [30], © Elsevier, 2013).*

It is expected that our efforts in formulating a simple tool will allow an accurate prediction of dissolved air concentration in DAF systems that can be used by water treatment practitioners and process engineers to periodically monitor the key parameters. A typical example is given below to illustrate the proposed correlation's simplicity of use.

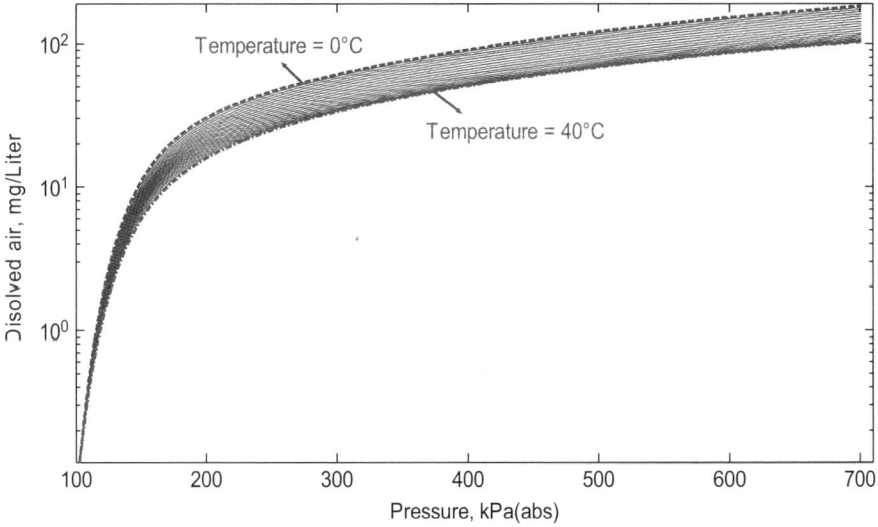

Figure 2.18 *The smoothness of results of the predictive tool for the estimation of dissolved air concentration in DAF systems (Reproduced with permission from [30], © Elsevier, 2013).*

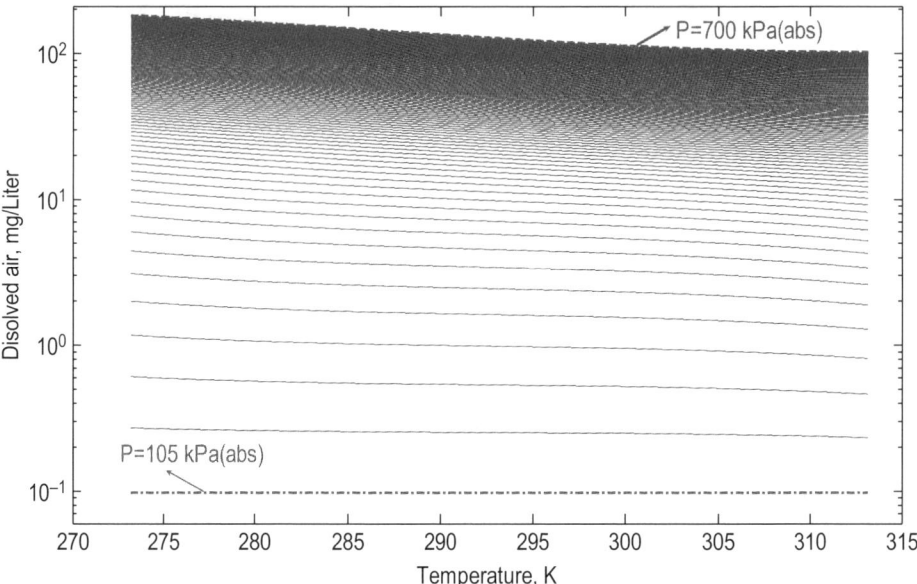

Figure 2.19 The smoothness of the predictive tool's results for the estimation of dissolved air concentration in DAF systems from another view point (Reproduced with permission from [30], © Elsevier, 2013).

Example A DAF plant is operating at 10% recycle with a saturator pressure of 500 kPa at 10 °C. Flocculated water enters the contact zone with a floc particle concentration (N_p) of 2000 particles/mL and a floc volume concentration (φ_p) of 2×10^{-6} L/L (2 ppm). Calculate the air mass concentration (C_b), bubble zone volume concentration (φ_b), and bubble number concentration (N_b) in the contact zone of the DAF tank and compare the concentrations of bubbles to floc particles. Assume the water temperature is 10 °C ($\rho_{air} = 1.20$ kg/m^3 = 1200 mg/L), the flocculated water has no oxygen deficit, so the air concentration in water is 24 mg/L, and the mean bubble diameter is 40 μm. Determine the mass of air in the recycle water.

Solution Calculate the mass concentration of air in the DAF tank. The dissolved air in the water through the saturator is calculated using the new predictive tool.

$$a = 6.4599$$
$$b - 1.26202 \times 10^3$$
$$c = 2.1006 \times 10^5$$
$$d = -1.7465 \times 10^7$$
$$x = 129.26 \text{ mg/L}$$

The mass of air in the recycle water is about 129.26 mg/L for 601 kPa (500 kPa gauge pressure plus 101 kPa atmospheric). Based on the air mass balance in the contact zone of the DAF tank, the concentration of air released can be calculated:

$$C_b = \frac{C_r - C_{fl}}{1+r} r = \frac{(129.26 - 24)}{1 + 0.1}(0.10) = 9.64 \text{ mg/L}$$

Then, the bubble volume concentration φ_b is calculated:

$$\varphi_b = \frac{C_b}{\rho_{air}} = \frac{9.64}{1200} = 8033 \times 10^{-6} \text{ L/L (8033 ppm)}$$

After that, the bubble number concentration N_b is calculated:

$$N_b = \frac{10^{12} \times 6\phi_b}{\pi d_b^3} = \frac{(10^{12})(6)(0.008033)}{\pi 40^3} = 2.4 \times 10^5 \text{ bubbles/mL}$$

Now we can compare the concentrations of bubbles to floc particles. The ratio of the concentration of bubbles to floc particles is calculated for the bubble number concentration and the bubble volume concentration:

$$\frac{N_b}{N_p} = \frac{2.4 \times 10^5}{2000} = 120$$

Because the ratio of bubbles to particles is high, there is a lot of opportunity for particle collision and attachment:

$$\frac{\phi_b}{\phi_p} = \frac{8033}{2} = 4017$$

Because the ratio of bubble volume to particle volume is high, the floc–bubble density is low, resulting in high rise velocities of the particle–bubble aggregate.

Unlike complex mathematical approaches for estimating dissolved air concentration in DAF systems, the proposed correlation is simple to use and would be an immense help to process engineers, especially those dealing with water and wastewater treatment operations. Additionally, the level of mathematical formulation associated with the estimation of dissolved air concentration in DAF systems can be easily handled by process engineers or water treatment practitioners who do not have any in-depth mathematical knowledge.

The example shown clearly demonstrates the simplicity and usefulness of the proposed tool. Furthermore, estimations are quite accurate, as evidenced from the comparisons with literature data (with average absolute deviation being less than 0.5%) and will help in attempting faster engineering and operations in the water treatment industry.

The proposed method has a clear numerical background, wherein the relevant coefficients can be retuned quickly if more data become available in the future. The solubility of air in water increases in a linear fashion with increasing pressure. In addition, the solubility of air in water is reduced at elevated temperatures, as shown by the proposed predictive tool. In DAF designs in tropical areas, or anywhere where raw-water temperatures routinely exceed 25 °C, it is sometimes necessary to adopt higher than normal design recycle rates to ensure sufficient air can be delivered to the process at warmer temperatures.

2.10 Granular-Media Filters

2.10.1 General

Granular media filtration is a process for the removal of suspended solids by passage of water through a porous medium. Filtration is commonly the final polishing step in conventional

Figure 2.20 *A granular-media filter system (Reproduced with permission from © Environmental Products and Solutions (EPS)).*

water treatment processes, designed to meet final treated water turbidity limits. Figure 2.20 shows a granular-media filter.

The filtration process results in a gradual accumulation of entrapped solids within the granular media, which require intermittent removal by means of a filter backwash cycle. This cycle typically comprises both air scour and water wash phases, to effectively loosen and flush out the retained solids to waste.

Granular-media filters are facilities to separate undissolved (free) oil and solid suspended matter from the wastewater by filtering the water through a bed of granular media such as sand. Granular-media filters do not remove phenol, H2S, soluble BOD, or other soluble constituents. They remove solids from the water stream by straining through the filter bed, and by other mechanisms such as sedimentation between the media particles and adsorption upon them.

As impurities are removed, bed pressure drop increases, thus, the filters should periodically be taken out of service and backwashed in the up-flow direction to reduce pressure drop. Backwashing should be accomplished by expanding or fluidizing the bed with filter influent or effluent at a relatively high flow-rate and injecting air to scour the particles. Occasionally, a detergent wash or steaming may be employed to more thoroughly clean the bed.

2.10.2 Filter Types and Applications

Granular-media filters are differentiated by:

1. mode of operation:
   - down-flow
   - up-flow
2. number of media:
   - monomedia
   - mixed-media:

i. dual-media (e.g., sand and anthracite);
   ii. tri-media (e.g., garnet, sand, and anthracite).
3. material of media:
   - sand
   - anthracite
   - garnet or ilmenite.
4. driving force of flow:
   - gravity
   - pressure.
5. depth of penetration:
   - shallow-bed
   - deep-bed.

Granular-media filters can remove oil and suspended solids from a wastewater with or without the use of chemicals. In chemical applications, a chemical flocculent is often added to the wastewater prior to granular-media filtration, to enhance oil removal.

Additionally, granular-media filters can be designed and operated as simple coalescers, without chemicals, causing the small oil droplets to coalesce into larger ones as they pass through the filter bed.

When filters are operated in this manner, the oil is subsequently removed from the water by gravity separation. This technique can be used on oil-field produced water, but should not be used for refinery or petrochemical plant wastewater. Use of gravity filters should be avoided due to high water demand and large bed areas.

When possible, the design basis contaminant removal efficiency for granular media filters should be determined by tests using the particular waste stream of interest. With an existing wastewater stream, pilot plant work to establish effluent purity, cycle time, pressure drop, and pressure drop build-up, chemicals, and dosages should precede design.

For grassroots and/or other installations where tests cannot be conducted, the typical contaminant removal efficiencies given in Table 2.6 can be used for granular-media filters.

Table 2.6 Typical contaminant removal efficiencies for granular-media filters.

Filter Location	API Separator Efficiency	Biological Efficiency
Residual oil level:	less than 20 mg/L 80 to 90% free residual	80 to 90% free
Susp. sol. removal:	80 to 90%	80 to 90%
5-day BOD removal:	30% (Based on the removal of insoluble BOD when the filter is used as a pre-treatment process; this BOD reduction should not be expected when the filter is used for polishing. Granular-media filters do not remove soluble BOD or other soluble constituents.)	–

Deep-bed, down-flow, sand filters produce about the same quality effluent as up-flow sand filters or down-flow, dual-media filters, but with a shorter cycle time. Down-flow filters can operate with greater flow-rate variation than up-flow filters, because the media will not fluidize at high flow-rates. However, deep-bed down-flow sand filters may not be feasible with flocculents because of the low solid retention capacity of these filters. Shallow-bed, down-flow, gravity sand filters are conventionally used for polishing in domestic water treatment.

In general, in refinery or petrochemical plants, the following types of filters are preferred for wastewater treating purposes. Utilizing any other type should be approved by the Company:

1. Deep-bed, pressure, down-flow, sand filter.
2. Deep-bed, pressure, up-flow, sand filter.

2.10.3 System Design Parameters

1. The feed system should be designed to minimize flow surges.
2. Multiple units should be provided to allow continuous operation at maximum design capacity with one unit out of service for backwashing.
3. Units should be designed for continuous service and uninterrupted operation for a period of minimum 2 years.
4. The type of filters used should be vertical cylindrical filters with dished head ends supported on four legs. However, in the case of utilizing horizontal type filters (upon approval of the Company), the effective filtration area should be the area of the interface between the underdrain systems and the bottom media.
5. All equipment should be suitable for unsheltered, outdoor installation for the specified climatic zone.
6. The type of operation (manual or automatic) backwash, as well as provision of automatic programmable controllers, should be specified by the company.
7. Headers for feed water distribution, effluent collection, backwash water, and air scour should be provided.
8. Sufficient storage capacity with all necessary accessories should be provided for backwashing purposes.
9. Deep-bed, pressure filters, both up-flow and down-flow, may be used either with or without chemical additions depending upon the quality requirements of the effluent water requested by the Company. However, utilization of the chemicals should be proposed by the vendor based on the effluent water quality and should be approved by the Company.

2.10.3.1 Down-Flow Sand Filters

Down-flow sand filters should consist of a bed of sand supported on several layers of gravel resting on a backwash distributor that supports the entire bed. The sands shall be 1 to 2 mm in diameter. The minimum and maximum depth of sand bed should be 1.2 m and 2 m respectively. The gravel layers should be graded by particle size, with the coarsest layer on the bottom.

The backwash distributor should be designed such as to support the gravel and allow proper simultaneous distribution of air and water during backwashing.

Down-flow filters offer a more positive protection against breakthrough than do up-flow filters. Down-flow filters also offer the advantage of not fluidizing during filtration, and thus can operate with a relatively large temporary increase in flow.

2.10.3.2 Up-Flow Sand Filters

An up-flow sand filter should consist of a bed of sand supported by several layers of gravel of various sizes, resting on a bed support/distributor. A hold-down grid should be located near the top of the bed to prevent fluidization during filtration. The use of open-top vessels is not acceptable. The filters should be comprised of pressure vessels containing sands of 1 to 2 mm in diameter.

The filter bed support should be designed such as to allow simultaneous addition and distribution of air and water during backwashing. The minimum and maximum depth of sand bed should be 1.2 m and 2 m respectively.

The media in up-flow sand filters should be fine sand (fine relative to the other material in the filter) and should consist of a layer of coarse sand and two layers of gravel, each of a different size. All of the media should be of the proper size and shape. The shape should approach spherical (should be at least 0.8), to allow effective movement of the grains during backwashing. The uniformity coefficient (U) should be 1.2 minimum.

2.10.3.3 Backwashing Considerations

1. The water used to backwash the filter must be clean. It can be a portion of the filtrate, or clean water from another source. When filtrate is used, a holdup tank shall be provided to retain the necessary quantity. Alternatively, if there are enough filters and the pressure is sufficient, the backwash water can be taken directly from the filter effluent.
2. Two backwash pumps (one operated and one spare) should be provided. The backwash water rate depends upon the size of the sand particles and the temperature of the water and should be advised by the Vendor. The recommended backwash water rate is 39 m^3/h per each m^2 of the filter bed.
3. A restriction orifice, sized for about 45 to 50 m^3/h per each m^2 of the filter bed, should be provided in the backwash water line of each filter to limit the quantity of backwash water in order to avoid washing the sand from the bed. One hand control valve with a local indicator should be provided at the backwash pump's discharge line.
4. Air should be injected during backwashing to create movement, tumbling, and rotation of the media to scour off the retained material. The air rate will be advised by the vendor. The air pressure should be at least 300 kPa (ga). No deliberate back pressure is maintained on the filter beds during air blowing or backwashing. If safety problems (explosivity) could be encountered during air blowing, due to the presence of light components, inert gas or nitrogen should be used in place of air.
5. If plant air is unavailable, air blowers (with 100% spare capacity) to supply air for air scour during the backwash operation should be provided.
6. The backwash storage tank should have adequate capacity for at least 20 minutes of operation of the backwash pumps at design throughput.

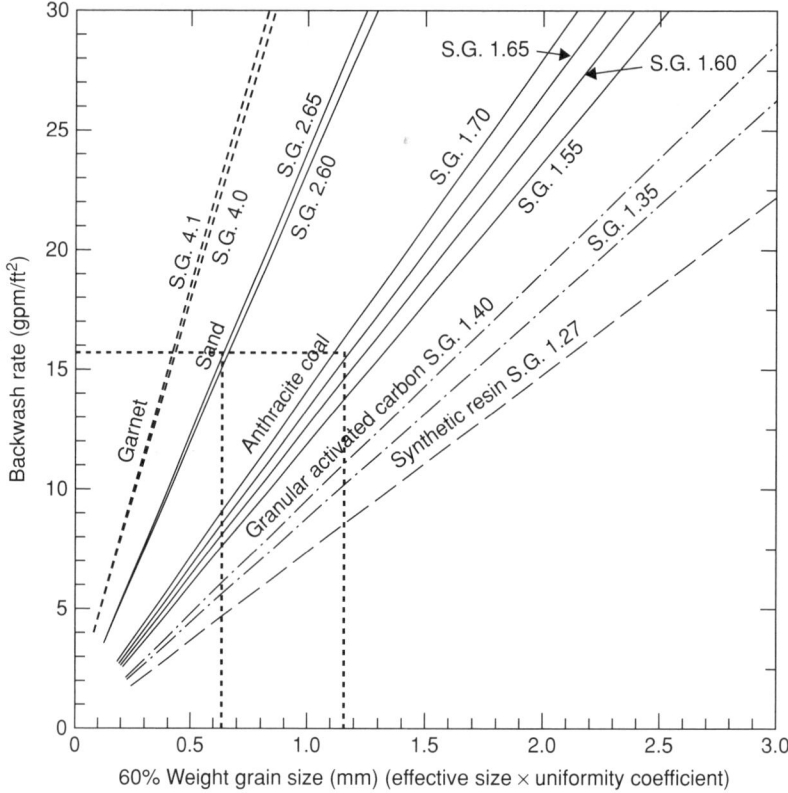

Figure 2.21 Estimation of backwash rate in a granular-media filter system at 20 °C (Reproduced with permission from © DDM Imagery).

7. When a bed is cleaned, the discharge from the backwash contains the bulk of the contaminants and should be sent to the sludge removal facilities.
8. If automatic backwashing is desirable and required by the Company, the automated backwash should be activated by a cycle timer, but should have overrides that activate the backwash procedure at high pressure drops across a filter and at low flowthrough ones. Additional overrides for high oil content in the effluent water and for high effluent turbidity should also be considered.
9. Figure 2.21 provides a rapid estimation of backwash water rate at 20 °C.

2.10.4 Cycle Time

On-stream filtration time depends upon the solid loading, oil content, and contaminant characteristics of the stream being filtered, and, when possible, should be determined by tests. Filtration run lengths on the order of 8 to 24 hours can be expected if chemicals are used; longer runs would be expected if chemicals are not used. Pressure drop may play a part in the retention of the oil and solids, so the flow to these filters should not be suddenly reduced or stopped and then increased or restarted, otherwise oil breakthrough or high pressure drops may ensue.

2.10.5 Vessels and Appurtenances

1. The filter vessel must be high enough to accommodate the required depth of filter media and an inlet distributor, and to allow a freeboard for bed expansion during backwashing. The manufacturer determines the space required under the inlet distributor. Filter media should be a minimum of 1.2 m for vertical vessels.

   The freeboard should be a minimum of two-thirds of the total bed depth, measured from the top of the filter media to the tangent line at the top of vessel, but not less than 1.2 meters.

   Filter media retainers should be provided above the bed of each filter vessel in up-flow filters to prevent fluidization of the bed during filtration.
2. Sand filters that are not subject to steaming should be designed for a minimum mechanical design temperature of 93 °C and a mechanical design pressure at least equal to the shut-off pressure of either the filter feed pumps or backwash water circulation pumps, whichever is greater. If steam is connected to the filter, the mechanical design pressure and temperature of the steam system should also be considered in addition to the above conditions. Pressure vessels should be designed and fabricated per ASME Code Section VIII Div. 1.
3. The filter bed should rest on a perforated baffle. The holes in the baffle should be covered with devices in order to retain the filter media in the bed. These devices should be equipped with slots or holes that are smaller than the particle dimension of the filter media.
4. The total required filtration area will be determined by dividing the flow-rate to be filtered by the filter hydraulic loading (allowable filtration rate per unit of the area with one unit out of service for backwashing). The maximum hydraulic loading for a deep-bed pressure, down-flow type filter should be 12.2 $m^3/h/m^2$ when one filter is in backwash operation.
5. The maximum media loss should be 5% of media volume per year.
6. Filter media traps should be provided on the outlet of each pressure filter unit to indicate loss of media in the event of underdrain failure. The maximum pressure drop through the trap should be 35 kPa (ga) when the unit is operating at maximum design flow-rate.
7. Units of 900 mm diameter and larger should have one 500 mm manhole in the top head and one of the same size below the media support. Units smaller than a 900 mm diameter should have one 250 mm diameter handhole located in the top head and another one below the media support. Manholes, of 500 mm (20 inch) minimum diameter, should be provided for access to the compartments of the units, as required. Manhole covers should be equipped with handling davits. In any case, access for removal of the that internal should be provided.
8. A filter media discharge connection should be provided close to the vessel bottom. It should be DN 150 (6 inch) as a minimum size and have a blind flange.
9. Drains should be provided to dewater each unit. The minimum size of the should be DN 50 (2 inch) for units up to 750 mm (30 inch) diameter and 75 mm (3 inch) for larger units.

10. Cleaning connections should be provided for each vessel.
11. A manual air relief valve for each vessel should be provided.
12. Sample valves should be placed on the inlet and the outlet of each vessel. Units operating above 65 °C should have a common sample cooler to cool the sample below 38 °C.
13. A pressure safety valve for each pressure vessel should be provided.
14. The filter vessels should be constructed of killed carbon steel with a minimum corrosion allowance of 3 mm. Where lining is required, the lining material should be compatible with the water quality and media abrasiveness. The Vendor's proposal for coatings and linings should be submitted to the Company for approval. Vessel interiors should be stainless steel.
15. Winterizing should be provided for filters if required by climatic conditions of the site.

2.10.6 Instrumentation and Controls

2.10.6.1 Pressure

1. A differential pressure gauge, with an adjustable high pressure alarm in the control room, should be furnished to monitor the DP across the combined inlet and outlet headers. Additional pressure gauges are required on the inlet and the outlet backwash headers for use during backwashing.
2. For each vessel, facilities should be provided to measure the pressure at the inlet and the pressure drop across the vessel during either filtration or backwashing operation. A high DP alarm in the control room should be provided for each vessel. These instruments will not reveal whether a filter is plugged, so individual flow instruments are also required.

2.10.6.2 Flow

Adjustable flow controls and flow indicators should be provided to control and indicate service, air scour or subsurface wash, backwash, and effluent flow from each vessel.

2.10.6.3 Level

Some means (preferably a sight glass) must be provided to detect the water level in the vessels when they are drained to the top of the sand prior to backwashing.

2.10.6.4 Temperature

Some means should be provided to measure and control the temperature of the backwash water.

2.10.6.5 Venting

Air vents should be DN 20 ($^3/_4$ inch) minimum size, piped to the nearest drain. When units are operated automatically, air vents should also operate automatically. In addition, an auxiliary, manually operated vent valve should be furnished. Automatically operated vent valves should be installed with a block valve to permit removal of the vent valve without having to shut down the unit.

2.10.6.6 Analyzers

If specified by the Company, the following in-line analyzers with indicators and alarms in the control room should be provided for the effluent water from the unit:

1. Turbidity analyzer (surface scatter type).
2. Oil analyzer.

2.10.6.7 Programmable Controller for Automatic Operation

If the provision of a programmable controller for automatic operation is specified by the Company, it should be provided in accordance with the following requirements:

1. The number of programmable controllers required should be specified.
2. The control mode (automatic/step/manual) should permit selection and control of every step within the sequence of operation.
3. The control system should provide for status selection (service/standby/power off) and identification.
4. Interlocks should be provided to prevent:
   - more than one unit from backwashing simultaneously; and
   - filter backwash, where the levels in either the clean or the dirty backwash storage tank are insufficient for complete backwash.
5. Some means to verify that a step has been completed within the specified time should be provided. If a step fails, an alarm should be activated and either the sequence stopped or another contingency step initiated.
6. Programmable controllers should have provisions for manual operation from a locally mounted solenoid cabinet in the event of a programmable controller malfunction.
7. Local solenoid cabinets containing solenoids for the manual operation of automatic valves should be located adjacent to the units. Operation of the solenoids should be by means of switches mounted outside the cabinet. Such switches should be protected by a hinged, weatherproof cover.
8. A remote interface should be provided in the control room. This interface should include graphics in the form of either a hard-wired graphics panel or a display presented on a video screen, as specified. The minimum graphics display for each unit shall be as follows:
   - Control mode (automatic/step/manual).
   - Unit status (service/standby/power off).
   - Programmable controller malfunction.
   - Indication of each step of the backwash sequence.
   - On/off indication of pumps and compressors.
   - Position of each automatic valve.
   - Valve malfunction during the sequence.
   - A schematic of the filter unit to display the status.
9. Programmable controller input and output circuits should be individually isolated and protected against transitions in field wiring affecting the internal logic.

3
Chemical Treatment

3.1 Introduction

Chemicals are used during wastewater treatment in an array of processes designed to expedite disinfection. These chemical processes, which induce chemical reactions, are called chemical unit processes and are used alongside biological and physical cleaning processes to achieve various water standards. There are several distinct chemical unit processes, including chemical coagulation, chemical precipitation, chemical oxidation and advanced oxidation, ion exchange, and chemical neutralization and stabilization, which can be applied to wastewater during cleaning.

3.1.1 Chemical Precipitation

Chemical precipitation is the most common method for removing dissolved metals from a wastewater solution containing toxic metals. To convert the dissolved metals into solid particle form, a precipitation reagent is added to the mixture. A chemical reaction, triggered by the reagent, causes the dissolved metals to form solid particles. Filtration can then be used to remove the particles from the mixture. How well the process works is dependent upon the kind of metal present, the concentration of the metal, and the kind of reagent used. In hydroxide precipitation, a commonly used chemical precipitation process, calcium or sodium hydroxide is used as the reagent to create solid metal hydroxides. However, it can be difficult to create hydroxides from dissolved metal particles in wastewater because many wastewater solutions contain mixed metals.

3.1.2 Chemical Coagulation

This chemical process involves destabilizing wastewater particles so that they aggregate during chemical flocculation. Fine solid particles dispersed in wastewater carry negative electric surface charges (in their normal stable state), which prevent them from forming larger groups and settling. Chemical coagulation destabilizes these particles by introducing positively charged coagulants that then reduce the negative particles' charge. Once the charge is reduced, the particles freely form larger groups. Next, an anionic flocculant is introduced to the mixture. Because the flocculant reacts against the positively charged mixture, it either neutralizes the particle groups or creates bridges between them to bind the particles into larger groups. After larger particle groups are formed, sedimentation can be used to remove the particles from the mixture.

3.1.3 Chemical Oxidation and Advanced Oxidation

With the introduction of an oxidizing agent during chemical oxidation, electrons move from the oxidant to the pollutants in the wastewater. The pollutants then undergo structural modification, becoming less destructive compounds. Alkaline chlorination uses chlorine as an oxidant against cyanide. However, alkaline chlorination as a chemical oxidation process can lead to the creation of toxic chlorinated compounds, and additional steps may be required. Advanced oxidation can help remove any organic compounds that are produced as a byproduct of chemical oxidation, through processes such as steam stripping, air stripping, or activated carbon adsorption.

3.1.4 Ion Exchange

When water is too hard, it is difficult to use it for cleaning and it often leaves a grey residue. (This is why clothing washed in hard water often retains a dingy tint.) An ion exchange process can be used to soften the water. Calcium and magnesium are common ions that lead to water hardness. To soften the water, positively charged sodium ions are introduced in the form of dissolved sodium chloride salt, or brine. Hard calcium and magnesium ions exchange places with sodium ions, and free sodium ions are simply released in the water. However, after softening a large amount of water, the softening solution may fill with excess calcium and magnesium ions, requiring the solution be recharged with sodium ions.

Ion exchange resins are classified according to their specific application as per Table 3.1.

3.1.4.1 Design Criteria for an Ion Exchange System

Design criteria for an ion exchange system should be based upon (API, 1988; ASTM, 1978; DIN, 1970; Chapman and Grammas, 1989):

- the required flow rate;
- the influent water quality,
- the desired effluent water quality;
- the exchange capacity and hydraulic characteristics of the exchanger;
- the period between regenerations;
- the type of operation – manual or automatic; and
- the flexibility required, that is the number of softener units.

Ion exchangers are not economically suitable for demineralizing waters containing more than 1000–2000 mg/kg of dissolved solids, except in a few specialized industrial applications. Table 3.1 shows the classification of ion exchange systems.

Table 3.1 Classification of ion exchange resins.

Type	Application	Ionic Form in the Ready-to-Use Condition	Regenerating Agent Aqueous Solution of
Cation exchange resins Strongly acidic	reduction of calcium ion concentration	Na	NaCl
Cation exchange resins Strongly acidic	reduction of salt content	H	HCl, H_2SO_4
Weakly acidic	reduction of hydrogen carbonate concentration	H	HCl, H_2SO_4, CO_2
Weakly acidic	reduction of heavy metal ion content	Na, H	$HCl, H_2SO_4,$ NaOH
Anion exchange resins Strongly basic	reduction of salt content;	OH	NaOH
Anion exchange resins Strongly basic	reduction of the content of certain ions, e.g., nitrate ions, sulfate ions	Cl, HCO_3	NaCl, $NaHCO_3$
Anion exchange resins Strongly basic	reduction of the organic substance content, e.g., humic acids	Cl, OH	NaCl, NaOH
Weakly basic	reduction of salt content	Free base	NaOH
Weakly basic	reduction of heavy metal ion content	Free base	NaOH
Weakly basic	reduction of the organic substance content, e.g., humic acids	Free base	NaOH

The process of ion exchange for softening waters is preferable to the precipitation process when one or more of the following conditions exist:

- less than 100 mg/kg of hardness expressed as calcium carbonate is present in the water;
- an extremely low dissolved solids content is required;
- only a limited volume of treated water is required.

The relative exchange capacity of cation exchangers and regenerative salt dosages will be as per Table 3.2.

Anion exchangers typically have an exchange capacity calculated as $CaCO_3$ of 27.4–57.2 g/L at a sodium dosage of 1.05–7 kg/kg removed. Interstate degasification in demineralization systems should be considered for flows over 22.7 m^3/h and alkalinity over 100 mg/kg.

When ultra-pure water is required, using the mixed bed demineralizer is recommended.

Table 3.2 Relative exchange capacity of cation exchangers.

Cation Exchanger	Nominal Exchange Capacity, g/L	Regenerative Salt Dosage	
		Volumetric kg/m³	Effective kg/kg Hardness Removed
Greensand	6.4	20.2	3.1
Processed greensand	12.6	39.5	3.1
Synthetic siliceous Zeolite	25.2	79.2	3.1
Resin, polystyrene	73.2	201.7	3.1
Resin, polystyrene	50.3	80	1.7

The demineralized water storage(s) should be designed in order to store the demineralized water produced and to cover the following uses:

- Make-up to deaerators.
- Process units.
- Regeneration of condensate treatment.

3.1.5 Chemical Stabilization

This process works in a similar fashion to chemical oxidation. Sludge is treated with a large amount of a given oxidant, such as chlorine. The introduction of the oxidant slows down the rate of biological growth within the sludge, and also helps deodorize the mixture. The water is then removed from the sludge. Hydrogen peroxide can also be used as an oxidant, and may be a more cost-effective choice.

3.2 Definition and Application

Chemical unit processes are those used for the treatment of wastewater in which change is brought about by means of or through chemical reaction. In the field of wastewater treatment, chemical unit processes are usually used in conjunction with the physical unit operations and the biological unit processes to meet treatment objectives.

Applications of chemical unit processes in wastewater treatment are presented in Table 3.3 below. In considering the application of the following chemical unit processes, the inherent disadvantages associated with most chemical processes (activated carbon adsorption is an exception), as compared with the physical unit operations, are that they are additive processes and cause an increase in the total dissolved solids concentration of the wastewater. If the treated wastewater is to be reused, this can be a significant factor. Another disadvantage of chemical unit processes is that they are all intensive in operating costs.

For additional chemical applications in wastewater collection, treatment, and disposal, reference can be made to Table 3.4.

Table 3.3 Applications of chemical unit processes in wastewater treatment.

Process	Application
Chemical precipitation	Removal of phosphorus and enhancement of suspended solids removal in primary sedimentation facilities used for physical-chemical treatment.
Chemical flocculation	Removal of free oil & suspended solids from wastewater.
Adsorption	Removal of organic material not removed by conventional chemical and biological methods. Also used for dechlorination of wastewater before final discharge for treated effluent.
Disinfection	Selective destruction of disease-causing organisms (can be accomplished in various ways).
Disinfection with chlorine	Selective destruction of disease-causing organisms. Chlorine is the most commonly used chemical.
Dechlorination:	Removal of total combined chlorine residual that exists after chlorination (can be accomplished in various ways).
Disinfection with chlorine: dioxide, bromine chloride, ozone or ultraviolet light	Selective destruction of disease-causing organisms.
Other chemical applications	Various other chemicals can be used to achieve specific objectives.

3.2.1 Activated Carbon Adsorption

Activated carbon has been used for a long time in water and wastewater treatment. The adsorption capacity of activated carbon is of special interest due to its large surface area. The surface properties of activated carbon depend upon both the initial material used to produce the carbon and the exact production procedure. Generally, granular carbon from bituminous coal has a small pore size, a large surface area, and the highest bulk density. Lignite carbon has the largest pore size, least surface area, and the lowest bulk density.

The ability of activated carbon to adsorb pollutants is a function of both the characteristics and concentration of the pollutants and the temperature. Generally, the amount of material adsorbed is determined as a function of the concentration at a constant temperature, and the resulting function is called an adsorption isotherm. The most common adsorption isotherm equations used in water and wastewater treatment are the Freundlich isotherm and Langmuir isotherm.

- **Freundlich isotherm:**

$$\frac{x}{m} = K_f C_e^{1/n} \tag{3.1}$$

Where:

x/m: amount of adsorbate adsorbed per unit of carbon

C_e: equilibrium concentration of adsorbate in solution after adsorption

K_f, n: empirical constants

Table 3.4 Applications of chemical unit processes in wastewater treatment.

Application	Chemicals used	Remarks
Collection		
Slime-growth control	Cl_2, H_2O_2	Control of fungi and slime-producing bacteria.
Corrosion control (H_2S)	Cl_2, H_2O_2, O_3	Control brought about by destruction of H2S in sewers.
Corrosion control (H_2S)	$FeCl_3$	Control brought about by precipitation of H_2.
Odor control	Cl_2, H_2O_2, O_3	Especially in pumping stations and long, flat sewers.
Treatment		
Grease removal	Cl_2	Added before preaeration.
BOD reduction	Cl_2, O_3	Oxidation of organic substances.
pH control	KOH, $Ca(OH)_2$, NaOH	–
Ferrous sulfate oxidation	Cl_2	Production of ferric sulfate and ferric chloride [$6(FeSO_4 \cdot 7H_2O) + 3Cl_2 \rightarrow 2FeCl_3 + Fe_2(SO_4)_3 + 42H_2O$].
Filter-ponding control	Cl_2	Residual at filter nozzles.
Filter-fly control	Cl_2	Residual at filter nozzles, used during fly season.
Sludge-bulking control	Cl_2, H_2O_2, O_3	Temporary control measure.
Digester supernatant oxidation	Cl_2	–
Digester and Imhoff tank foaming control	Cl_2	–
Ammonia oxidation	Cl_2	Conversion of ammonia to nitrogen gas.
Odor control	Cl_2, H_2O_2, O_3	
Oxidation of refractory organic compounds	O_3	
Disposal		
Bacterial reduction	Cl_2, H_2O_2, O_3	Plant effluent, overflows, and stormwater.
Odor control	Cl_2, H_2O_2, O_3	

Note: Cl_2 = chlorine, H_2O_2 = hydrogen peroxide, O_3 = ozone, KOH = potassium hydroxide, $Ca(OH)2$ = calcium hydroxide, NaOH = sodium hydroxide.

- **Langmuir isotherm:**

$$\frac{x}{m} = \frac{abC_e}{1 + bC_e} \quad (3.2)$$

Where:

x/m: amount adsorbate adsorbed per unit of carbon

C_e: equilibrium concentration of adsorbate in solution after adsorption

a, b: empirical constants

Table 3.5 Chemicals used in wastewater treatment.

Chemical	Formula	Molecular Mass	Density, (kg/m³)	
			Dry	Liquid
Alum	$Al_2(SO_4)_3 \cdot 18H_2O$	666.7 594.3	960–1200	1250–1280 (49%)
	$Al_2(SO_4)_3 \cdot 14H_2O$		960–1200	1330–1360 (49%)
Ferric chloride	$FeCl_3$	162.1	–	1345–1490
Ferric sulfate	$Fe_2(SO_4)_3$	400 454	–	1120–1153
	$Fe_2(SO_4)_3 \cdot 3H_2O$			
Ferrous Sulfate (copperas)	$FeSO_4 \cdot 7H_2O$	278	993–1057	–
Lime	$Ca(OH)_2$	56 as CaO	560–800	–

Activated carbon can be classified into two types: powdered activated carbon (PAC), which has a diameter of less than 200 mesh, and granular activated carbon (GAC), which has a diameter greater than 0.1 mm. GAC is usually used in a fixed bed column while PAC is added to the effluent from secondary biological treatment to serve as a polishing process. Another application of PAC is the direct addition of PAC to the aeration basin of the activated-sludge process.

3.3 Chemical Precipitation

Chemical precipitation in wastewater treatment involves the addition of chemicals to alter the physical state of dissolved and suspended solids and to facilitate their removal by sedimentation. Chemical processes, in conjunction with the various physical operations, can be used in the secondary treatment of untreated wastewater, including the removal of either nitrogen or phosphorus, or both. The most common chemicals used in wastewater treatment are listed in Table 3.5. The degree of clarification obtained depends on the quantity of chemicals used and the care with which the process is controlled. It is possible to obtain a clear effluent by chemical precipitation, substantially free from matter in suspension or in the colloidal state. Around 80 to 90% of the total suspended matter, 40 to 70% of the BOD_5, 30 to 60% of the COD, and 80 to 90% of the bacteria can be removed by chemical precipitation. In comparison, when plain sedimentation is used, only 50 to 70% of the total suspended matter and 30 to 40% of the organic matter settles out.

3.4 Chemical Flocculation

To meet the requirements of potable, industrial, and agricultural water, the immediate need is to treat wastewater, particularly the sewage sludges and slimes from the municipal and chemical oil and gas industrial effluents respectively. These effluents are very undesirable and unsafe. The removal of contaminants from wastewater is a must before they can be reused.

The removal of contaminants from these effluents involves the processes of flocculation and coagulation. Colloidal particles in nature normally carry charges on their surface, which lead to the stabilization of the suspension. Through addition of some chemicals, the surface property of such colloidal particles can be changed or dissolved material can be precipitated so as to facilitate the separation of solids by gravity or filtration. The conversion of the stable state dispersion to the unstable state is termed destabilization and the processes of destabilization are coagulation and flocculation. Often the terms coagulation and flocculation are used synonymously despite there being a subtle difference between the two. If destabilization is induced through charge neutralization by addition of inorganic chemicals, the process is called coagulation. On the other hand, the process of forming larger agglomerates of particles in suspension or of small agglomerates already formed as a result of coagulation through high molecular weight polymeric materials is called flocculation. No substantial change of surface charge is accomplished in flocculation. The agglomerates formed by coagulation are compact and loosely bound, whereas the flocs are of larger size, strongly bound, and porous in the case of flocculation.

3.4.1 Definition and Applications

Chemical flocculation can be used to remove suspended matter and free oil from water. The coagulating agent hastens the settling of suspended material and often allows removal of solids not separated by conventional sedimentation.

Chemical flocculation consists of the following four steps:

1. **Coagulation**
   The formation of flocculent particles by adding chemicals. This is sometimes termed flash-mixing or rapid mixing (Figure 3.1)
2. **Flocculation**
   The provision of retention time with gentle agitation to allow the floc particles or precipitate, associated with the impurities, to increase in size by agglomeration. This step is sometimes called slow-mix (Figure 3.2).
3. **Sedimentation**
   Settling of the floc by gravity (Figure 3.3).

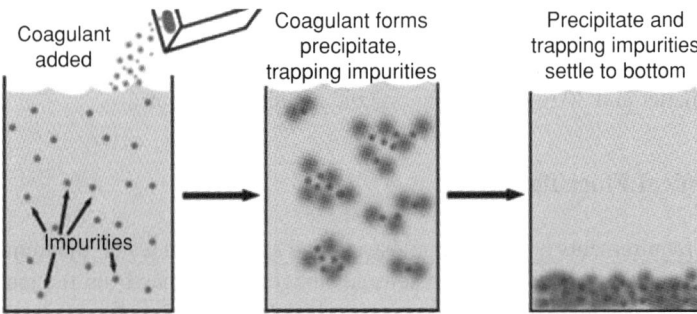

Figure 3.1 A schematic for coagulation process (Reproduced with permission from [40], © Springer, 2013).

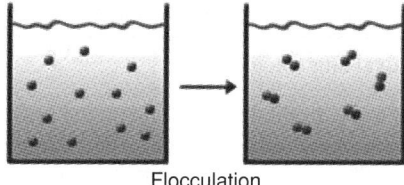
Flocculation

Figure 3.2 A schematic for the flocculation process (Reproduced with permission from [40], © Springer, 2013).

4. **Removal of the settled material**
   After coagulation, flocculation, and settling, the floc is removed as a sediment or sludge.

Chemical flocculation units normally serve to protect a downstream biological oxidation unit from excessive influxes of oil. Other potential uses for chemfloc are:

- To protect downstream carbon adsorption columns from excessive oil and solids.
- To increase the cycle time of downstream deep-bed sand or mixed media filters.
- To reduce the oil content of API separator effluent prior to final discharge.
- To reduce color and turbidity.
- To remove stabilized, fine oil droplets that are ineffectively removed by gravity separation.
- To remove some COD and BOD associated with the non-dissolved material (usually the BOD of the water is hardly affected by the chemical flocculation process).
- To reduce the concentration of sulfides (by using iron salts).

A chemical flocculation unit should consist of at least of the following major items:

- chemical storage, metering, and addition;
- flash-mixing;
- flocculation;
- clarification;
- sludge handling, treatment, and disposal.

Design contaminant removal efficiencies for a chemical flocculation unit are shown in Table 3.6.

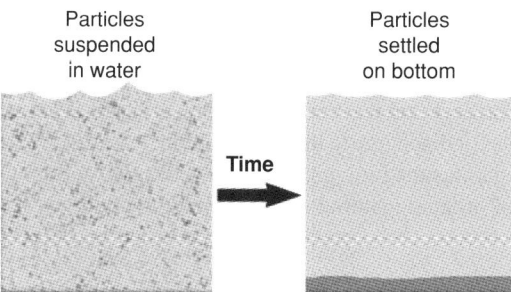

Figure 3.3 A schematic for the sedimentation mechanism in wastewater treatment (Reproduced with permission from [40], © Springer, 2013).

Table 3.6 *Design contaminant removal efficiencies for chemical flocculation units.*

Contaminant	Removal Efficiencies
Oil	5–25 mg/L residual (depending on the pre-treating facilities of the feed), measured by extraction and infrared analysis
Suspended Solids	50 to 80% removal with a minimum residual of 20 mg/L
Hydrogen Sulfide	10% removal
Phenol	10% removal
5-Day Bod	10% removal (based on removal of insoluble BOD)

3.4.2 Design Considerations

1. A chemical flocculation unit operating on refinery or petrochemical plant wastewater should generally be preceded by pre-treatment and by flow equalization. The type of pre-treatment and the amount of flow equalization vary with the waste and with the installation, and must be studied for each case. Pre-treatment should include at a minimum sulfide and ammonia stripping facilities, and settling in an API type or other gravity settler to remove oil and settleable solids.
2. Pumping of the flocculated water or subjecting it to any other turbulence should be avoided.
3. The flocculation chamber should contain a slow-speed mixer which gently agitates the fluid. The mixer should be a variable speed, mechanical agitator type.
4. High temperatures can upset the clarifier, so a high temperature alarm in the control room of the sewer water should be provided to warn of excessive influent temperature. In addition, provision of high temperature alarms on critical streams into the sewer should be taken into consideration. pH swings can also upset the clarifier and, therefore, meters to measure both alkalinity and free mineral acidity, or pH meters with attendant alarms on the sewer water, should be provided.
5. The alarms should allow diversion to avoid excessive swings in temperature, acidity, alkalinity, or pH.
6. The chemical addition equipment should be designed with sufficient flexibility to allow the addition of acids, alkali, coagulant, and coagulant aid over the range of rates.
7. Special attention should be paid to covering the flocculation chamber if required by either air pollution regulations or the relevant Company's instructions.

3.4.3 Clarifier

Secondary circular clarifiers for wastewater (Figure 3.4) are designed to provide a high quality effluent suitable for discharge to the environment or further treatment. The secondary clarifier effectively separates the biological floc and colloidal solids to produce wastewater with very low levels of organic material and suspended matter.

3.4.3.1 Design Considerations

Solids from the aeration basin settle out in the secondary clarifiers. Some of the solids are recirculated back to the aeration basin to increase the rate of organic decomposition. Solids

Figure 3.4 Secondary circular clarifier (Reproduced with permission from © Hickory wastewater treatment).

are also pumped out from the secondary clarifiers to control the age of the bugs in the aeration basin.

The treated wastewater runs over the edges of secondary clarifiers. A part of the settled sludge is returned into aeration tanks, where is mixed with "fresh" primary treated wastewater and bio-oxidation process continues.

The settled sludge goes for further treatment – anaerobic decomposition in controlled conditions with biogas (methane) production.

1. A minimum of two clarifiers should be provided. Each clarifier should be designed to operate under the following conditions:
   - Total retention time should not be less than 3 hours, including recycling.
   - Clarifier overflow (rise) rate should not be greater than 2.4 L/day/cm^2.
   - Oil 5–25 mg/L residual (depending on the pre-treating facilities of the feed), measured by extraction and infrared analysis.
   - Suspended solids 50 to 80% removal with a minimum residual of 20 mg/L.
   - Hydrogen sulfide 10% removal.
   - Phenol 10% removal.
   - 5-day BOD 10% removal (based on removal of insoluble BOD).
   - The system should be designed for a recycle flow rate equal to 100% of the fresh feed to the clarifier.
   - Sludge withdrawal should be designed based on the continuous handling of a flow-rate equal to 100% of the fresh feed.
2. The basins should preferably be circular in shape and of reinforced concrete construction with a bottom slope of 8%. However, rectangular or square basins can also be used depending upon space availability and the Company's approval.
3. The minimum side water depth should be 4.5 m. At least 600 mm freeboard above the water depth should be provided.

92 Waste Management in the Chemical and Petroleum Industries

4. The unit must be capable of effective operation at all rates from 40% of design to 100% of design.

3.4.3.2 Clarifier Components

1. A conical hood should be included in the center of the circular clarifiers to separate the flocculation chamber from the clarification chamber.
2. The outlet weirs should be level to promote even distribution. Level weirs and the capability of being leveled after installation should be specified. Submerged weirs should be specified where freezing problems are anticipated. Multiple weirs can be employed to reduce the weir rate in both rectangular and circular clarifiers. In circular clarifiers, multiple weirs should be of the radial type and should be at least equivalent in length to a circumferential weir. Weirs should be arranged to uniformly collect the effluent over a substantial portion of the surface to promote full utilization of the settling area and to reduce short-circuiting.
3. The clarified water flows over the weir(s) into a collecting trough or launder which leads it to the outlet pipe. Peripheral effluent launders should be flat-bottomed and sized such as to result in a velocity of around 0.6 m/s.
4. Sludge scrapers should be provided to move the settled floc and contaminants to an outlet at the bottom of the tank. The scrapers in circular clarifiers should move the sludge to a central outlet in the cone at the bottom of the tank. Sludge scrapers in circular units should be rotated from a central shaft. The rotary mechanism should operate at a relatively low speed (top speed of around 2.5 cm/s) to avoid disturbance of the settled solids. Circular sludge scrapers should be designed for continuous operation and constant speed drive.
5. The sludge should be removed from the clarifiers by the hydrostatic pressure in the tank to a pit outside of the clarifier.

    Sludge discharged from the clarifier is usually intermittent and should be controlled by an automated blow-off valve. Provisions should be made in the sludge outlet for backflushing with water under pressure prior to sludge discharge.

    The frequency and duration of the backflush and blow-off can be automatically controlled by cycle timers. The sludge withdrawal rate should be frequently regulated by a manually set timing sequence. The setting of this should be based on the amount of floc being formed and the amount of silt and dirt entering the chemfloc system. The accumulated sludge in the sludge sump should be recycled back by the sludge pumps to the clarifier influent. Around 2 to 5% of the sludge should be routed to the aerobic digester, sludge thickener, and/or sludge basin.
6. The clarifier should be equipped with an oil skimmer, in conjunction with an oil moving device. When the chemfloc unit is followed by a biox plant, it is particularly important to avoid sending large slugs of oil to the biox unit. The oil moving device usually consist of slowly moving blades, which direct settled oil to the skimmer for removal. In circular clarifiers, the oil moving device should be rotated by a common shaft with the sludge scraper. In rectangular clarifiers, the oil moving device should be of the continuous, flight scraper type.

3.4.4 Chemical Addition Systems

3.4.4.1 Type of Chemicals

1. **pH adjustment**
   Either a basic material or an acid should be consistently added for pH adjustment of the wastewater. Liquid caustic is normally used to raise pH; sulfuric acid is often used to lower it. Acid is normally added with a metering pump, and the acid addition rate is regulated by the pump speed or pump stroke, which is set by the pH controller. A pH meter/controller to control the addition of the basic or acidic material should be provided.
2. **Coagulants**
   Alum is the most commonly used coagulant. It can be used as a solid or water solution. If the alum solution requires dilution, it should be diluted to the optimum concentration (determined by testing) for the water to be treated. Dilution water should be added when the alum solution is being unloaded, in order to achieve good mixing. Alum is corrosive, and the piping and equipment in alum solution service should be fabricated of corrosion resistant materials.
3. **Coagulant aids**
   Polyelectrolytes are the most commonly used coagulant aids. Jar tests should be performed to determine the most suitable form of polyelectrolyte (cationic, acidic, or neutral) for a particular waste and contaminant. Activated silica may also be used as a coagulant aid. Utilizing activated silica is not recommended due to its extremely short shelf-life. Facilities should be provided for adding the coagulant aid to both the flash-mix and slow-mix (flocculation chamber) sections.

3.4.4.2 Polyelectrolyte Handling and Mixing System

Polyelectrolyte handling and mixing system requirements should be at the minimum as per the following clauses. The requirements outlined below are also applicable to any other chemical handling system (where required) and/or liquid polyelectrolyte plus activator.

1. The polyelectrolyte should be used as a coagulant aid at the influent of the dissolved air flotation unit and clarifier.
2. The equipment and facilities should receive the input plant water from the plant source and the polyelectrolyte from the concentrated polyelectrolyte containers for ratio balancing and transfer as separate streams to the tank and should also include a control system and electrical terminals. The design concentration of the dilute polyelectrolyte solution should be 1% mass (using liquid) and 0.2% mass (using dry powder).
3. Tanks should be completed with gauge glass, a hinged cover, and level control connections for drain and pump suction.
4. The control system should be provided to sequence water and polyelectrolyte flush cycles to prevent undispersed polyelectrolyte from entering the system. Safety circuits should be provided to monitor normal operating conditions and to shut down the unit in the event of a component failure. A low level switch should also be provided on all tanks to stop the relevant pumps when the tank experiences low level conditions.

5. The system should be located inside a suitable shelter or building.
6. Suitable material should be used for all equipment and piping.
7. The system should include but not be limited to the following equipment:
    - water pumps(one operating and one spare);
    - polyelectrolyte metering pumps (one operating and one spare);
    - polyelectrolyte solution tank with minimum capacity of 3 m^3;
    - activator supply tank with minimum capacity of 2 m^3;
    - water tank with a minimum capacity adequate for 24 hours;
    - polyelectrolyte inline static mixer(s);
    - polyelectrolyte dosing screen conveyor;
    - polyelectrolyte concentrate tank with minimum capacity of 60 liters (60 L);
    - calibration pots on the activator metering pumps suction line;
    - polyelectrolyte injection pumps to the clarifier (one operating and one spare) including a calibration pot on the suction line;
    - polyelectrolyte injection pumps to the DAF unit (one operating and one spare) including a calibration pot on the suction line;
    - polyelectrolyte premix water pumps to the DAF unit (one operating and one spare);
    - all necessary instrumentation including a local control panel.

3.5 Disinfection

Primary, secondary, and even tertiary treatment cannot be expected to remove 100% of the incoming waste load and, as a result, many organisms still remain in the waste

Table 3.7 Characteristics of an ideal chemical disinfectant.

Characteristics	Remarks
Toxicity to microorganisms	Should have a broad spectrum of activity at high dilutions.
Solubility	Must be soluble in water or cell tissue.
Stability	Loss of germicidal action on standing should be low.
Non-toxic to higher forms of life	Should be toxic to organisms and non-toxic to man and other animals.
Homogeneity	Solution must be uniform in composition.
Interaction with extraneous material	Should not be absorbed by organic matter.
Toxicity at room temperature	Should be effective in environmental temperature range.
Penetration	Should have the capacity to penetrate surfaces.
Non-corrosive and non-staining	Should not disfigure metals or stain clothing.
Deodorizing ability	Should deodorize while disinfecting.
Detergent capacity	Should have cleansing action to improve effectiveness of disinfectant.
Availability	Should be available in large quantities and be reasonably priced.

stream. To prevent the spread of water-borne diseases and to minimize public health problems, the destruction of pathogenic organisms in wastewaters is required. While most of these microorganisms are not pathogens, pathogens must be assumed to be potentially present.

Thus, whenever wastewater effluents are discharged to receiving waters that may be used for water supply, or any other purposes, the reduction of bacterial numbers to minimize health hazards is a very desirable goal. When disease-causing organisms are not 100% destroyed during the processing, disinfection should be used. In wastewater treatment, disinfection should be accomplished through use of chemical agents, physical agents, mechanical means, and radiation.

3.5.1 Chemical Agents

Chemical agents that should be used as a disinfectant include: alcohols, iodine, chlorine and its compounds, bromine, ozone, dyes, soaps and synthetic detergents, quaternary ammonium compounds, hydrogen peroxide, and various alkalis and acids. The characteristics of an ideal chemical disinfectant are shown in Table 3.7.

3.5.2 Mechanical Means

Typical removal efficiencies for various treatment processes are listed in Table 3.8:

Table 3.8 Removal or destruction of bacteria by different treatment processes.

Process	Percent Removal
Coarse screens	0–5
Fine screens	10–20
Grit chambers	10–25
Plain sedimentation	25–75
Chemical precipitation	40–80
Trickling-filters	90–95
Activated sludge	90–98
Chlorination or treated sewage	98–99
Fine screens	10–20
Grit chambers	10–25

3.6 Chlorination

Chlorination is by far the most common method of wastewater disinfection and is used worldwide for the disinfection of pathogens before discharge into receiving streams, rivers, or oceans. Chlorine is known to be effective in destroying a variety of bacteria, viruses, and protozoa, including Salmonella, Shigella, and Vibrio cholera. Today, wastewater chlorination is widely practiced to reduce microbial contamination and potential disease risk for exposed populations.

Figure 3.5 *Chlorine contact chamber (Reproduced with permission from © Hickory wastewater treatment).*

Figure 3.5 shows a chlorine contact chamber. The contact chamber is maze shaped to allow the chlorine to completely mix with the water. Extra chlorine residual is removed by sulfur dioxide. Chlorination plays a key role in the wastewater treatment process by removing pathogens and other physical and chemical impurities. Chlorine's important benefits to wastewater treatment are listed below:

- disinfection;
- controlling odor and preventing septicity;
- aiding scum and grease removal;
- controlling activated sludge bulking;
- controlling foaming and filter flies;
- stabilizing waste activated sludge prior to disposal;
- foul air scrubbing;
- destroying cyanides and phenols;
- ammonia removal.

3.6.1 Application

The principal uses of chlorine and its compounds are listed in Table 3.9.

3.6.2 Chlorine Dosages

Typical chlorine dosages for disinfection are shown in Table 3.10.

3.6.3 Design Considerations

1. Each chlorinator unit should be furnished complete, including a chlorinator, any necessary manifold valving for the chlorinator cylinders, an ejector, a chlorinator flow-rate

Table 3.9 Chlorination applications in wastewater collection and treatment.

Application	Dosage Range mg/L	Remarks
Collection		
Slime-growth control	1–10	Control of fungi and slime-producing bacteria.
Corrosion control (H_2S) [Per mg/liter of H_2S]	2–9	Control brought about by destruction of H_2S in sewers.
Odor control [Per mg/liter of H_2S]		Especially in pump stations and long flat sewers.
Treatment		
Grease removal	2–10	Added before preparation.
BOD reduction, [Per mg/liter of BOD, destroyed]	0.5–2	Oxidation of organic substance.
Ferrous sulfate oxidation, $6FeSO_4, 7H_2O + 3Cl_2$ $\rightarrow 2FeCl_3 + 2Fe_2(SO_4)_3 + 42H_2O$	Based on reaction stoichiometry	Production of ferric sulfate and ferric chloride.
Filter-pounding control	1–10	Residual at filter nozzles.
Filter-fly control	0.1–0.5	Residual at filter nozzles, used during fly season.
Sludge-bulking control	1–10	Temporary control measure.
Digester supernatent oxidation	20–140	–
Digester and Imhoff tank	2–15	–

   indicator and adjusting valve, a strainer, and all necessary solution feed and vacuum tubing, valves, adapters, check valves, etc.
2. Separate housing for the chlorination system with forced ventilation that has start-up bottom at the room entrance should be supplied.
3. Chlorine cylinders with handling equipment should be provided.
4. A space heater should be provided for the room.
5. The chlorinators should be manually operated on an intermittent push-button start-up basis, with all vacuum gauges and the rotameter located on the front of the panel. The rate of chlorine feed should remain constant until manually changed.

Table 3.10 Typical chlorine dosages for disinfection.

Effluent From	Dosage Range (mg/L)
Untreated wastewater (prechlorination)	6–25
Primary sedimentation	5–20
Chemical-precipitation	2–6
Trickling-filter plant	3–15
Activated-sludge plant	2–8
Multi-media filter following activated-sludge plant	1–5

6. The chlorine feed-rate, when set, should be maintained within less than 3% accuracy. Operation should be continuous and intermittent.
7. The chlorinator should be equipped with an ejector assembly of the chlorine water solution type utilizing water pressure to produce a vacuum in the chlorine system. The ejector should be constructed of corrosion resistant plastic and the ejector nozzle should be of the fixed throat orifice type. The ejector nozzle should be precisely engineered for the given hydraulic conditions such that the chlorinator will operate efficiently over the full range of its operating capacity. The ejector should contain a diaphragm type check valve to prevent water from entering the chlorinating system when the water flow is shut down.
8. The chlorinator should be constructed of materials suitable for wet and dry chlorine gas service. All springs used in the chlorinator should be of hastelloy C.
9. The cabinet should be made of plastic fiberglass, floor mounted and self-supporting, and suitable for indoor installation.
10. All necessary chlorine valves, flexible connectors, fittings, etc. and any additional items as required for an operable unit should be provided. The basic instruments at minimum should be as follows:
    - a gas inlet pressure gauge;
    - an outlet vacuum gauge;
    - a rota-meter (easily visible for flow adjustment from the front of unit);
    - a spring loaded diaphragm actuated pressure relief valve (for operation if pressure becomes excessive or when excessive air develops);
    - a chlorine gas shut-off valve for interrupted water flow or loss of vacuum.
11. The chlorinator should be provided with a heated trap.
12. The chlorinator should be provided with an indicator in the cabinet to show when the chlorine supply is exhausted or shut-off.
13. The following accessories should be included:
    - header valves and double unions at each chlorinator inlet;
    - auxiliary valves, flexible connections, and header valves at each chlorine container;
    - an in-line chlorine pressure switch with adjustable contacts.

4
Biological Treatment

Biological treatment – the use of bacteria and other microorganisms to remove contaminants by assimilating them – has long been a mainstay of wastewater treatment in the chemical process industries. Because they are effective and widely used, many biological treatment options are available today.

They are, however, not all created equal, and the decision to install a biological treatment system requires much thought. When considering biological wastewater treatment for a particular application, it is important to understand the sources of the wastewater generated, the typical wastewater composition, the discharge requirements, events, and practices within a facility that can affect the quantity and quality of the wastewater, and pre-treatment ramifications.

4.1 Theory

The objectives of biological treatment of wastewater are to coagulate and remove the non-settleable colloidal solids and to stabilize the organic matter. For domestic wastewater, the major objective is to reduce the organic content and, in many cases, nutrients such as nitrogen and phosphorus.

In many locations, the removal of trace organic compounds that may be toxic is also an important treatment objective. For industrial wastewater, the objective is to remove or reduce the concentration of organic and inorganic compounds. Pre-treatment is required, because many of the compounds in industrial wastewater are toxic to microorganisms.

Depending on the type of wastewater being treated, as well as its level of biodegradability and toxicity, a series of physical, chemical, and biological processes are combined, in order to remove different forms of carbon, nitrogen, and phosphorus present in the water, among other elements, in order to achieve the maximum possible efficiency in the treatment. Compared with physical and chemical processes, biological processes are considered to be efficient and cost effective; in most cases they will be the treatment of preference for a diverse range of wastes.

All biological treatment reactors are designed using mass balance calculations. These are calculated for each of the "electron donor" constituents of interest that are entering or leaving the system, along with their rates of depletion or production within the system, as well as the flow-rates crossing a defined volume, which are measured as mass per unit volume per unit time. Reactors are designed to operate with high microorganism concentrations; but their overall rate of metabolism is controlled by the limiting substrate concentration.

Monod kinetics (Monod, 1949) are most commonly used to model the relationship between the residual concentration of substrate that limits the microbial specific growth rate and the maximum specific growth rate of biomass obtained as a function of the substrate utilization. This is given by the equation:

$$\mu = \mu_{max} \frac{S}{K_S + S} \qquad (4.1)$$

Where:

μ = specific growth rate (d^{-1})
μ_{max} = maximum growth rate at saturation concentration of growth limiting substrate (d^{-1})
S = substrate concentration (mg/L)
K_s = half saturation constant (mg/L) which is the concentration of limiting substrate at which the specific growth rate equals one-half of the maximum specific growth rate ($\mu = \mu_{max}/2$).

The specific growth rate μ corresponds to the change in biomass per day (related to the amount of biomass present) and is a function of the substrate concentration. The maximum bacterial specific growth rate relates to the maximum specific substrate utilization rate and occurs at high substrate concentrations. Hence, when cells grow rapidly in the presence of non-limiting substrate concentrations, they make their maximum investment of energy in cell synthesis. However, when the substrate concentration (electron donor) is limited, a larger portion of the energy obtained from the substrate oxidation must be used for cell maintenance (Gray, 1990).

A loss term not explicitly included in the standard Monod equation (equation 4.2) is endogenous decay. Endogenous decay includes cell mass losses derived from the oxidation of internal storage products for energy used for cell maintenance, cell death and lysis, and predation by organisms higher in the food chain (Gray, 1990). As the endogenous decay may also affect the specific growth rate μ, in practice endogenous decay is also commonly considered by engineers in the Monod equation as follows:

$$\mu = \mu_{max} \frac{S}{K_S + S} - K_d \qquad (4.2)$$

Where:

μ = specific growth rate (d^{-1})
μ_{max} = maximum growth rate at saturation concentration of growth limiting substrate (d^{-1})
S = substrate concentration (mg/L)
K_s = half saturation constant (mg/L) which is the concentration of limiting substrate at which the specific growth rate equals one-half of the maximum specific growth rate ($\mu = \mu_{max}/2$)
K_d = endogenous decay

The ratio of the amount of biomass produced to the amount of substrate (electron donor) consumed is defined as the biomass yield coefficient Y and is expressed as mass (or mole) of organism produced per mass (or mole) of substrate consumed. As a definitive stoichiometric relationship exists between the substrate removed and the observed biomass yield, the expression of yield can also be combined with the Monod equation to give the rate of substrate utilization as follows:

$$\frac{ds}{dt} = -\frac{\mu X}{Y} \tag{4.3}$$

Where:

s = concentration of substrate (mg/L)
t = time (d)
μ = specific growth rate (d^{-1})
Y = yield coefficient
X = concentration of microorganisms (mg/L)

In summary, the performance and efficiency of biological processes are directly linked to the dynamics of substrate utilization and microbial growth. Therefore, the effective operation of biological systems requires an understanding of the types of microorganisms that are involved, the specific reactions that they perform, and their nutritional needs and kinetics.

In addition, knowledge and control of environmental conditions, such as temperature, pH, dissolved oxygen, and other relevant factors that affect organisms of interest, is also essential. By taking into account all these factors, organic and inorganic chemicals may then be effectively removed from water and thus the best treatment possible may be implemented.

4.1.1 Biological Activated Carbon Process

Biological activated carbon (BAC) is a modification of conventional biological processes. In the process, activated carbon is added to the biological tank. The functions of activated carbon are adsorbency and attaching media for bacterial growth.

The advantages of biological activated process are widely proved from many researches, namely:

- More stable system during shock loadings.
- Reduction of refractory compounds.
- Color and ammonia removal.
- Improved settleability.

Normally, bacteria are very sensitive to organic loading. Adding Granular Activated Carbon (GAC) in biological treatment will reduce the impact of shock loading by its adsorption property. If biofilm is generated in the BAC system, the adsorbate adsorbed in the activated carbon can be degraded by the microorganisms attached to the surface of the activated carbon through the biofilm process. It thereby extends the time for activated carbon regeneration. Furthermore, some compounds, which are non-biodegradable, can be adsorbed on activated carbon and subsequently metabolized by attached microorganisms.

One of the important considerations in the quantity of activated carbon added is sludge age. Equation 4.4 (Metcalf and Eddy, 2003) illustrates the dependence of GAC and mixed liquor GAC suspended solid dosage on solid residence time and hydraulic retention time:

$$X_p = \frac{X_i \cdot \theta_c}{\theta} \quad (4.4)$$

Where:

X_p: equilibrium GAC mixed liquor suspended solid content mg/L
X_i: GAC dosage, mg/L
θ_c: solid residence time, d
θ: hydraulic retention time, d

The quantity of activated carbon added is different from one process to the next, but the higher the sludge age, the better the organic removal per unit of carbon, which results in greater removal efficiency.

4.1.2 Biokinetic Theoretical Model

The effectiveness of biological treatment is governed by the growth of microorganisms. The following reviews are basis equations concerned with bacterial growth, substrate utilization rate, and yield coefficient in growths both with and without inhibition.

4.1.2.1 Growth Without Inhibition

Growth of bacterial can be expressed into two ways: as numbers and mass. The kinetic consideration of bacterial growth is very important to control the growth rate of microorganism. Experimentally, the effect of a limiting substrate or nutrient on bacterial growth is usually defined adequately using two well-known Monod equations:

- **Specific growth rate:**

$$\frac{dX}{dt} = \mu X \quad (4.5)$$

$$\mu = \mu_{max} \frac{S}{K_S + S} \quad (4.6)$$

Where:

X = Biomass concentration (mg/L)
S = Growth-limiting substrate concentration (mg/L)
μ_{max} = Maximum growth rate (d^{-1})
μ = Specific growth rate (d^{-1})
K_S = Half-velocity constants (mg/L)

From equation 4.6, one can see that growth rate of bacterial is limited even the substrate is available.

- **Substrate Utilization Rate:**

Substrate Utilization Rate means the amount of biomass produced per 1 mg of substrate removed per day and can be expressed as:

$$U = \frac{(S_0 - S)}{\theta \cdot X} \quad (4.7)$$

Where:

U = Substrate utilization rate (mg substrate removed/mg MLSS.day)
S_0 = Initial substrate concentration (mg/L)
X = Biomass concentration (mg/L)
S = Final substrate concentration (mg/L)
θ = Hydraulic retention time (day)

- **Growth Yield Coefficient (Y):**

In biological treatment, a portion of substrate is converted to new cells and the relation between bacterial growth rate and substrate consumption is represented by the Growth Yield Coefficient.

$$\frac{dX}{dt} = -Y \frac{dS}{dt} \quad (4.8)$$

Where:

Y = Growth yield coefficient (g SS/g substrate removed.day)
X = Biomass concentration (mg/L)
S = Substrate concentration (mg/L)

- **Growth with Inhibition**

In complex or particular wastewater, some compounds that inhibit biological treatment are present. Thus, this leads to a change in growth kinetics. Some researchers have developed models for growth kinetics in these cases. Table 4.1 shows some kinetic models for growth with inhibitory substances.

Table 4.1 Kinetic models for inhibition growth (Han and Levenspiel, 1988).

Equations	Equation number	Model name
$\mu = \mu_{max} \left(1 - \frac{I}{K_I}\right) \frac{S}{S + K_S}$	4.9	Ghose and Tyagi
$\mu = \mu_{max} \left(1 - \frac{I}{K_I}\right)^{0.5} \frac{S}{S + K_S} = \mu_{obs} \frac{S}{S + K_S}$	4.10	Bazua and Wilke
$\mu = \mu_{max} \left(1 - \frac{I}{K_I}\right)^{n} \frac{S}{S + K_S} \mu_{obs} \frac{S}{S + K_S}$	4.11	Han and Levenspiel

Where:

I = Concentration of inhibitor
K_i = Critical inhibitor concentration above which reactions stop
μ_{max} = Maximum specific growth rate at zero concentration of inhibitor (I = 0)
μ_{obs} = Observed maximum specific growth rate at certain concentration of inhibitor

Table 4.2 Growth models for inhibitor as substrate.

Equations	Equation number
$\mu = \mu_{max} \dfrac{S}{K_s + S}\left(1 - \dfrac{S}{S_c}\right)$	4.12
$\mu = \mu_{max} \dfrac{S}{(K_s + S)\left(1 - \dfrac{S}{K_i}\right)}$	4.13
$\mu = \mu_{max} \dfrac{S}{K_s + S + \dfrac{S^2}{K_i}}$	4.14

K_i = Critical inhibitor concentration above which reactions stop
μ_{max} = Maximum specific growth rate at zero concentration of inhibitor (I = 0)
μ_{obs} = Observed maximum specific growth rate at certain concentration of inhibitor
S_c = Critical substrate concentration above which reactions stop

In the case where the substrate is the inhibitor at a critical point, some researchers have studied a model for the inhibition of growth on acetic acid as substrate (Table 4.2).

The primary purpose of wastewater purification is not cell multiplication and biomass production but removal of organic compounds from solution. Combining equations 4.5, 4.7, and 4.8 leads to:

$$-\frac{dS}{dt}\frac{1}{X} = \frac{\mu_{max}}{Y_{obs}}\frac{S}{K_s + S} \qquad (4.15)$$

$$r_x = r_{x,m}\frac{S}{K_s + S} \qquad (4.16)$$

$r_x = -(dS/dt)/X$ is actual substrate removal rate (1/time)
$r_{x,m} = \mu_{max}/Y_{obs}$ is maximum substrate removal rate (1/time)

Equation 4.16 shows the dependence of the substrate removal rate on the actual concentration of a single component substrate and their relative value with K_s. In order to control the operation of a treatment system and thus produce good efficiency from a practical point of view, it is important to know the biokinetic data of a particular wastewater. And one of the most easy-to-use methods is respirometry.

4.2 Biological Treatment Processes

The major biological processes used for wastewater treatment are identified in Table 4.3. All the biological processes used for the treatment of wastewater, as reported in Table 4.3, are derived from processes occurring in nature.

By controlling the environment of the microorganisms, the decomposition of wastes is sped up. Regardless of the type of waste, the biological treatment process consists of controlling the environment required for optimum growth of the microorganisms involved.

Table 4.3 Major biological treatment processes used for wastewater treatment.

Type	Common Name	Use
	Aerobic processes	
Suspended-growth	Activated-sludge process Conventional (plug-flow) Complete-mix Slop aeration Pure oxygen Sequencing batch reactor Contact stabilization Extended aeration Oxidation ditch Deep tank (27.5 m) Deep shaft	Carbonaceous BOD removal (nitrification)
	Suspended-growth nitrification	Nitrification
	Aerated lagoons	Carbonaceous BOD removal (nitrification)
	Aerobic digestion Conventional air Pure oxygen	Stabilization, carbonaceous BOD removal
Attached-growth	Tricking-filters Low-rate High-rate	Carbonaceous BOD removal, nitrification
	Roughing filters	Carbonaceous BOD removal
	Rotating biological contactors	Carbonaceous BOD removal (nitrification)
	Packed-bed reactors	Carbonaceous BOD removal (nitrification)
Combined suspended-and attached-growth processes	Activated biofilter process, trickling-filter solids-contact process, biofilter activated-sludge process, series trickling-filter activated-sludge process	Carbonaceous BOD removal (nitrification)
	Anoxic processes	
Suspended-growth	Suspended-growth denitrification	Denitrification
Attached-growth	Fixed-film denitrification	Denitrification
	Anaerobic processes	
Suspended-growth	Anaerobic digestion Standard rate, single-stage High-rate, single-stage, two-stage	Stabilization, carbonaceous BOD removal
	Anaerobic contact process	Carbonaceous BOD removal
	Upflow anaerobic sludge-blanket	Carbonaceous BOD removal
Attached-growth	Anaerobic filter process	Carbonaceous BOD removal, waste stabilization (denitrification)
	Expanded bed	Carbonaceous BOD removal, waste stabilization

(Continued)

Table 4.3 (Continued)

Type	Common Name	Use
Combined aerobic, anoxic, and anaerobic processes		
Suspended-growth	Single- or multi-stage processes, various proprietary processes	Carbonaceous BOD removal, nitrification, denitrification, phosphorus removal
Combined suspended-and attached-growth	Single- or multi-stage processes	Carbonaceous BOD removal, nitrification, denitrification, phosphorus removal
Pond processes	Aerobic ponds	Carbonaceous BOD removal
	Maturation (tertiary) ponds	Carbonaceous BOD removal (nitrification)
	Facultative ponds	Carbonaceous BOD removal
	Anaerobic ponds	Carbonaceous BOD removal (waste stabilization)

Note: Major uses are presented first; other uses are identified in parentheses.

The principal applications of these processes are for:

- removal of the carbonaceous organic matter in wastewater, usually measured as BOD, total organic carbon (TOC), or chemical oxygen demand (COD);
- nitrification;
- denitrification;
- phosphorus removal;
- waste stabilization.

Biological processes are affected by environmental conditions. The environmental conditions can be controlled by pH and/or temperature regulation, nutrient or trace element addition, oxygen addition or exclusion, and proper mixing. Biological processes are classified by the oxygen dependence of the primary microorganisms responsible for waste treatment. The processes are usually referred to as aerobic, anaerobic, or facultative.

4.2.1 Major Differences in Aerobic and Anaerobic Treatment

Before we go into discussion of the various aerobic biological treatment processes, it is important to briefly discuss the terms aerobic and anaerobic. Aerobic, as the title suggests, means in the presence of air (oxygen); while anaerobic means in the absence of air (oxygen). These two terms are directly related to the type of bacteria or microorganisms that are involved in the degradation of organic impurities in a given wastewater and the operating conditions of the bioreactor.

Therefore, aerobic treatment processes take place in the presence of air and utilize those microorganisms (also called aerobes) that use molecular/free oxygen to assimilate organic impurities i.e., convert them in to carbon dioxide, water, and biomass. Anaerobic treatment processes, on other hand, take place in the absence of air (and thus molecular/free oxygen) by those microorganisms (also called anaerobes) which do not require air (molecular/free oxygen) to assimilate organic impurities. The final products of organic assimilation in anaerobic treatment are methane and carbon dioxide gas and biomass.

Table 4.4 summarizes the major differences in these two types of processes. From this summary, it can be concluded that it is not anaerobic or aerobic treatment, but a combination of the two types of technologies, that give an optimum configuration for those wastewater treatment applications where organic impurities are at a relatively higher concentration.

4.2.2 Aerobic Processes

An aerobic environment is achieved by the use of diffused or mechanical aeration. The reactor contents are referred to as the mixed liquor. The temperature dependence of the biological reaction-rate constant is very important in assessing the overall efficiency of a biological treatment process.

In aerobic photosynthetic ponds, the oxygen is supplied by natural surface aeration and by algal photosynthesis. Only a portion of the original waste is actually oxidized to low-energy compounds, such as NO_3, SO_4, and CO_2; the remainder is synthesized into cellular material.

The most important inorganic compound is ammonia, because its presence in the plant effluent can stimulate the lowering of the dissolved oxygen in the receiving stream through the biological process of nitrification. In nitrifical, ammonia is biologically oxidized to nitrite. The nitrite is then oxidized by another group of microorganisms to nitrate. Nitrate is the final oxidation state of the nitrogen compounds, and as such represents a stabilized product.

There are multitudes of aerobic biological treatment processes and technologies in literature and practice; such as activated-sludge, trickling-filter, and aerobic stabilization ponds. The activated-sludge process is used almost exclusively in large units. However, for the purpose of this section, the following four biological treatment technologies are described.

4.2.2.1 Conventional Activated-Sludge Process (ASP) System

This is the most common and oldest biotreatment process used to treat municipal and industrial wastewater. Typically wastewater after primary treatment, i.e., suspended impurities removal, is treated in an activated-sludge process based biological treatment system comprising an aeration tank followed by a secondary clarifier. The aeration tank is a completely-mixed or a plug-flow (in some cases) bioreactor where a specific concentration of biomass is maintained along with sufficient dissolved oxygen (DO) concentrations (typically 2 mg/L) to effect biodegradation of soluble organic impurities measured as biochemical oxygen demand (BOD_5) or chemical oxygen demand (COD).

Figure 4.1 shows a conventional activated-sludge process (ASP) system. The aeration tank is provided with a fine bubble diffused aeration pipework at the bottom to transfer the required oxygen to the biomass and also to ensure completely mixed reactor. A roots type air blower is used to supply air to the diffuser pipework. The aerated mixed liquor from the aeration tank overflows through gravity to the secondary clarifier unit, to separate out the biomass and allow clarified, treated water flow to the downstream filtration system for finer removal of suspended solids. The separated biomass is returned to the aeration tank

Table 4.4 Major differences in aerobic and anaerobic treatment.

Parameter	Conventional Activated-Sludge Process (ASP) System	Cyclic Activated-Sludge System	Integrated Fixed Film Activated-Sludge (IFAS) System:	Membrane Bioreactor (MBR)
Treated effluent quality	Meets specified discharge standards with additional filtration step	Meets/exceeds specified discharge standards without additional filtration step	Meets/exceeds specified discharge standards with additional filtration step	Exceeds specified discharge standards without additional filtration step. Very good for recycling provided TDS level permits
Ability to adjust to variable hydraulic and pollutant loading	Average	Very good	Very good	Very good
Pre-treatment requirement	Suspended impurities e.g., oil & grease and TSS removal	Suspended impurities e.g. oil & grease and TSS removal	Suspended impurities e.g., oil & grease and TSS removal	Fine screening for suspended impurities like hair and almost complete oil & grease removal
Ability to cope with ingress of oil	Average	Good	Average	Poor & detrimental to membrane
Secondary clarifier requirement	Needed	Aeration basin acts as clarifier	Needed	Clarifier is replaced by membrane filtration
Complexity to operate & control	Simple, but not operator friendly	Operator friendly	Operator friendly	Requires skilled operators
Reliability & proveness of technology	Average	Very good	Very good	Limited references in industrial applications
Capital cost	Low	Low	High	Very high
Operating cost	Low	Low	High	Very high
Space requirement	High	Low	Average	Low

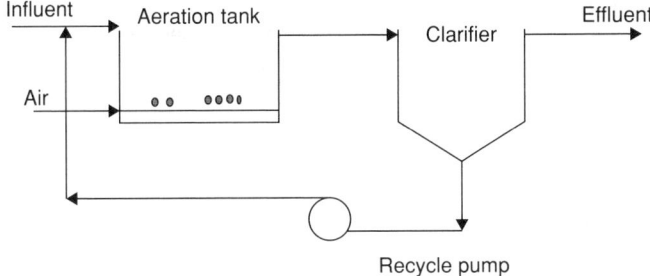

Figure 4.1 Conventional activated-sludge process system.

by means of a return activated-sludge (RAS) pump. Excess biomass (produced during the biodegradation process) is wasted to the sludge handling and dewatering facility.

4.2.2.2 Cyclic Activated-Sludge System (CASS^{TM})

The Cyclic Activated-Sludge System (CASS^{TM}), as the name suggests, is one of the most popular sequencing batch reactor (SBR) processes employed to treat municipal wastewater and wastewater from a variety of industries including refineries and petrochemical plants.

The cyclic activated-sludge system incorporates a high level of process sophistication in a configuration that is cost and space effective and offers a methodology that has operational simplicity, flexibility, and reliability that is not available in conventionally configured activated-sludge systems. Its unique design provides an effective means of control of filamentous sludge bulking, a common problem with conventional processes and other activated-sludge systems.

The system design is such that the sludge return rate causes an approximate daily cycling of biomass in the main aeration zone through the selector zone. No special mixing equipment or formal anoxic mixing sequences are required to meet the effluent discharge objectives. The basin configuration and mode of operation enables combined nitrogen and phosphorous removal mechanisms to take place through a simple "one-shot" control of the aeration.

Important reasons for choosing CASS^{TM} over the conventional constant volume activated-sludge aeration and clarifier process include:

- Operates under continuous reduced loading through simple cycle adjustment.
- Operates with feed-starve selectivity, So/Xo operation (control of limiting substrate to microorganism ratio), and aeration intensity to prevent filamentous sludge bulking and ensures endogenous respiration (removal of all available substrate), nitrification, and denitrification together with enhanced biological phosphorus removal.
- Simultaneous (co-current) nitrification and denitrification by variation of aeration intensity.
- Tolerates shock loads caused by organic and hydraulic load variability. The system is easily configured and adjusted for short-term diurnal and long-term seasonal variations.
- Inherent ability to remove nutrients without chemical addition, by controlling the oxygen demand and supply.

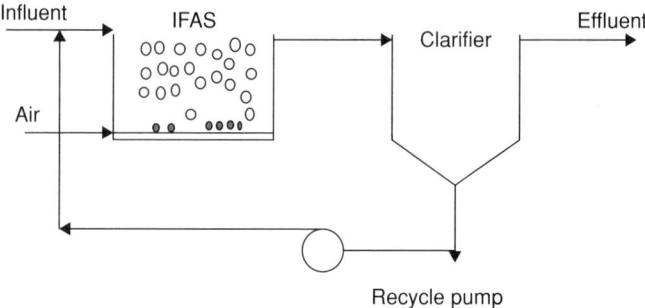

Figure 4.2 Integrated fixed-film activated system (IFAS).

4.2.2.3 Integrated Fixed-Film Activated-Sludge (IFAS) System

There are several industrial installations where two-stage biological treatment comprising a stone or plastic media trickling-filter (also known as packed-bed biotower) followed by an activated-sludge process based aeration tank, followed by a secondary clarifier, are in operation.

Another modification of the above configuration that has been implemented in newer industrial wastewater treatment systems is the fluidized media bioreactor (also known as the moving bed bioreactor (MBBR)) in lieu of a biotower followed by activated-sludge process. In some industries (e.g., refineries and petrochemical plants, where the existing wastewater treatment system was a single stage conventional activated-sludge process based on an aeration tank and clarifier unit), that underwent capacity expansion and/or faced stricter discharge regulations, up-grading of the activated-sludge process by the addition of fluidized biomedia has been implemented to meet these requirements. This hybrid process of fluidized media and activated-sludge process taked place in a single aeration tank and is known as the integrated fixed-film activated sludge (IFAS) process. Figure 4.2 shows a schematic of this system.

The common advantages of all of the above described configurations are as follows:

- Fixed-film media provides additional surface area for biofilm to grow on and degrades the organic impurities that are resistant to biodegradation or may even be toxic to some extent.
- The overall efficiency of the two-stage biotreatment system is better than the activated-sludge process alone.
- Fixed-film processes are more effective in the nitrification of the wastewater than the activated-sludge process.
- The overall footprint for a fixed-film process based system is smaller than the activated-sludge process system.
- Due to less sludge wastage, the sludge handling and dewatering facility is smaller compared to the activated-sludge process.

4.2.2.4 Membrane Bioreactor (MBR)

The Membrane Bioreactor (MBR) is the latest technology for biological degradation of soluble organic impurities. MBR technology has been in extensive use for treatment of

Figure 4.3 Aerobic basin (Reproduced with permission from © AECOM).

domestic sewage, but for industrial waste treatment applications, its use has been somewhat limited or selective. The MBR process is very similar to the conventional activated-sludge process, in that both have mixed liquor solids in suspension in an aeration tank (Figure 4.3). The difference in the two processes lies in the method of separation of biosolids. In the MBR process, the biosolids are separated by means of a polymeric membrane based on a microfiltration or ultrafiltration unit, as against the gravity settling process in the secondary clarifier in the conventional activated-sludge process. Therefore, the advantages of the MBR system over the conventional activated-sludge system are obvious as listed below:

1. Membrane filtration provides a positive barrier to suspended biosolids such that they cannot escape the system, unlike gravity settling in the activated-sludge process, where the biosolids continuously escape the system along with clarified effluent and sometimes a total loss of solids also occurs due to process upsets causing sludge-bulking in the clarifier. As a result, the biosolids concentration measured as MLSS/MLVSS can be maintained at 3 to 4 times higher in an MBR process ($\sim$ 10 000 mg/l) in comparison to the activated-sludge process ($\sim$2500 mg/l).
2. Due to the above aspect of MBR, aeration tank size in the MBR system can be one-third to one-fourth of the size of the aeration tank in an activated-sludge system. Further, instead of a gravity settling based clarifier, a much more compact tank is needed to house the membrane cassettes in the case of a submerged MBR and skid mounted membrane modules in the case of a non-submerged, external MBR system.
3. Thus, the MBR system requires only 40–60% of the space required for the activated-sludge system, therefore significantly reducing the concrete work and overall footprint.
4. Due to membrane filtration (micro/ultrafiltration), the treated effluent quality in the case of the MBR system is far superior compared to conventional activated-sludge, so the

treated effluent can be directly reused as cooling tower make-up or for gardening etc. Typical treated water quality from MBR system is:
- $BOD_5 < 5$ mg/L
- Turbidity < 0.2 NTU.

4.2.3 Anaerobic Waste Treatment

Anaerobic waste treatment involves the decomposition of organic and/or inorganic matter in the absence of molecular oxygen. The major application is in the digestion of the concentrated sewage sludges and in the treatment of some industrial wastes. Another application of anaerobic treatment is in anaerobic lagoons or ponds. The usual mode of operation of an anaerobic waste treatment unit receiving a concentrated sewage sludge is by use of a complete-mix reactor system with minimum cellular recycling for the purpose of heating and mixing the tank contents.

4.2.4 Aerobic, Anaerobic (Facultative) Waste Treatment

Stabilization ponds of facultative bacteria are known as aerobic–anaerobic ponds. Such ponds have an aerobic upper layer and an anaerobic bottom layer. In practice, oxygen is maintained in the upper layer by the presence of algae or by the use of surface aerators. Where surface aerators are used, algae are not required.

4.3 Activated-Sludge Units

Activated-sludge units are multi-chamber reactor units that make use of (mostly) aerobic microorganisms to degrade organics in wastewater to produce a high-quality effluent. To maintain aerobic conditions and to the keep the active biomass suspended, a constant and well-timed supply of oxygen is required. Activated-sludge systems (refer to Figures 4.4 and 4.5) normally make use of bar screens and/or comminutors, grit chambers, primary settling tanks, secondary settling tanks, and digesters, which are operated in the same manner as those of trickling-filter systems. They differ from the trickling-filter systems in that they make use of an aeration tank instead of a trickling-filter.

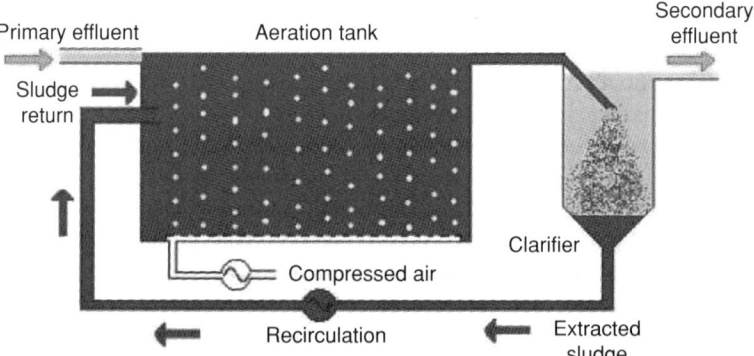

Figure 4.4 *Activated-sludge system (example 1).*

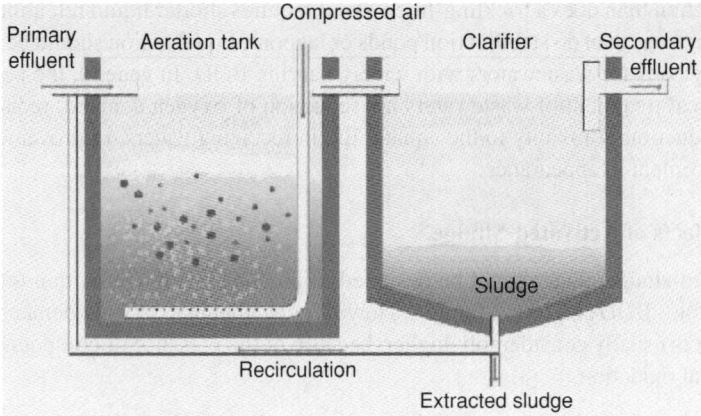

Figure 4.5 Activated-sludge system (example 2).

Different configurations of the activated-sludge process can be employed to ensure that the wastewater is mixed and aerated (with either air or pure oxygen) in an aeration tank. The microorganisms oxidize the organic carbon in the wastewater to produce new cells, carbon dioxide, and water.

Although aerobic bacteria are the most common organisms, aerobic, anaerobic, and/or nitrifying bacteria along with higher organisms can be present. The exact composition depends on the reactor design, environment, and wastewater characteristics. During aeration and mixing, the bacteria form small clusters, or flocs. When the aeration stops, the mixture is transferred to a secondary clarifier where the flocs are allowed to settle out and the effluent moves on for further treatment or discharge. The sludge is then recycled back to the aeration tank, where the process is repeated.

Compressed air is continually diffused into the sewage as it flows through the aeration tank. This provides both a source of oxygen for the aerobic bacterial floc that forms in the tank and the turbulence necessary to bring the waste and the bacteria into contact. Aerobic bacteria attack the dissolved and finely divided suspended solids not removed by primary sedimentation. Some of the floc is removed with the sewage that flows out of the aeration tank and is carried into the secondary settling tank. Here the floc settles to the bottom of the tank, and is later pumped back into the aeration tank. The liquid portion then flows over a weir at the surface of the settling tank to be chlorinated and released to a receiving stream.

To achieve specific effluent goals for BOD, nitrogen, and phosphorus, different adaptations and modifications have been made to the basic activated sludge design. Aerobic conditions, nutrient-specific organisms (especially for phosphorus), recycling design, and carbon dosing, among others, have successfully allowed activated-sludge processes to achieve high treatment efficiencies.

4.3.1 Applications

The activated-sludge process and its modifications, the extended-aeration activated-sludge process and the oxygen activated-sludge process, are used to decompose organic matter and stabilize soluble organic wastes in water. An activated-sludge unit achieves more

BOD reduction than does a trickling-filter, and it requires shorter liquid retention times and smaller plot areas than do stabilization ponds or lagoons. Equalization should be considered to stabilize loads for wastewaters with widely varying BOD. In general, the benefits from the biological treatment of wastewaters are reduction of oxygen demand, reduction of oil content, reduction of toxicity to the aquatic life in receiving waters, improvement in odor, and improvement in appearance.

4.3.2 Effects of Activated-Sludge

An activated-sludge plant should be designed to produce an effluent with a total (soluble plus insoluble) BOD_5 of 10 to 20 mg/L. However, the chemical oxygen demand (COD) of the effluent is usually considerably higher, because of the presence of compounds resistant to biological oxidation.

1. The residues from an activated sludge system can include the following:
    - **Soluble BOD, this is composed of:**
        - Any of the soluble BOD_5 in the wastewater not removed by the process. The concentration of these residues decreases with retention time in the aeration basin.
        - Organic byproducts produced by the microorganisms and not metabolized. These residues increase with retention time.
        - Organic material produced by the decomposition of cells during endogenous respiration (auto-oxidation, oxidation by microorganisms of material from bacterial cells). A residual BOD_5 of 5 to 15 mg/L will persist from these residues as long as active cells are present in the system.
    - **Biological solids**
      Although the biological solids (cells) are removed in the settler following aeration, the settling step is not perfect and some solids overflow with the clarifier effluent. These solids often hold some oil. In addition, these solids will contribute a BOD on the order of 0.3 to 0.7 mg BOD_5/mg suspended solids depending on the age of the sludge.
    - **Non-biodegradable residues**
      Some soluble wastes are not biodegradable. These will exhibit a COD or TOC, but not a BOD. In addition, some solid metabolic byproducts that are non-biodegradable are produced.
    - **Non-biological organic suspended material**
      Some of the insoluble organic material originally present in the wastewater may carry through with the settled effluent. This unstabilized material could include oil, plastics, rubber, etc.
    - **Inorganic, non-volatile suspended matter**
      Most of the inorganic suspended matter (e.g., silt) will settle with the floc and will be purged with wasted sludge. Some, however, may not settle and so may appear as non-volatile suspended solids in the clarifier effluent.
2. **Free oil:**
   An activated-sludge unit is designed to remove soluble organic compounds and not free oil. Free oil above about 50 mg/L must be first removed in a pre-treatment step. The small amount of free oil remaining after pre-treatment (5 to 20 mg/L) will be further reduced in the activated-sludge unit, primarily by adsorption onto the floc.

3. **Ammonia**

A small amount of ammonia (equal to 1 to 3% of the BOD_5 removed) can be removed from water by activated-sludge treatment in the usual case where nitrification does not occur. Thus, activated-sludge treatment is not a practical way to remove ammonia from wastewater due to too small removal efficiency.

4.3.3 Feed Composition

Certain materials in the feed and other environmental factors affect the operation of an activated-sludge plant. Table 4.5 gives criteria for pre-treatment of activated-sludge feed. In Table 4.5, shock loads as well as continuous loads should be considered. Moreover the values given are for satisfactory operation of the activated-sludge unit. Pre-treatment may be required to reach more restrictive values to meet discharge requirements.

Pre-treatment of feed to an activated-sludge or extended-aeration plant should be provided when the feed is outside the limits shown in Table 4.5.

In addition to feed pre-treatment, a holding basin to which the biox influent can be diverted in the event of toxic shocks or organic shocks is recommended.

4.3.3.1 Oil Content

Pre-treatment of the wastewater to remove free oil should be provided if the oil content of the wastewater to be treated in an activated-sludge system is over 50 mg/L.

4.3.3.2 Nutrients

Any nutrient not already present in sufficient quantity should be supplied. Ammonia is most readily assimilated by bacteria in the form of nitrogen. The source of ammonia can be either free ammonia in the wastewater or ammonia formed in the activated-sludge unit by the biodegradation of organic nitrogen compounds such as MEA (mono-Ethanolamine) and DEA (di-Ethanolamine). Both refinery and petrochemical plant wastewaters are usually deficient in phosphorus, although phosphorates enter the wastewater from boiler and cooling tower blow-down and from spent phosphoric acid catalyst from polymerization units.

The need for nitrogen and phosphorus should be determined by pilot plant testing. If this has not been done, an analysis of the feed wastewater is required. Based on this analysis, the contractor should provide nutrient addition facilities, if necessary, to result in a ratio of N:P:BOD metabolized of 5:1:100, and in addition, provide for residual concentrations of 0.1 mg/L or the phosphate phosphorus (PO_4—) and 0.5 mg/L ammonia nitrogen in the cffluent.

4.3.3.3 Toxic and Inhibitory Materials

The presence of toxic materials in the wastewater can seriously impair the performance of an activated-sludge unit and reduce its efficiency. Toxicants either kill microorganisms or inhibit their growth. The effect depends on concentration, contact time, and the environment (e.g., pH).

A number of toxic materials are potentially present in refinery and petrochemical plant wastewaters, including zinc, chromium, copper, cyanide, sulfide, biocides, furfural, and certain other organic compounds. In addition, highly alkaline or acid discharges may

Table 4.5 Criteria for pre-treatment of activated-sludge feed.

Constituent	Limiting Concentration	Appropriate Pre-treatment	Remarks
Oil	50 mg/L	Dissolved-air flotation, granular-media filtration, chemical flocculation	Limiting concentration varies with the solids retention time
Sulfides	10 mg/L max	Stripping, precipitation with iron salts	—
Ammonia	500 mg/L max.	Stripping, dilution	—
Phenols	500 mg/L max.	Stripping	Although the activated-sludge unit will operate satisfactorily within the indicated range, the size will be affected by the indicated parameter. Determine if there is an economic incentive to provide pre-treatment to adjust this parameter in order to reduce the cost of the activated-sludge or extended-aeration plant. The concentration given applies if the concentration is relatively constant. A lesser concentration can only be tolerated if it occurs as a non-steady variation.
Chlorides	100% sea water max	Dilution, de-ionization	Although the activated-sludge unit will operate satisfactorily within the indicated range, the size will be affected by the indicated parameter. Determine if there is an economic incentive to provide pre-treatment to adjust this parameter in order to reduce the cost of the activated-sludge or extended-aeration plant. The concentration given applies if the concentration is relatively constant. A lesser concentration can only be tolerated if it occurs as a non-steady variation. De-ionization and ion exchange are seldom practical for refinery or petrochemical plant wastewater.
Suspended Solids	150 mg/L max	Sedimentation, granular-media filtration, chemical floc, DAF Unit	—

Copper	1 mg/L max	Precipitation, ion exchange	De-ionization and ion exchange are seldom practical for refinery or petrochemical plant wastewater
Heavy Metals	1 to 10 mg/L max.	Precipitation, ion exchange	De-ionization and ion exchange are seldom practical for refinery or petrochemical plant wastewater
Dissolved Salts	30 000 mg/L max	Dilution, stream segregation	Although the activated sludge unit will operate satisfactorily within the indicated range, the size will be affected by the indicated parameter. Determine if there is an economic incentive to provide pre-treatment to adjust this parameter in order to reduce the cost of the activated sludge or extended-aeration plant. The concentration given applies if the concentration is relatively constant. A lesser concentration can only be tolerated if it occurs as a non-steady variation.
Cyanides	5 mg/L max	Stripping, oxidation (ozone, alkaline chlorination)	—
Thiocyanides Organic Load Variation (Based on 4 hour composite;	30 mg/L max. max of 2 to 4	Retention, equalization	Depends upon the wastewater
pH	6.5 to 8.5	Neutralization	
Alkalinity	Max of 0.2 kg alkalinity as CaCO$_3$ per 0.4 kg BOD removed	Neutralization	
Free Mineral Acidity	Zero free mineral acidity, maximum 10 to 40 °C or 46 to 60 °C	Neutralization	
Temperature		Cool, dilute, heat	Although the activated sludge unit will operate satisfactorily within the indicated range, the size will be affected by the indicated parameter. Determine if there is an economic incentive to provide pre-treatment to adjust this parameter in order to reduce the cost of the activated-sludge or extended-aeration plant. Avoid operation in the 40 to 46 °C temperature range

be toxic. Besides being toxic, sulfides preferentially consume oxygen and can create an anaerobic environment.

Table 4.5 gives guidance on the limiting concentrations for certain materials. Heavy metals, in general, inhibit the activated sludge process. Dissolved heavy metal ions (cadmium, cobalt, copper, dichromate, ferric iron, lead, manganese, nickel, silver, vanadium, and zinc), above certain concentrations, impede biological metabolism. The effect of any particular concentration of these depends on the environment; therefore, it is not possible to give exact concentration limits for most of these heavy metals. Microorganisms can tolerate only 1 mg/L or less of some heavy metals (cadmium, copper, silver), but other metal ions (ferric iron, manganese, nickel, vanadium) have been successfully handled at 10 mg/L with little or no loss in efficiency.

Organic compounds that are toxic to the microorganisms will inhibit their utilization of other materials. Hence, intermittent slugs of many organic compounds or sudden, large changes in their concentration can disrupt the biological process. When feasible, laboratory studies to evaluate toxic limits should be made before wastes from a new process are introduced into an existing activated-sludge system.

4.3.3.4 Organic Compounds

Organic compounds differ in their biodegradability; some are readily degraded by the microorganisms normally present, some are degraded if the organisms have become acclimatized to the compound, and others are practically non-biodegradable. Table 4.6 lists organic compounds that are biodegradable by acclimated organisms, and compounds generally resistant to biological degradation.

4.3.3.5 Organic Shocks

An organic shock is a sudden increase in the organic content of the incoming wastewater. Organic shocks can be caused by a number of materials including phenols, alcohols, detergents, solvents, MEA, DEA, hydrocarbons, etc. The concentration of each component in the feed to an activated-sludge unit should not vary by more than 50% per hour. In any case, retention or equalization holdup should be provided to avoid any organic shock.

4.3.3.6 Salt Content

Rapid changes in salt concentration can devitalize the microorganisms; salt concentrations about 3000 mg/L (ppm) TDS (Total Dissolved Solids) retard sludge settling and any concentration of dissolved solids reduces the solubility of oxygen in water and increases the rate of transfer of oxygen to water. In any case, shock loads of salt, as from occasional releases of salt-water ballast, should be avoided.

4.3.3.7 Suspended Solids

High concentrations of influent suspended solids decrease efficiency by reducing the fraction of active biological solids, create an excessive oxygen demand, and may result in a sludge less amenable to dewatering.

Table 4.6 *Relative biodegradability of certain organic compounds.*

Biodegradable Organic Compounds	Compounds Generally Resistant to Biological Degradation
Acrylic acid	Diethanolamine
Aliphatic acids	Ethers
Aliphatic alchohols (normal, iso, secondary)	Ethylene chlorohydrin
Aliphatic aldehydes	Insoluble oil
Aliphatic esters	Isoprene
Aromatic amines	Methyl vinyl ketone
Benzaldehyde	Morpholine
Dichlorophenols	Polymeric compounds
Ethylene glycol	Selected hydrocarbons
Ketones	Aliphatics
Linear alkyl benzene sulfonates	Aromatics
Methacrylic acid	Alkyl-aryl groups
Methyl methacrylate	Tertiary aliphatic alcohols
Monochlorophenols	Tertiary benzene sulfonates
Monoethanolamine	Tetrapropylene and other polypropylene benzene
Nitriles	Sulfonates
Phenols	Trichlorophenols
Primary aliphatic amines	Triethanolamine
Styrene	
Vinyl acetate	

Note: Some compounds can be degraded biologically only after an extended period of acclimation.

4.3.3.8 pH, Alkalinity, and Acidity

The pH of the aeration basin determines which microorganisms predominate in the system and also affects the rate of reaction. Facilities to adjust the pH of the feed to the activated-sludge unit should be provided. The pH of the aeration basin should be kept in the range of 6.5 to 8.5, the optimum pH range is 7 to 8. An aeration basin pH outside the range of 5 to 9 may result in destruction of microorganisms.

4.3.3.9 Temperature

The temperature in the aeration basin affects the reaction rate in the aeration basin and the settling characteristics of the sludge. Either of two ranges of temperature can be maintained in the aeration basin.

The growth of a different type of microorganism predominates in each of these ranges: mesophilic microorganisms grow in the range of 10 to 40 °C and thermophilic microorganisms grow in the range of 46 to 60 °C. An activated-sludge plant should not be designed to operate in the 40 to 46 °C temperature range. Winterization facilities should be foreseen in cold climates. Temperature changes should be kept smaller than 10 °C per hour so as not to retard performance. An activated-sludge plant should not be designed to operate below 15 °C.

4.3.3.10 Oxygen Concentration

The design value for dissolved oxygen concentration should be a minimum of 2 mg/L. This should be sufficient to maintain a minimum DO (Dissolved Oxygen) residual of 0.5 mg/L throughout the aeration chamber with normal fluctuations in incoming BOD and COD load. If shock BOD loads are expected, provision should be made for more than 2 mg/L dissolved oxygen, to avoid a DO of less than 0.3 mg/L during or after the shock load period.

4.3.4 Process Design

The biological oxidation unit should utilize the completely-mixed recycle activated-sludge process of treatment. For the design of the unit, influent wastewater specifications shall be based on either daily average or maximum figures extracted from the specified conditions. However, the unit should be checked for the expected maximum concentrations of the contaminants in the influent water in order to meet the guaranteed effluent wastewater quality. The effluent from the aeration basin should be routed to the activated sludge clarifier.

A portion of the activated sludge from the clarifier underflow should be recycled to the aeration basin and mixed with the influent waste. A portion of the clarifier underflow should be wasted to the aerobic sludge digester to maintain the proper concentration of mixed liquor suspended solids in the aeration basin. Floatable solids from the scum trough in the clarifier should flow to a scum pit. Effluent from the clarifier should flow to a surge tank and be pumped to the deep bed filters for final treatment.

Effluent from the clarifier should contain 10 mg/L(ppm) maximum of dissolved BOD_5 and 25 mg/L(ppm) maximum of total dissolved solids unless otherwise specified.

4.3.5 Design Considerations

4.3.5.1 Biological Oxidation Basin(s)

Adequate numbers of aeration basins should be provided based on the optimum size of the basin and fresh feed design flow-rate. Adequate hours of retention time should be provided by each aeration basin. Basin walls shall have a 1.5 to 1 slope and a freeboard of not less than 500 mm. The basins should be surrounded by guard rails to prevent personnel from falling into the aeration basins.

The inlet line should allow the influent to be added to the aeration basin both at the head end of the basin and under each aerator. The recycled sludge should enter at the head end. The aeration basins should be designed so that the air is disengaged from the microbial floc and the skimming of foam, scum, or floating sludge is minimized in the discharge. An overflow weir and baffles extended 500 mm below and 300 mm above the water surface should be provided.

4.3.5.2 Mechanical Aerators

Low speed mechanical aerators should be provided for each aeration basin to supply the oxygen required. Their number will be determined based on the total oxygen requirement. The circles of influence of adjacent aerators should just touch but not overlap. The aerators should be of the easily-adjusted, variable weir platform mounted type driven by a gear box consisting of a stationary base, ball race, and internal gear. The latter should be provided with a mounting rim for attaching the drive shaft.

The internal gear should be rotated by a pinion, keyed to a first stage speed reducer, connected to an electric motor. The aerobic sludge digester aerator should be of the floating type. Each unit should be complete with an impeller shaft and assembly and all the necessary anchor bolts and accessories. Each unit should be ready for easy on-site assembly requiring only the bolting of the base plate to the appropriate platform or poles. A platform with a vertical access ladder should be provided to the aerobic sludge digester.

The drive assembly should be designed for continuous duty and heavy shock loads. The impeller assembly should be of suitable design to provide oxygen transfer and mixing, withstand design loads, and be easily removable from drive assembly. Each impeller should be statically balanced prior to shipment. The impeller shaft should be of suitable material and of proper size to withstand all operating and static stresses.

4.3.5.3 Nutrient Feed Systems

The nutrient feed systems should be designed to provide a sufficient amount of nitrogen and phosphorous to sustain growth of the biological mass in each aeration basin. The nutrients should be injected at a single point up-stream of the aeration basin(s). Each nutrient feed system should include a chemical solution tank, an agitator, and two chemical feed pumps (one as a spare). The chemical solution tank capacity should be based on at least 24 hours of operation.

The tank should be equipped with a level indicator, sight glass, bag breaker, the necessary pump controls, and an agitator. The pumps should have facilities to permit adjustment of capacity from 0 to 100% of the maximum specified. The pump accessories should include a coupling guard and strainer.

4.3.5.4 Aerobic Sludge Digester

The aerobic sludge digester should be of an adequate size to provide a minimum of 16 days retention time based on a design input of at least 3% of sludge withdrawal rate from the clarifiers. Floating type low-speed mechanical surface aerators should be provided. The tank should be equipped with a platform (including ladder) and poles to fix the mechanical aerators. The Vendor should specify the oxygen requirement for the aerobic sludge digester.

4.3.5.5 Control and Instrumentation

The design of the biological oxidation units should incorporate the necessary instrumentation for a minimum amount of operator observance. Alarms should be installed on all critical water levels and for over-torque condition on the clarifiers. DO (Dissolved Oxygen) and TOC (Total Organic Content) analyzers should be provided at appropriate locations in the biological oxidation unit. The following instrumentation should be specified as the minimum requirement:

- A flow controller (indicating or recording) on sludge recycling.
- A flow controller (indicating or recording) on sludge wastage.
- An indicator or recorder for dissolved oxygen in the aeration basin.
- An indicator or recorder for TOC of feed and effluent.
- A recorder of the pH of the feed.
- An indicator or recorder of the pH in the aeration basin.
- An indicator of the temperature of the influent and aeration basin.
- An indicator or recorder of dissolved oxygen in the final effluent.

4.3.5.6 Supplemental Dissolved Oxygen Using Hydrogen Peroxide

Hydrogen peroxide has been used to reduce Biological Oxygen Demand (BOD) and Chemical Oxygen Demand (COD) in refinery wastewaters for many years.

This includes:

- Supply of supplemental Dissolved Oxygen (DO) when biological treatment systems experience temporary overloads or equipment failure. Figure 4.6 shows a typical aeration basin with supply of supplemental dissolved oxygen.
- Pre-digestion of wastewaters that contain moderate to high levels of compounds that are toxic, inhibitory, or recalcitrant to biological treatment.

As indicated by these examples, hydrogen peroxide can be used as a standalone treatment or as an enhancement to existing physical or biological treatment processes, depending on the situation.

The BOD/COD removal efficiency of aerobic biological treatment processes depends on a number of factors including (but not limited to): influent BOD/COD loading, temperature,

Figure 4.6 Aerobic basin with supplemental dissolved oxygen system (Reproduced with permission from © DDM Imagery).

nutrient levels, and DO concentrations. Many biological treatment facilities use hydrogen peroxide to supplement DO levels when oxygen limited conditions in aeration basins result in poor BOD/COD removal.

These conditions can be brought about by unexpected peaks in influent BOD/COD loading; seasonal variations in BOD/COD loading; and hot weather – which reduces the efficiency of oxygen transfer by mechanical aeration equipment (i.e., O_2 solubility decreases as temperature increases). These conditions may or may not be accompanied by filamentous bulking.

When hydrogen peroxide is used to supplement DO, it is metered directly into the aeration basin of a biological treatment system to provide an immediate source of dissolved oxygen. The conversion of hydrogen peroxide to DO in an activated-sludge mixed liquor proceeds according to the following reaction:

$$\text{Catalase Enzyme: } 2\,H_2O_2 \rightarrow O_2 + 2\,H_2O$$

Catalase enzyme is a natural decomposition catalyst for hydrogen peroxide, and is found in all activated-sludge mixed liquors, being produced by most aerobic organisms. Because this enzymatic decomposition of hydrogen peroxide is very rapid, the oxygen supplied by hydrogen peroxide is immediately available for uptake by the aerobic organisms.

The above reaction shows that two parts of hydrogen peroxide will yield one part of dissolved oxygen. Therefore, the amount of hydrogen peroxide required to oxygenate the wastewater is surprisingly small. In actual practice, the requirement may be higher due to side reactions with oxidisable compounds.

4.4 Trickling-Filters

A trickling-filter (see Figures 4.7 and 4.8) is a fixed bed, biological filter that operates under (mostly) aerobic conditions. Pre-settled wastewater is "trickled" or sprayed over the filter. As the water migrates through the pores of the filter, organics are degraded by the biomass covering the filter material.

The trickling-filter is filled with a high specific surface-area material such as rocks, gravel, shredded PVC bottles, or special pre-formed filter material. A material with a specific surface area between 30 and 900 m^2/m^3 is desirable. The filter is usually 1–3 m

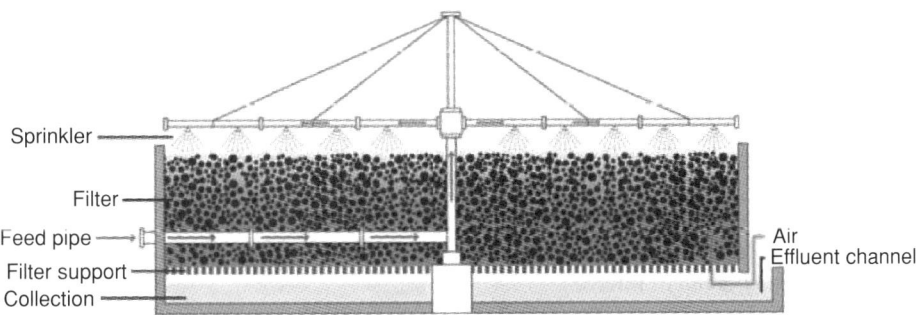

Figure 4.7 Trickling-filter.

Figure 4.8 Trickling-filter (Reproduced with permission from © DDM Imagery).

deep but filters packed with lighter plastic filling can be up to 12 m deep. Pre-treatment is essential to prevent clogging and to ensure efficient treatment. The pre-treated wastewater is "trickled" over the surface of the filter. Organisms that grow in a thin biofilm over the surface of the media oxidize the organic load in the wastewater to carbon dioxide and water while generating new biomass.

The incoming wastewater is sprayed over the filter with the use of a rotating sprinkler. In this way, the filter media goes through cycles of being dosed and exposed to air. However, oxygen is depleted within the biomass and the inner layers may be anoxic or anaerobic.

The ideal filter material has a high surface to volume ratio, is light, durable and allows air to circulate. Whenever it is available, crushed rock or gravel is the cheapest option. The particles should be uniform such that 95% of the particles have a diameter between 7 and 10 cm. Both ends of the filter are ventilated to allow oxygen to travel the length of the filter. A perforated slab that allows the effluent and excess sludge to be collected supports the bottom of the filter.

The bed consists of crushed rock or slag (1–2 m deep) through which the sewage is allowed to percolate. The stones become coated with a zoogloea film (a jelly-like growth of bacteria, fungi, algae, and protozoa), and air circulates by convection currents through the bed. Most of the biological action takes place in the upper 0.5 m of the bed. Depending on the rate of flow and other factors, the slime will slough off the rocks at periodic intervals or continuously, whenever it becomes too thick to be retained on the stones. A secondary settling basin is necessary to clarify the effluent from the trickling-filter. The overall reduction of BOD for a complete trickling-filter system averages around 80–90%.

4.4.1 Trickling-Filter Process Design

The first process design approach to use fundamental principles was published by Velz in 1948. His equation expressed BOD removal as a first order function of filter depth:

$$\ln\left(\frac{L_e}{L_o}\right) = -kd \qquad (4.1)$$

L_o = BOD of filter influent
L_e = BOD at any depth in filter
d = filter depth, m
k = Coefficient

Schultz modified the Velz equation to account for the hydraulic loading rate (gpm/ft² or m³/m²/hr), and Germain later applied Schultz's formula to plastic trickling-filter media.

$$\ln\left(\frac{L_e}{L_o}\right) = -kd/q^n \tag{4.2}$$

L_o = BOD of filter influent
L_e = BOD at any depth in filter
d = filter depth, m
k = Coefficient,
q = Hydraulic dosage of wastewater (not including re-circulation)
n = Exponent characteristic of filter media

4.4.1.1 Wetting Rates

The overall application rate of wastewater to the trickling-filter, including recirculation, expressed as gpm/ft² (or m³/m²/hr) of the filter area, is known as the "Wetting Rate." The desired wetting rate ranges from 0.05 gpm/ft² (0.12 m³/m²/hr) to a maximum of 3 gpm/ft² (7.32 m³/m²/hr), but is more typically in the range of 0.25 to 1 gpm/ft² (0.61 to 2.44 m³/m²/hr) for BOD removal systems and 0.75 to 2 gpm/ft² (1.83 to 4.88 m³/m²/hr) for nitrification trickling-filters.

If the average wetting rate is too low, the water may not penetrate the depth of the filter bed uniformly. It may channel away from some areas and leave damp unwetted areas that can act as incubators for pests like filter flies and snails (in nitrification towers). Also, biological populations not continuously wetted and fed by wastewater become ineffective. Those areas of the filter tower will not be available to provide effective treatment of wastewater during periods of higher flow. Semi-dry biomass can also putrefy and create odor problems. Recycling of treated wastewater is an effective method of keeping all areas and depths of the trickling filter biologically active when the influent flow is too low for proper wetting.

4.4.1.2 Instantaneous Wetting Rate

From the work of the German wastewater treatment industry, a term has been developed that identifies the Instantaneous Application Rate. This term is the SpülKraft Rate, or SK Rate, that gives the units of mm of water per pass of the distributor arms.

$$SK = -kd\left(\frac{q+R}{a \times n}\right)\left(\frac{1000}{6}\right) \tag{4.3}$$

SK: Flushing intensity, mm/pass
$q+R$: Influent and recycle hydraulic loading, $(m^3/m^2.h)$
a: Number of arms
n: Distributer, RPM

Hydraulically-driven rotary distributors in the normal operating mode usually rotate at a rate of 1 revolution per 0.75 to 1.5 minutes and have two or four arms. The SK Rate may be in the range of 0.3 to 0.5 mm per pass in rock filters and from 5 to 30 mm per pass in more modern filters.

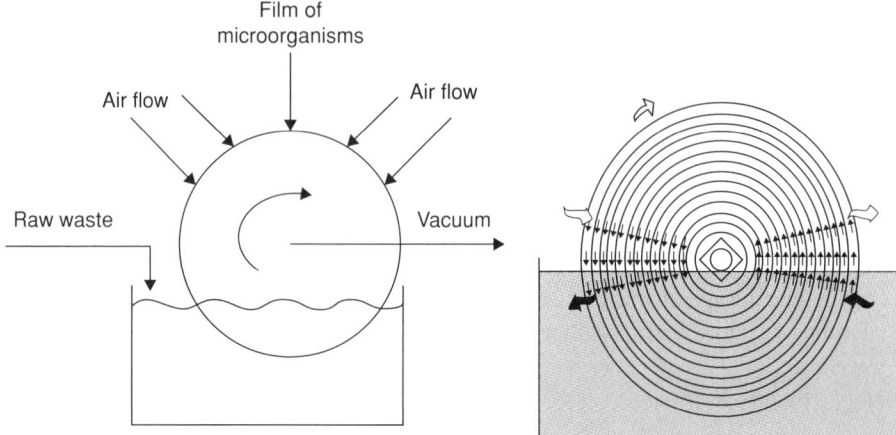

Figure 4.9 Rotating biological contactor.

If the recycling capacity is minimal, and the operator has the ability to slow the rotation speed of the distributor, it is possible to compensate somewhat for low wetting rates by using higher SK values. Higher SK values will provide more complete penetration of the filter media depth and keep the bulk of the filter wetted.

Short cycle times of dryness between flushing will not be as detrimental to the biomass as a general starvation for water in pockets of media that are bypassed at low wetting rates.

4.5   Rotating Biological Contactor System

Rotating biological contactor systems (refer to Figures 4.9, 4.10, and 4.11) normally make use of bar screens and/or comminutors, grit chambers, primary settling tanks, secondary tanks, and digesters, which are operated in the same manner as those of trickling-filter systems. The rotating biological contactor (RBC) is a simple, effective method of providing secondary wastewater treatment. The system consists of biomass media, usually plastic, that is partially immersed in the wastewater.

As it slowly rotates, it lifts a film of wastewater into the air. The wastewater trickles down across the media and absorbs oxygen from the air. A living biomass of bacteria, protozoa, and other simple organisms attaches and grows on the biomass media. The organisms then remove both dissolved oxygen and organic material from the trickling film of wastewater. Any excess biomass is sloughed off as the media is rotated through the wastewater. This prevents clogging of the media surface and maintains a constant microorganism population. The sloughed off material is removed from the clear water by conventional clarification. The RBC rotates at a speed of 1 to 2 rpm and provides a high degree of organic removal.

4.6   Sewage Oxidation Ponds

Sewage oxidation ponds are 0.8–1.2 m in depth, and may be used singly, in parallel, or in a series following primary treatment (see Figures 4.12 and 4.13). Their use is

Biological Treatment 127

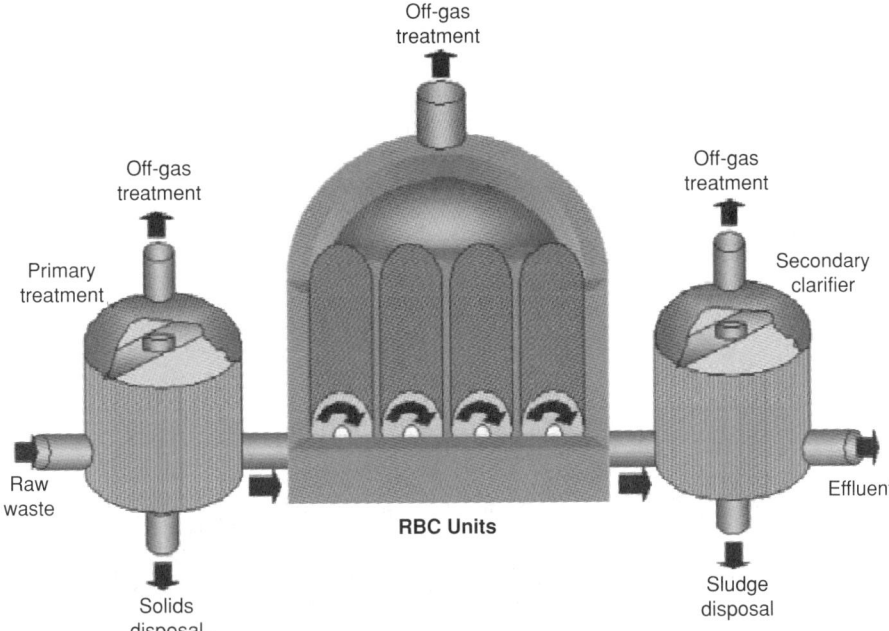

Figure 4.10 Rotating biological contactors preceded by pre-treatment and followed by secondary sedimentation.

particularly suited to locations with available land and warm climates. Their ability to absorb shock loads and ease of operation and maintenance make them desirable treatment units. Biological life in ponds uses the organic and mineral matter in the sewage for food to produce more stable products.

The products often stimulate abundant growth of algae and other vegetation. Solution of oxygen from the atmosphere, and the ability of vegetation to produce oxygen when exposed

Figure 4.11 Rotating biological contactor (Reproduced with permission from © DDM Imagery).

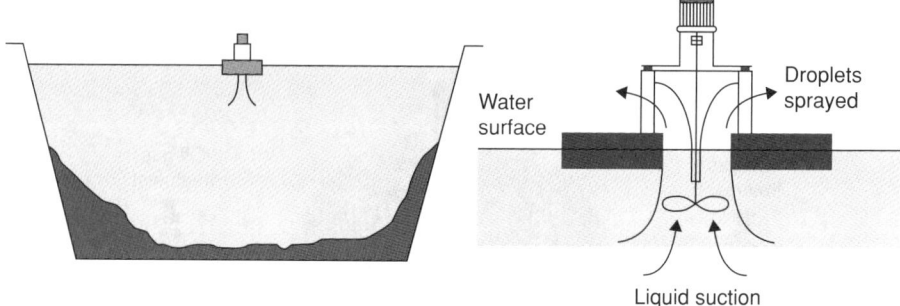

Figure 4.12 Schematic for sewage oxidation ponds.

Figure 4.13 Sewage oxidation ponds (Reproduced with permission from © DDM Imagery).

to sunlight, help maintain aerobic conditions. The lagoons will develop an odor similar to fresh water ponds in wooded areas. Allowable loading can vary from 125–2000 persons per hectare depending upon the location. Where complete treatment is to be provided by ponding, the cells are known as raw sewage lagoons, with depths of 1–1.5 m and reduced loading.

5

Wastewater Treatment in Unconventional Oil and Gas Industries

The aim of this chapter is to identify issues associated with the proper environmental management of waste formation waters from exploration and production activities in unconventional oil and gas industries.

Further, it will outline treatment and disposal options to produce and manage formation wastewater to meet or exceed environmental performance and responsible care requirements.

5.1 Background

Exploration and extraction of petroleum hydrocarbons – specifically natural gases (coal seam gas [CSG], shale gas etc.) – is not a new industry worldwide. However the resource potential has started gaining attention in the past two decades as more conventional resources have become depleted.

Marcellus shale is of commercial interest because it is an organic-rich shale containing high quantities of natural gas. Thus, there is enormous interest in commercial exploitation of the shale formation for natural gas production using drilling technologies such as hydraulic fracturing or directional drilling.

Because shales ordinarily have insufficient permeability to allow significant fluid flow to a well bore, most shales are not commercial sources of natural gas. Shale gas is one of a number of unconventional sources of natural gas; others include coal bed methane, tight sandstones, and methane hydrates. Shale gas areas are often known as resource plays (as opposed to exploration plays).

Figure 5.1 Used frack water from a drilling operation around a shale gas drilling rig (Reproduced with permission from © Kate Ausburn).

Shale has low matrix permeability, so gas production in commercial quantities requires fractures to provide permeability. Shale gas has been produced for years from shales with natural fractures; the shale gas boom in recent years has been due to modern technology in hydraulic fracturing (fracking) to create extensive artificial fractures around well bores.

The market-changing growth of shale gas production in recent years has resulted in the emergence of pressing environmental and water management challenges. Treating the large quantities of water used in the hydraulic fracturing process is a major concern for operators. Fracking, or hydraulic fracturing, is the process of drilling sideways into subterranean shale and blasting it open with millions of gallons of water to release natural gas. Figure 5.1 shows typical used frack water from drilling operation around a shale gas drilling rig.

The technology innovation targeted toward wastewater treatment from shale gas operations has escalated rapidly. After a well is drilled, water mixed with various chemicals and sand is forced down the well to fracture the shale and release the gas. The tainted water that returns to the surface is called "flow back."

Disposal of the tainted water is a relatively new and evolving industry. In common with other young technology markets, we observe transfer and adaptation of technologies from other industries – in this case from utilities water treatment and the desalination space. As the market matures, we can expect increased development of water treatment technologies specific to shale gas extraction.

In the coal seam gas industry, there are wide variations of water production rates from coals in any basin. Ease of dewatering any well depends on the coal's permeability, interference with other wells or mines, and links to an aquifer or meteoric waters. Past mining in the area, even though presently inactive, may have depleted water in the seams.

Quantity and chemical content are the two important considerations for waters produced from coal seams. Some treatment at the surface is necessary regardless of the disposal method. TDS, oxygen content, and suspended solids must be controlled for discharge into surface streams and conformity to the environmental regulations. Chloride content, TDS, particulate matter, and formation compatibility must be established for water disposal by well injection.

Besides establishing disposal requirements, water composition gives some insight into the permeability of the formation. Bicarbonate ions exist in larger concentrations in those formations having meteoric waters continually replenishing the coal seams.

Chloride ions occur in greater concentration in those more stationary coal bed waters. Therefore, the anion HCO_3^- is indicative of a good permeability in the coals and of continuity in the seams that allows water circulation along an uninterrupted path. The Cl^- anion, on the other hand, suggests a discontinuity in the seams or a lack of permeability that leaves the waters uncirculated.

Generally, the produced water is composed of dissolved and dispersed oil components, dissolved formation minerals, production chemicals, dissolved gases (including CO_2 and H_2S) and produced solids. There is a wide variation in the level of its organic and inorganic composition due to geological formation, the lifetime of the reservoir, and the type of hydrocarbon produced.

5.1.1 Dissolved and Dispersed Hydrocarbon Components

Dispersed and dissolved components are a mixture of hydrocarbons including BTEX (benzene, toluene, ethylbenzene, and xylene), PAHs (polyaromatic hydrocarbons), and phenols. BTEX, phenols, aliphatic hydrocarbons, carboxylic acid, and low molecular weight aromatic compounds are classified as dissolved hydrocarbons, while less-soluble PAHs and heavy alkyl phenols are present in produced water as dispersed liquid hydrocarbons (Stephenson, 1989). Dissolved and dispersed oil content in produced water is dangerous to the environment and its concentration can be very high in some fields.

5.1.2 Dissolved Mineral

Dissolved inorganic compounds or minerals are usually high in concentration and classified as cations and anions, naturally occurring radioactive materials, and heavy metals. Cations and anions play a significant role in the chemistry of produced water. Na^+ and Cl^- are responsible for salinity, ranging from a few milligrams per litre to ~300 000 mg/l. Cl^-, SO_4^{2-}, CO_3^{2-}, HCO_3^-, Na^+, K^+, Ca^{2+}, Ba^{2+}, Mg^{2+}, Fe^{2+}, and Sr^{2+} affect conductivity and scale-forming potential. Typical oilfield produced water contains heavy metals in varied concentrations, depending on the formation geology.

^{226}Ra and ^{228}Ra are the most abundant naturally occurring radioactive elements present in the produced water. The radioactivity of the produced water results primarily from radium that is co-precipitated with barium sulfate (scale) or other types of scales. The concentration of barium ions in produced water could give a strong indication of radium isotopes present in it.

5.1.3 Production Chemicals

Production chemicals can be pure compounds or compounds containing active ingredients dissolved in a solvent or a co-solvent, and are used for inhibition of corrosion, scale deposition, bacterial growth, and emulsion breaking in order to improve the separation of oil and water. These chemicals enter produced water in traces or sometimes significant amounts and vary from platform to platform. Active ingredients partition themselves into all phases present depending on their relative solubilities in liquid hydrocarbons, gas, or water. The fate of these chemicals is difficult to determine because some active ingredients are consumed within the process.

5.1.4 Produced Solids

Produced solids include clays, precipitated solids, bacteria, carbonates, sand and silt, corrosion and scale products, proppant, formation solids, and other suspended solids (ASTM, 1978). Their concentrations vary from one platform to another.

5.1.5 Dissolved Gases

The major dissolved gases in produced water are carbon dioxide, oxygen, and hydrogen sulfide. They are formed naturally, by the activities of bacterial or by chemical reactions in the water.

Currently, oil and gas operators treat produced water via one or more of the following options:

1. Avoid production of water: water fractures are blocked by polymer gel or downhole water separators, although this option is not always possible.
2. Inject into formations: produced water may be injected back to its formation or into other formations. This option often requires transportation of water, and treatment to reduce fouling and bacterial growth. In the long term, the stored produced water may pollute the underground waters.
3. Discharge to the environment: produced water may be discharged to the environment as long as it meets onshore and offshore discharge regulations.
4. Reuse in oil and gas industry operations: minimally treated produced water may be used for drilling and workover operations within the petroleum industry.
5. Application in beneficial uses: produced water may be consumed for irrigation, wildlife consumption and habitat, industrial water, and even drinking water. However, beneficial uses of produced water may involve significant treatment. (See Figure 5.2.)

5.2 Toxicity Limitations of Coal Bed Water

Coal bed waters are regulated to specify the following chemical contents and conditions:

- Dissolved oxygen (DO).
- Biochemical oxygen demand (BOD).
- Iron.
- Manganese.
- Total dissolved solids (TDS).

The first four conditions are dependent on adequate oxygen being added to the waters from the coal beds before the waters can be disposed of in surface streams. Dissolved oxygen must be input to the produced waters because waters from coal seams are devoid of oxygen.

Figure 5.2 *A holding pond at Woodview, west of Casino, used by a coal seam gas company to store wastewater (Reproduced with permission from © Green-left, NSW, Australia).*

Upon aerating the waters at the surface, iron and manganese are oxidized and precipitate as solids. Additionally, aerating supplies the oxygen for BOD; about 1.2 g oxygen is required for 1 g BOD. Therefore, supplying oxygen is a primary requirement for surface treatment of produced waters, and 5 mg/L of O_2 is required for waters discharged from the treating process.

Table 5.1 summarizes important characteristics of oxygen in water.

Note from Table 5.1 that oxygen solubility in water is naturally limited to 7.6–11.3 mg/L under ordinary conditions. As temperature or chloride content increases, oxygen solubility is further diminished. In the hot months, therefore, O_2 solubility in Alabama's surface waters decreases, and supplying or maintaining the oxygen in produced waters becomes more difficult.

Despite the freedom of surface disposal, strict regulations are imposed on the treatment, disposal, and monitoring of waters in surface streams. A series of treating ponds in any producing field of the basin serve as staging points for the treatment process.

Table 5.1 *Oxygen in coal bed waters.*

Parameter	Value
Oxygen in water from coal seam	0
Oxygen required in discharged waters	≥5 mg/L
Oxygen for ferrous ion oxidation	1 mg/L O_2 per 7 mg/L Fe_2+
Oxygen for manganous ion oxidation	1 mg/L O_2 per 3 mg/L Mn_2+
Per 1.0 g of BOD	1.2 g O_2
7.6 to 11.3 mg/L O_2 dissolves in H_2O	at 50 to 86°F w/o Cl

The first treatment must be to provide oxygen to the waters collected in the ponds using one of three methods:

1. **Spraying**. The surface area of the water is increased to absorb oxygen from the air.
2. **Agitating mechanically**. The surface area of water in contact with air is renewed constantly. The new surface absorbs more oxygen; the concentration gradient is increased for more rapid absorption of the oxygen.
3. **Pumping air beneath the water surface**. A primary objective in the initial treatment is to transfer oxygen to the water to feed the growth of microorganisms that degrade organic matter in the water.

   The preceding methods increase surface areas of water exposed to air to enhance absorption of oxygen. Then agitation is supplied to bring bacteria, oxygen, and organic matter into contact.

Bates, *et al.* gave the rate of oxygen transfer across the gas and liquid films of the liquid–air interface in equation 5.1.

$$dC/dt = k_L a(C_s - C) \tag{5.1}$$

Where:

C = concentration of oxygen
C_s = saturation concentration of oxygen in water
a = area of interface per unit volume of liquid
t = time
k_L = proportionality constant

Biochemical oxygen demands (BOD) result from the bacteria that degrade organic compounds in the water. It is this oxygen that must be supplied to the bacteria to degrade the mass of organic matter in the water within a time frame.

Microorganisms increase their activity exponentially with temperature so a standard temperature must be set. The standard is 20 °C. BOD_5 – the biochemical oxygen demand over a 5-day test period at 20 °C – must not exceed 30 mg/L in the disposal waters.

The organic constituents of the produced coal seam waters that the microorganisms feed upon may come from organic compounds of the coal or from decaying organic matter in the treatment ponds.

However, the greatest amounts of organic matter that must be biodegraded come from fracturing fluids expelled from formation. Consequently, the fracturing fluids place a heavy periodic demand for oxygen on the waters.

After fracturing, these stimulation fluids continue to be returned for several months during production of methane. Hydroxypropyl guar and other water-soluble polymers used in fracturing are the main culprits.

To correct the heavy surge of BOD_5 upon start-up of a field, hydrogen peroxide is added to supply oxygen demand as a temporary fix to a problem that could require large capital expenditure for more facilities. The start-up BOD_5 demand results from the return of stimulation fluids used to fracture wells in the field.

Removal of manganese and iron is more straightforward than TDS removal. The oxidized manganese and iron forms precipitate in the holding ponds. Their oxidation is much faster at a higher pH, which should be maintained above 7.2. Manganese content in the effluent waters must be less than 2 mg/L as a monthly maximum. Total iron must be less than a 3.0 mg/L monthly maximum.

The TDS is the most troublesome chemical content of produced waters and the most damaging to plant life. Sodium chloride is the main constituent of the dissolved solids.

Considerable environmental concern exists with regard to the potential impact of discharged untreated formation waters from exploration and extraction operations on aquatic ecosystems, groundwater resources, and soils. Other jurisdictions have experienced problems with certain disposal methods for these types of wastewater. In order to understand the potential environmental impacts of formation waters, their physical/ chemical properties must first be understood. To this end a brief description of formation waters is given below.

Formation water is the naturally occurring water that is contained within the geological formation itself. In order to extract the methane/natural gas resource within the formation, often large volumes of water must be pumped to the surface. The quantity and quality of the formation water can both be problematic. Depending on the formation, each well can produce in the range of 3 to 100 liters of water per minute, which equates to 4500 to 140 000 liters per day per well.

Over time, the water volume should decrease as the gas volumes increase. Formation water has been found to contain high levels of chlorides, arsenic, iron, barium, manganese, and may even contain naturally occurring radioactive materials (NORMS). Chlorides in some cases have been found to be in excess of four times the level of concentrations found in the ocean.

Discharge of these waters into streams that are not saline would be lethally toxic to aquatic life. Discharge onto the land can have an adverse effect on soil structure and negatively impact vegetative growth. Sodium-induced dispersion can cause reduced infiltration, reduced hydraulic conductivity, and surface crusting of soils.

Excess salinity in soil water can decrease water available to plants and cause stress to vegetation. Livestock or humans cannot utilize water with high levels of chlorides due to cellular disruption. High levels of arsenic and barium can be lethally toxic to animal and aquatic life. Elevated levels of iron and manganese can cause accelerated bacterial growth in fresh water systems and reduce ground and surface water quality.

Other jurisdictions have permitted the discharge of untreated formation waters directly into surface waters. Pollution caused by these discharges has been common. Many of these jurisdictions have since amended their regulations and policies to prohibit direct discharge of untreated formation waters into any surface water body. Discharge of highly saline formation water has led to serious pollution of surface waters. In the meantime, leakage of lagoons holding highly saline mine formation waters may led to serious contamination of a potable aquifer.

5.3 Shale Gas and Coal Seam Gas Produced Water, Treatment and Disposal

In general, four techniques are possible to dispose of produced coal bed waters: (1) well injection, (2) discharge into surface streams, (3) land application, and (4) membrane processes. All coal seam gas (coal bed methane) produced water that is sent through a treatment facility is discharged into drainage. Thus water treatment is generally used when the well field is in close proximity to a large river or stream. While water treatment is sometimes used farther away from the drainage, it then requires a large scale pumping and piping infrastructure to move the water to the treatment facility and then to the discharge point.

Formation waters in most places typically cannot meet these requirements without prior treatment before discharge. The following is a list of commercially available treatment technologies currently being used by the Industry. This list does not represent a hierarchy of options; it is expected that the generators of the waste will choose the treatment option that is best suited to the characteristics of the formation waters they are dealing with and within what is economically achievable for the project.

The Industry is also not limited to the technologies listed in this chapter. As technology advances, and more is learned about the potential environmental effects of formation waters, it can be expected that new technologies will emerge and standards will change.

5.3.1 Evaporation Pond

An evaporation pond is an artificial pond that requires a relatively large area of land and is designed to efficiently evaporate water by solar energy. They are designed either to prevent subsurface infiltration of water or the downward migration of water depending on produced water quality.

It is a favorable technology for warm and dry climates because of the potential for high evaporation rates. Evaporation ponds are typically economical and have been employed for the treatment of produced water on-site and off-site.

Ponds are usually covered with netting to prevent potential problems to migratory waterfowl caused by contaminants in the produced water. All water is lost to the environment when using this technology, which is a major setback when water recovery is an objective for water treatment.

5.3.2 Surface Stream Disposal

A flow diagram of the surface treating facilities for coal bed waters is given in Figure 5.3. The waters pass through an aeration pond, a sedimentation pond, and possibly a storage pond before being disposed of through a diffuser into the surface stream.

Aeration of the "dead" coal bed waters that have no dissolved oxygen is an important step in processing because the introduction of oxygen into the water has multiple beneficial effects. Foremost of the benefits is the oxidation and resulting precipitation of the suspended solid iron and manganese.

Oxidation of the two metals in the aeration and sedimentation ponds removes these metals from the waters and from further consideration. (After several years, the solid sediments must be removed from the bottom of the pond.) Volatile organic matter is lost during aeration in the holding pond. Also, aeration adds dissolved O_2 and decreases biological oxygen demand by as much as 50–90%. The aeration process, however, decreases neither the chloride content nor the TDS of the coal bed waters.

In case of emergencies, the storage and transfer pond in Figure 5.3 is designed to hold waters temporarily, primarily during the low flow of surface streams in the dry summer months.

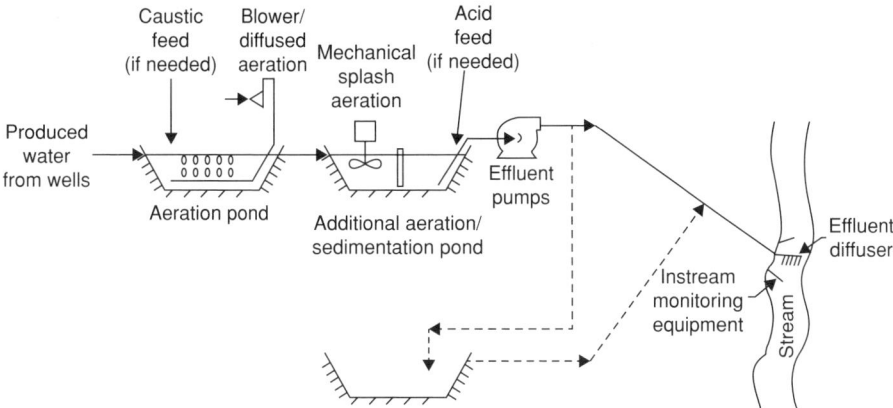

Figure 5.3 Surface treatment of coal bed waters.

Chloride removal requires a more expensive and higher technology treatment, specifically, ion exchange, reverse osmosis, or electrodialysis; evaporation is also feasible. Consequently, the most difficult problem for any type of surface disposal lies with meeting the TDS specifications, primarily for chlorides. The options are to remove chlorides at considerable expense or to dilute the water to acceptable chloride concentration levels. Upon addition of production waters, the content of the natural stream becomes the sum of innate TDS and Cl^- plus TDS and Cl^- added from CBM waters.

Timing of CBM water disposal in streams must allow for the prescribed dilution, a feat dependent upon the highly variable seasonal surface stream flow. Four parameters are necessary to develop the surface disposal plan: (1) the water quality of the produced and natural streams; (2) the well start-up schedule; (3) the projected flow history of the well; and (4) the natural stream's capacities.

$$Q_s = Q_e \frac{C_e - C_m}{C_m - C_s} \tag{5.1}$$

Where:

C_m = in-stream quality limitation
C_s = background stream concentration
Q_s = minimum surface stream natural flow to accommodate, cfs
Q_e = effluent from coal bed methane wells, bpd
C_e = effluent water concentration of TDS

Environmental regulations set C_m. Assimilative capacities of streams can therefore be calculated. If consistent units of C are used, if Q_e is given in units of BWPD, and if Q_s is desired in units of cubic feet per second, the relationship becomes that of equation 5.2.

$$Q_s = 6.5 \times 10^{-5} Q_e \frac{C_e - C_m}{C_m - C_s} \tag{5.2}$$

Example The "A" Creek in the month of August has an average minimum monthly flow rate of 12 cu ft/sec (184,814 bbl/D). Inherent background TDS in its waters amounts to 15 mg/L. Government regulations limit raising the TDS to a maximum of 200 mg/L. Coal bed methane wells in the adjacent field produce waters having an average TDS content of 1800 mg/L. What maximum volumetric rate of the produced waters in BWPD could be disposed of in "A" Creek in August?

Solution:

$$Q_e = Q_s \frac{C_m - C_s}{C_e - C_m}$$

Where:

Q_e = coal bed well effluent, b/d
Q_s = 12 cfs
C_e = 1800 mg/L
C_m = 200 mg/L
C_s = 15 mg/L

Therefore, during the low-flow period of August (assuming an average river flow rate for the month) 21 369 BWPD of the coal bed produced waters could be disposed in "A" Creek without exceeding the total dissolved solids' limit of the stream. September, October, and November would be the low-flow months, so the flow allowable in August would be expected to decrease further in the subsequent three months.

5.3.3 Ion Exchange

Ion exchange is conventionally used to describe the purification, separation, and decontamination of aqueous and other ion-containing solutions with use of solid ion exchangers. Commercial ion exchange is used in many locations for formation water treatment. The technology is capable of treating water of varying quality, which makes it one of the more flexible treatment options.

The finished water quality is high enough that it can be used for irrigation, livestock, or watering in some jurisdictions. A waste brine is also produced which, depending on the chemistry, must be disposed of at an approved facility. Ion exchange has proven to be economically viable under appropriate conditions.

5.3.4 Membrane Filtration Technology

Membranes are microporous films with specific pore ratings, which selectively separate a fluid from its components. There are four established membrane separation processes: microfiltration (MF), ultrafiltration (UF), reverse osmosis (RO), and nanofiltration (NF). RO separates dissolved and ionic components, MF separates suspended particles, UF separates macromolecules, and NF is selective for multivalent ions. MF and UF can be used as a standalone technology for treating industrial wastewater, but RO and NF are usually employed in water desalination. Membrane technology operates two types of filtration process, cross-flow filtration or dead-end filtration, which can be pressure (or vacuum)-driven systems.

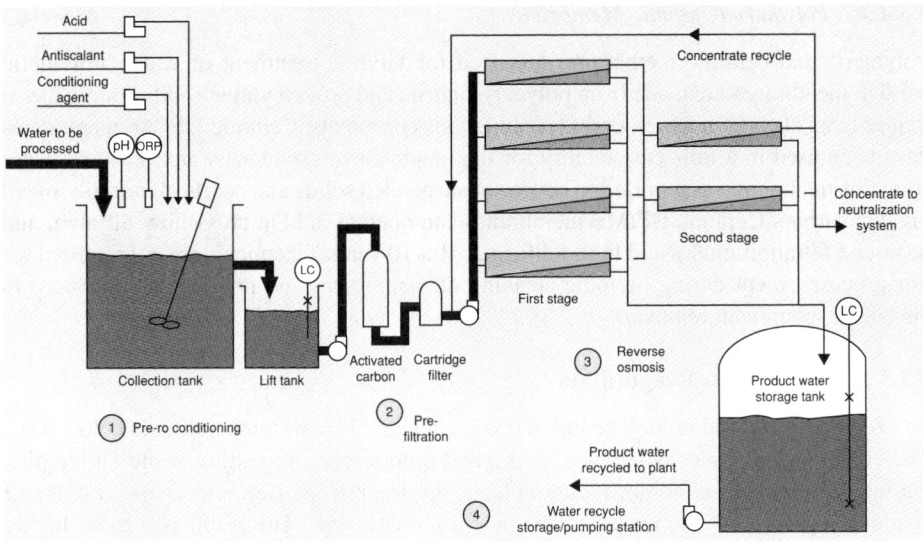

Figure 5.4 A typical reverse osmosis (RO) treatment process.

5.3.4.1 Reverse Osmosis

Reverse osmosis is a separation process that uses pressure in excess of osmotic pressure to force a solvent through a membrane that retains the solute on one side and allows the purified solvent to pass to the other side.

The process is used most often for desalination but is also effective in removal of all particles larger than one angstrom. Pre-treatment is required as membranes cannot be backwashed because of the design that allows water to flow through in only one direction.

Waste brine is typically piped to a storage pond for further evaporation and eventual disposal at an approved facility. This process has been effectively utilized worldwide for various types of water and wastewater treatment. It is energy consumptive due to the high pressures required to force the solvent through the membrane. Figure 5.4 shows a typical reverse osmosis treatment process.

5.3.4.2 Microfiltration/Ultrafiltration

MF has the largest pore size (0.1–3 μm) and is typically used for the removal of suspended solids and turbidity reduction. It can operate in either cross-flow or dead-end filtration. UF pore sizes are between 0.01 and 0.1 μm. They are employed in the removal of color, odor, viruses, and colloidal organic matter. UF is the most effective method for liquid hydrocarbon removal from produced water in comparison to traditional separation methods, and it is more efficient than MF for the removal of hydrocarbons, suspended solids, and dissolved constituents from oilfield produced water. Both MF and UF operate at low trans-membrane pressure (1–30 psi) and can serve as a pre-treatment to desalination but cannot remove salt from water.

5.3.4.3 Polymeric/Ceramic Membranes

Polymeric and ceramic membranes are used for UF/MF treatment of water. Polymeric MF/UF membranes are made from polyacrylonitrile and polyvinylidene and ceramic membranes from clays of nitrides, carbides, and oxides of metals. Ceramic UF/MF membranes have been used in a full-scale facility for the treatment of produced water. Product water from this treatment was reported to be free of suspended solids and nearly all non-dissolved organic carbon. Ceramic UF/MF membranes can operate in both cross-flow filtration and dead-end filtration modes and have a lifespan of >10 years. Chemicals are not required for this process except during periodic cleaning of membranes and pre-coagulation (used to enhance contaminant removal).

5.3.5 Freeze–Thaw Evaporation

Freeze–thaw evaporation may be utilized when the ambient air temperature is below 0 °C. The contaminated water is sprayed or dripped onto a freezing pad to create an ice pile. During sub-freezing conditions, runoff from the ice pile will elevate concentrations of chemical constituents compared to the original wastewater. The result is a more highly concentrated solute that will be need to be stored for later disposal at an approved facility. Once temperatures increase to allow for thaw conditions, the melt water can be suitable for beneficial uses such as irrigation.

There are also implications for soil contamination associated with this method of treatment, but it may be feasible in jurisdictions with large land areas that are isolated from population bases.

5.3.6 Adsorption

Adsorption is generally utilized as a polishing step in a treatment process rather than as a standalone technology, since adsorbents can be easily overloaded with organics. It has been used to remove manganese, iron, total organic carbon (TOC), BTEX, oil, and more than 80% of heavy metals present in produced water. There are a variety of adsorbents, such as activated carbon, organoclays, activated alumina, and zeolites.

5.3.7 Chemical Oxidation

Chemical oxidation destroys organic contaminants that are either dissolved in water or present in free phase. Chemical oxidants are introduced to the wastewater to react with such organic compounds as amines, phenols, chlorophenols, cyanides, halogenerated aliphatic compounds, and mercaptans. Chemical oxidation can be combined with other treatment technologies to maximize clean-up results.

Treated water quality is high. A waste brine is also produced which, depending on the chemistry, must be disposed of at an approved facility. The efficiency of the process is reduced with reduced water quality. Capital and operating costs are typically high.

Chemical oxidation is an established and reliable technology for the removal of color, odor, COD, BOD, organics, and some inorganic compounds from produced water.

Chemical oxidation treatment depends on oxidation/reduction reactions occurring together in produced water, because free electrons cannot exist in solution. Oxidants commonly used include ozone, peroxide, permanganate, oxygen, and chlorine. The oxidant mixes with contaminants and causes them to break down. The oxidation rate of this technology depends on the chemical dose, type of the oxidant used, raw water quality, and contact time between oxidants and water.

5.3.8 Filtration

Filtration technology is extensively used for the removal of oil and grease and TOC from produced water. Filtration can be accomplished by the use of various types of media such as sand, gravel, anthracite, walnut shell, and others. Walnut shell filters are commonly used for produced water treatment.

This process is not affected by water salinity and may be applied to any type of produced water. Media filtration technology is highly efficient for the removal of oil and grease, and efficiencies of more than 90% have been reported. Efficiency can be further enhanced if coagulants are added to the feed water prior to filtration. Media regeneration and solid waste disposal are setbacks to this process.

5.3.9 Constructed Wetlands

Constructed wetlands are artificially constructed wastewater treatment systems that utilize wetland vegetation to promote natural chemical and biological processes to achieve effective treatment of wastewater. There are two principle categories of constructed wetland systems: free water surface and subsurface flow.

Constructed wetland systems are very effective in removing nutrients such as nitrogen and phosphorous. Depending on the vegetation incorporated, wetlands can also remove organic or inorganic contaminants given sufficient retention time.

Efficiencies of the process have been shown to be in the 95% range but are highly dependent on the characteristics of the wastewater and the type of vegetation that is utilized.

Constructed wetland technology is considered a new technology with respect to formation water treatment. However, this technology has been successfully employed for treatment of coal mine discharge waters which have a similar chemistry to that of formation waters. Use of an existing natural wetland or a wetland that has been created as part of a compensation plan is strictly prohibited.

5.3.10 Electrodialysis/Electrodialysis Reversal

Electrodialysis (ED) and ED reversal (EDR) are mature electrochemically driven desalination technologies. These processes involve separation of dissolved ions from water through ion exchange membranes. They use a series of ion exchange membranes containing electrically charged functional sites arranged in an alternating mode between the anode and the cathode to remove charge substances from the feed water (Figure 5.5). If the membrane is positively charged, only anions are allowed to pass through it. Similarly, negatively charged membranes allow only cations to pass through them. EDR uses periodic reversal of polarity to optimize its operation.

5.3.11 Deep Well Injection at Dedicated Onshore Sites

Disposal of wastes through injection down wells into pores, fissures, and caverns in underground strata has been a favored strategy in many areas. These disposal sites often take

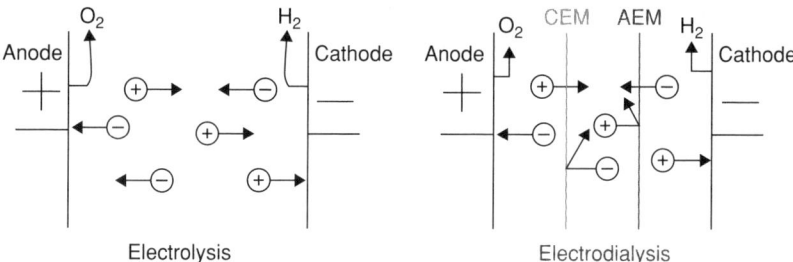

Figure 5.5 Comparison of electrolysis and electrodialysis (CEM, cation exchange membrane; AEM, anion exchange membrane).

advantage of exploratory wells that did not find commercial quantities of natural gas or oil, and therefore represent a non-performing asset to the drilling company, or have been evacuated of the oil or gas resource.

Using the site for waste disposal therefore allows some cost recovery. The technical rationale for using deep well disposal is that the materials are deposited in impermeable strata which are hydraulically isolated from potable aquifers and surface waters.

In addition, since the particular fluids used in drilling were considered safe enough to use during the original construction of the well, it should be reasonable to assume that more of these materials placed permanently will not pose additional risk.

The arguments against this form of disposal are also persuasive. Injection is a form of dumping since no treatment has occurred or is likely to occur at depth. This results in an area that is permanently impacted and withdrawn from potential beneficial use. There is also additional risk if the underlying geology is not perfectly understood, since leakage may occur and monitoring is difficult and expensive to install.

Since many locations do not have an extensive history of land-based petroleum exploration and therefore do not have preexisting wells or land areas that are impacted by drilling fluids, it does not appear reasonable at this time to establish deep well injection as a disposal option for formation waters.

5.3.12 Biological Aerated Filters

Biological aerated filters (BAF) are a class of biological technologies that consist of permeable media that use aerobic conditions to facilitate biochemical oxidation and removal of organic constituents in polluted water. Media is not more than 4 inches in diameter to prevent clogging of pore spaces when sloughing occurs. BAF can remove oil, ammonia, suspended solids, nitrogen, chemical oxygen demand (COD), biological oxygen demand (BOD), heavy metals, iron, soluble organics, trace organics, and hydrogen sulfide from produced water. It is most effective for produced water with chloride levels below 6600 mg/l.

This process requires upstream and downstream sedimentation to allow the full bed of the filter to be used. Removal efficiencies of up to 70% of nitrogen, 80% of oil, 60% of COD, 95% of BOD, and 85% of suspended solids have been achieved with BAF treatment.

Solids disposal is required for the accumulated sludge in sedimentation basins and can account for up to 40% of the total cost of this technology.

5.3.13 Macro-Porous Polymer Extraction Technology

Macro-porous polymer extraction (MPPE) is one of the best available technologies and best environmental practices for produced water management on offshore oil and gas platforms. It is a liquid–liquid extraction technology where the extraction liquid is immobilized in the macro-porous polymer particles. These particles have a diameter of ~1000 μm, pore sizes of 0.1–10 μm, and porosity of 60–70%. Polymers were initially designed for absorbing oil from water but later applied to produced water treatment in 1991. Table 5.2 shows a summary of different produced water membrane treatment technologies.

5.3.14 Thermal Technologies

Thermal treatment technologies of water are employed in regions where the cost of energy is relatively cheap. The thermal separation process was the technology of choice for water desalination before the development of membrane technology. Multistage flash (MSF) distillation, vapor compression distillation (VCD), and multieffect distillation (MED) are the major thermal desalination technologies. Hybrid thermal desalination plants, such as MED–VCD, have been used to achieve higher efficiency. Although membrane technologies are typically preferred to thermal technologies, recent innovations in thermal process engineering make thermal processes more attractive and competitive in treating highly contaminated water.

5.3.14.1 Multistage Flash

The MSF distillation process is a mature and robust technology for brackish and sea water desalination. Its operation is based on evaporation of water by reducing the pressure instead of raising the temperature. Feed water is pre-heated and flows into a chamber with lower pressure where it immediately flashes into steam. Water recovery from MSF treatment is ~20% and often requires post-treatment because it typically contains 2–10 mg/L of TDS. A major setback in operating MSF is scale formation on heat transfer surfaces, which often means this process requires the use of scale inhibitors and acids. Overall costs vary depending on the size, site location, and materials of construction. Its energy requirement is between 3.35 and 4.70 kWh/bbl.

Globally, MSF market share has significantly decreased due to competition of membrane technologies, but it is a relatively cost-effective treatment method with a plant life expectancy of more than 20 years, and can be employed for produced water treatment.

5.3.14.2 Multieffect Distillation

The MED process involves application of sufficient energy to convert saline water to steam, which is condensed and recovered as pure water. Multiple effects are employed in order to improve the efficiency and minimize energy consumption. A major advantage of this system is the energy efficiency gained through the combination of several evaporator systems. Product water recovery from MED systems are in the range of 20–67% depending on the type of the evaporator design employed. Despite the high water recovery from MED systems, it has not been extensively used for water production like MSF because of scaling problems associated with old designs. Recently, falling film evaporators have been introduced to improve heat transfer rates and reduce the rate of scale formation.

Table 5.2 Comparison of produced water membrane treatment technologies (Igunnu and Chen, 2012).

Method	Reverse Osmosis (RO)	Nanofiltration (NF)	Polymeric Microfiltration/ Ultrafiltration (MF/UF) Membrane	Ceramic Microfiltration/ Ultrafiltration Membrane
Disadvantages	(1) It is highly sensitive to organic and inorganic constituents in the feed water (2) Membranes cannot withstand feed temperatures in excess of 45 °C	(1) It is highly sensitive to organic and inorganic constituents in the feed water (2) Membranes cannot withstand feed temperatures in excess of 45 °C. (3) It requires several backwashing cycles	(1) Membrane requires periodic cleaning (2) Waste generated during backwash and cleaning processes require disposal/recycling or further treatment	(1) Irreversible membrane fouling can occur with significant amount of iron concentration in feed water (2) Membrane requires periodic cleaning (3) Waste generated during backwash and cleaning processes require disposal/ recycling or further treatment
Advantages	(1) It has high pH tolerance (2) System can be operated automatically leading to less demand for skilled workers (3) Energy costs can be reduced by implementing energy recovery subsystems (4) It performs excellently for produced water treatment with appropriate pre-treatment (5) It does not require concentrate treatment as the brine generated is usually disposed into sea (6) Product water recovery in SWRO is between 30 and 60%, and between 60% and 85 in BWRO	(1) It has high pH tolerance (2) System can be operated automatically leading to less demand for skilled workers (3) Energy costs can be reduced by implementing energy recovery subsystems (4) It does not require solid waste disposal (5) Water recovery between 75 and 90%	(1) Product water is free of suspended solids (2) Product water recovery range from 85 to 100%	(1) Product water is totally free of suspended solids (2) It can be operated in cross-flow or dead-end filtration mode (3) Product water recovery range from 90 to 100% (4) Ceramic membranes have a longer lifespan than polymeric membranes

Life cycle	3–7 years	3–7 years	7 years or more	>10 years
Pre/Post-Treatment	Extensive pre-treatment is required to prevent fouling of membrane. Product water may require remineralization or pH stabilization to restore SAR values	Extensive pre-treatment is required to prevent fouling of membrane. Product water may require remineralization to restore SAR values	Cartridge filtration and coagulation are usually used as a pre-treatment. Post-treatment may be required for polishing depending on the product water	Cartridge filtration and coagulation are usually used as a pre-treatment. Post-treatment may be required for polishing depending on the product water
Overall cost	Capital costs of BWRO vary from $35 to $170/bpd and operating costs are $0.03/bbl. Capital costs of SWRO vary from $125 to $295/bpd and operating costs are $0.08/bbl	Capital cost range from $35 to $170/bpd. Operating cost is $0.03/bbl.	Capital costs depend on feed water quality and size of the polymeric membrane system. Approximate capital cost is $0.02–$0.05/bpd. Approximate Operation and Maintenance costs $0.02–$0.05/bpd	
Chemical use	Caustic and scale inhibitors are required to prevent fouling. NaOH, H_2O_2, Na_2SO_4, H_3PO_4, HCl, or Na4EDTA are required for cleaning the system	Caustic and scale inhibitors are required to prevent fouling. NaOH, H_2O_2, Na_2SO_4, HCl, or Na4EDTA are required for cleaning the system	Ferric chloride, polyaluminium chloride and aluminium sulphate are common coagulants used for pre-coagulation. Acids, bases and surfactants are used in the cleaning process	Ferric chloride, polyaluminium chloride and aluminium sulphate are common coagulants used for pre-coagulation. Acids, bases and surfactants are used in the cleaning process

(Continued)

Table 5.2 (Continued)

Method	Reverse Osmosis (RO)	Nanofiltration (NF)	Polymeric Microfiltration/ Ultrafiltration (MF/UF) Membrane	Ceramic Microfiltration/ Ultrafiltration Membrane
Energy consumption	RO use electrical energy for its operation. SWRO requires 0.46–0.67 KWh/bbl if energy recovery device is integrated. BWRO require less energy than equivalent SWRO system. BWRO requires 0.02–0.13 kWh/bbl of energy to power the system's pumps	It uses electrical energy and its energy requirement is less than that required in RO systems. NF system requires 0.08 Kwh/bbl to power its high-pressure pump	Not available	Not available
Feasibility	This is a robust technology for seawater desalination and has been employed in produced water treatment. For this technology to be effective in produced water treatment, extensive pre-treatment of feed water is necessary. Several pilot studies failed due to poor pre-treatment and insufficient system integration	This technology is used for water softening and removal of metals from wastewater. It is specifically efficient for feed water containing TDS ranging from 500 to 25 000 mg/L. NF is a poor technology for produced water treatment and is inappropriate as a standalone technology	Applicable to water with high TDS and salt concentrations and also has the potential to treat produced water however it is extensively used in the municipal water treatment	Ceramic membranes have been used to treat oilfield produced water and extensively used in other industrial water treatments. They are applicable to all types of produced water irrespective of their TDS and salt concentrations, but produced water with high concentrations may be problematic

5.3.14.3 Vapour Compression Distillation

The VCD process is an established desalination technology for treating seawater and RO concentrate. Vapor generated in the evaporation chamber is compressed thermally or mechanically, which raises the temperature and pressure of the vapor. The heat of condensation is returned to the evaporator and utilized as a heat source. VCD is a reliable and efficient desalination process and can operate at temperatures below 70 °C, which reduces scale formation problems. Table 5.3 summarizes various produced water thermal treatment technologies.

5.4 Re-Thinking Technologies for Safer Facing

With water treatment predicted to increase, the advance of innovative and ground-breaking technologies will grow to meet the industry's need. Below are a few key companies that are working to revolutionize facing through innovative water treatment processes:

- Some companies have developed a high-energy electrocoagulation technology that addresses heavy metals, biological matter, and hydrocarbons, but is limited to areas where salt levels are moderate.
- A group of companies lead in oxidation technologies. They use catalyzed UV to achieve many of the same results, but it also removes metals. Though it's still an early-stage start-up, it ranks as a potential big player.
- Some researchers use propane to fracture gas wells. Shell, Husky, and others are testing the technology.
- Some environmental active companies have deployed a proven mobile wastewater treatment system in the field with great success.

As companies set out to revolutionize the industry with new water treatment solutions, it looks like the most cost-effective treatment systems could be based on a mobile platform.

Mobile wastewater treatment systems allow drilling companies to operate off the grid, which is a valuable time- and money-saving strategy. Being mobile just makes a lot of sense in an industry where job sites are constantly moving.

Another solution is on the brink of revolutionizing the industry. Some companies are working together to commercialize a geographic information system (GIS) that will help predict – and prevent – ecological harm from drilling operations.

The system will enable the formulation of land-use benchmarks to assist in the optimal placement of wells, roads, gathering lines, and other necessary infrastructure.

The point is, with unfavorable public opinion, companies are searching for ways to improve the drilling process by building faster, better, and cheaper technologies.

There are numerous types of other treatment methods that seem to be marketed as viable treatment methods for CBM water prior to discharge that include the following.

1. CBM–WTS is marketed as a CBM water treatment technology that is designed to treat specific feed specifications of numerous incoming chemical compositions and chemical removal of sodium ions, with throughput ranging from 50 gallons per minute to over 1000 gallons per minute, thus producing a secondary waste of salt concentrate which must be stored and disposed of.

Table 5.3 Comparison of produced water thermal treatment technologies (Igunnu and Chen, 2012).

Method	Multistage Flash (MSF)	Multieffect Distillation, (MED)	Vapor Compression Distillation Technology	Mult Effect Distillation–Vapor Compression Hybrid	Freeze–Thaw Evaporation
Disadvantages	(1) Low product water recovery usually between 10 and 20% (2) It is not flexible for varying water flow-rates (3) Scaling and corrosion can be a problem	(1) Typically low product water recovery usually between 20 and 35% (2) It is not flexible for varying water flow-rates (3) Scaling and corrosion can be a problem (4) High level of skilled labor required	(1) Typically low product water recovery is usually around 40 (2) It is not flexible for varying water flow-rates (3) Scaling and corrosion can be a problem (4) High level of skill required to operate system	(1) Not applicable to produced water wells point source (2) Being a hybrid design, it requires very highly skilled labor	(1) Cannot treat produced water with high methanol concentration (2) Moderate product water quality containing around 1000 mg/LTDS (3) Can only work in winter timeand in places with below freezing temperatures (4) A significant amount of land is required (5) It generates secondary waste streams

Advantages			
(1) It requires less rigorous pre-treatment and feed condition compared with membrane technologies
(2) It has a significantly longer lifespan
(3) MSF system can withstand harsh conditions
(4) It can easily be adapted to highly varying water quality
(5) Cost of labour is cheaper than using membrane technology
(6) Good for high TDS produced water treatment
(7) Product water quality is high with TDS levels between 2 mg/l and 10 mg/l.	(1) It requires less rigorous pre-treatment and feed condition compared with membrane technologies
(2) It has a long lifespan.
(3) Energy requirement is cheaper than using MSF.
(4) It can easily be adapted to highly varying water quality
(5) Cost of labor is cheaper than using MSF or membrane technology
(6) Good for high TDS produced water treatment
(7) Product water quality is high
(8) It does not require special concentrate treatment
(9) Product water recovery of up to 67% can be achieved using stacked vertical tube design	(1) Applicable to all types of water and water with high TDS 40 000 mg/L.
(2) It is a smaller unit compared with MSF and MED
(3) It has high ability to withstand harsh conditions
(4) It does not require special concentrate treatment
(5) Pre-treatment is less rigorous compared with membrane treatment	(1) It has high product water quality
(2) Excellent treatment technology for produced water with high TDS and zero liquid discharge
(3) System can withstand harsh condition	(1) Excellent for zero liquid discharge
(2) It requires low skilled labour, monitoring and control
(3) It is highly reliable and can be easily adapted to varying water quality and quantity

(Continued)

Table 5.3 (Continued)

Method	Multistage Flash (MSF)	Multieffect Distillation, (MED)	Vapor Compression Distillation Technology	Mult Effect Distillation–Vapor Compression Hybrid	Freeze–Thaw Evaporation
Life cycle	Typically 20 years but most plants operate for more than 30 years	Typically 20 years	Typically 20 years but may operate for more years	Typically 20 years but may be longer if made of materials with high corrosion resistance	Expected lifespan is 20 years
Pre/Post Treatment	Pre-treatment is done to remove large suspended solids. This requires screens and rough filtration. Product water stabilization is required because of its low TDS	Pre-treatment is done to remove large suspended solids similar to MSF. This requires screens and rough filtration. Product water stabilization is required because of its low TDS	Pre-treatment and post-treatments are required in order to avoid fouling and because of low TDS level in product water, respectively	It requires a less rigorous pre-treatment compared with membrane technologies. Lime bed contact post-treatment is required because of low TDS of product water	It requires minimal pre- and post-treatment depending on product water quality and discharge standards
Overall cost	Capital costs vary between $250 and $360 per bpd. Operating costs are $0.12/bbl and total unit costs are $0.19/bbl	Overall cost is lesser than in MSF. Capital costs ranges from $ 250 to $330 per bpd. Operating costs are 0.11/bbl and total unit costs are $ 0.16/bbl	Capital costs of vapour compression for sea water desalination ranges from $140 to 250 per bpd depending on various factors. Operating costs are 0.075/bbl and total unit costs are $0.08/bbl for seawater desalination	Capital cost is $250 per bbl per day. Operation costs depend on the amount of energy consumed	It depends on location

150

Chemical use	EDTA, acids, and other antiscaling chemicals are used to prevent scaling. pH control is also necessary to prevent corrosion	Scale inhibitors are required to prevent scaling. Acid, EDTA, and other antiscaling chemicals are required for cleaning and process control	Scale inhibitors and acids are required to prevent scaling. EDTA and other antiscaling chemicals are required for cleaning and process control. Corrosion is prevented by pH control	Scale inhibitors are required to prevent scaling. Acids, EDTA, and other antiscaling chemicals are required for cleaning and process control. Corrosion is prevented by pH control	None
Energy consumption	Electrical energy required ranges from 0.45 kWh/bbl to 0.9 kWh/bbl. Thermal energy required is estimated at 3.35 kWh/bbl. Overall energy required for MSF ranges from 3.35 to 4.73 kWh/bbl.	MED requires both thermal and electrical energy types. Electrical energy consumed is approximately 0.48 kWh/h/bbl and power consumption is 1.3–1.9 kWh/bbl	VCD requires both thermal and electrical energy. For desalination, power energy consumption is around 1.3 kWh/bbl [53]. Electricity consumption is 1.1 kWh/bbl for mechanical vapor compression (MVC) and to achieve zero-liquid discharge energy demand is 4.2–10.5 kWh/bbl	It uses both thermal and electrical energy. Power consumption for desalination is around 0.32 kWh/bbl To achieve zero-liquid discharge energy consumption is around 4.2–10.5 kWh/bbl	It uses electrical energy, but data are not available

(Continued)

151

Table 5.3 (Continued)

Method	Multistage Flash (MSF)	Multieffect Distillation, (MED)	Vapor Compression Distillation Technology	Mult Effect Distillation–Vapor Compression Hybrid	Freeze–Thaw Evaporation
Feasibility	This is a mature and robust desalination technology that can be employed for produced water treatment. MSF is applicable to all types of water with high TDS range up to 40 000 mg/L	This is a mature and robust desalination technology that can be employed for produced water treatment. MED is applicable to all types of water and a wide range of TDS	This is a mature and robust seawater desalination technology. It is applicable to all types of wastewater with TDS level greater than 40 000 mg/L. Various enhanced VCD have been applied in produced water treatment	A mature desalination technology that has been employed in produced water treatment. It is usually employed for treating water with high TDS. In future product, water quality may be increased. For example, product water recovery of 75% was achieved by GE using a brine concentrator and analyzer	This is a mature and robust technology for produced water treatment. It does not require infrastructure. This process requires favorable soil conditions, a significant amount of land and a substantial number of days with temperatures below freezing

2. Submerged combustion is marketed as a water treatment technology for the CBM industry that can be utilized to heat liquids in order to treat wastewater. This system is touted for the mining industry and pulp and paper industry for wastewater treatment, ethanol and glycol reduction, and many more industrial applications.
3. Catalyx Fluid Solution (CFS) is marketed as a water treatment technology for the CBM industry.
4. Utilizing a continuous counter-current ion exchange (CCIX) method for removing sodium and other cations from CBM produced water. This CCIX system is based on the Higgins Loop technology.

    Produced water containing high Na levels enters into the adsorption zone within the Higgins Loop where it makes contact with a strong acid cation resin which loads Na^+ ions in exchange for hydrogen ions. The "loop" extracts 95% of the cations. Concurrent with adsorption and in the lower section of the Higgins Loop, Na-loaded resin is regenerated with hydrochloric acid and a small, concentrated spent brine stream is produced.

    Regenerated resin is rinsed with water prior to re-entering the adsorption zone in order to remove acid from its pores. As resin in the adsorption zone becomes loaded with Na, the flows to the Higgins Loop are momentarily interrupted to allow "pulsing" the resin bed through the loop in the opposite direction of the liquid flow. Liquid flows restart after the resin pulsing is complete.
5. There are operator owned chemical treatment plants. While it is possible to alter the chemistry of sodic water by adding calcium and magnesium, this does not eliminate or reduce sodium, but changes the ratio of sodium to other salts, thus decreasing the sodium adsorption ratio (SAR). The net result is more saline water with the sodium salt still dissolved in the water. This approach is not likely to work with CBM product water because the added calcium will combine with carbonate from the CBM water and precipitate out as calcium carbonate (lime). To make this process work, CBM product water must be de-gassed of carbonate by addition of acid, or additional calcium must be made available in the soil by acidification from sulfur additions.

5.5 Water Treatment for Oil Sands Mining

To produce oil from oil sands fields, a huge amount of water is used. Specifically, water is used to separate bitumen from the sand at a high water to oil ratio – it is estimated that around 2 to 4.5 barrels of fresh water are used to produce a barrel of synthetic crude oil.

The water is used for in situ recovery of bitumen, i.e., water is converted to steam to heat up the bitumen underground and then it is pumped to the surface through wells. Wastewater is usually alkaline, slightly brackish, and toxic to aquatic biota due to high concentrations of organic acids. Strong regulation, i.e., a zero discharge policy, has lead to mostly water recycling as practiced by the oil sands producers. The other alternative is to tap underground aquifers hosting non-potable, saline water for bitumen recovery.

5.5.1 Recycling and Water Treatment Options

Despite the aggressive recycling of wastewater by oil sands producers for reuse in the recovery process, repeated extraction cycles have resulted in a deterioration of water quality that can upset the bitumen extraction process through scaling, fouling, increased corrosivity, and interference with extraction chemistry.

Furthermore, with a zero discharge policy, oil sands producers have to treat and properly dispose millions of cubic meters of toxic process water and tailings, which are currently held in large tailings ponds. As reclamation depends on natural detoxification of process waters in tailings ponds, its viability is uncertain.

To meet the concerns over wastewater management in general, and ensure a supply of recycled water for bitumen processing in particular, oil producers could avail themselves of various water treatment technologies. These include:

- adsorption;
- membranes – microfiltration and ultrafiltration;
- nanofiltration and reverse osmosis;
- biological treatment;
- advanced oxidation; and
- constructed wetlands.

Adsorbents are used to remove a wide variety of pollutants associated with oilfield produced waters, especially organic carbon compounds, oil and grease, and heavy metals. Adsorbents can be activated carbon, natural organic matter, zeolites, clays, and synthetic polymers.

Micro and ultrafiltration are pressure-driven membrane processes that reject particles as small as 0.1 micrometer and 0.01 micrometer respectively. Lab and pilot scale studies using membranes to treat produced waters have shown over 90% oil rejection with permeate concentrations of less than 20 ppm.

But on a wider scale, problems such as fouling and membrane durability can occur. Nanofiltration has the potential for partial demineralization, softening, and removal of soluble organic compounds from produced water as it can reject divalent ions, dissolved organic matters, pesticides, and other macromolecules.

Reverse osmosis, on the other hand, requires the feedwater to be forced against a concentration gradient though a semi-permeable membrane. Studies have showed that using reverse osmosis, rejection of hardness typically exceeds 98% (residual concentration of <2 ppm $CaCO_3$) and rejection of monovalent ions exceeds 90%.

Other projects have also used reverse osmosis to convert produced water from an oilfield to fresh water for agricultural and potable use.

Biological treatment (the use of microorganism to remove contaminants), in one pilot study, has been shown to remove 98–99% of total petroleum hydrocarbons from produced water, but the microorganisms have to get used to the saline conditions before a certain treatment efficiency (>95%) can be achieved.

Advanced oxidation could be either photocatalytic oxidation or sonochemical oxidation. The lab-scale experiments have shown that photocatalytic oxidation can decompose organic and inorganic pollutants in oilfield produced water. The degradation rates are dependent on the efficient adsorption of pollutants into the catalyst, which could then be affected by the feedwater pH.

Sonochemical oxidation involves the formation and collapse of bubbles when ultrasound is applied to a liquid i.e., produced water. This collapse of the microbubbles produces cavities (cavitation) of high temperature and pressure that can break or destroy particles or molecules. Used with an oxidant such as hydrogen peroxide, the cavitation process in some studies has been demonstrated to degrade phenols, organic acids, and polyaromatic hydrocarbons.

Constructed wetlands, since their earlier use to treat municipal and stormwater, have become popular in the oil sector. The processes of pollutant removal in wetlands include sedimentation, adsorption, denitrification, photo-oxidation, plant uptake, and volatilization. Wetlands have proven to be effective in removing contaminants in produced water but treatment performance is variable, as shown in various pilot and full-scale studies, i.e., 55–85% of BOD_5, 53–86% for COD, 54–94% for oil and grease, and 10–94% for phenols.

Table 5.4 summarizes the different treatment processes and their potential applications to treat produced water from the oil sands mining.

This section reveals that most water treatment technologies have shown the potential to treat – i.e., de-oil, soften, detoxify, and demineralize – oilfield-produced water on a small-scale basis, but their effectiveness has yet to be proven at a bench or pilot levels.

It is suggested that further preliminary studies are needed on pre-treatment requirements, performance on target pollutants, energy consumption and costs, among others, to be able to compare these to existing/conventional technologies.

There is an urgent need to further understand how these technologies will perform given the unique physical and chemical nature of produced water from oil sands mining on a larger scale.

5.5.2 Oily Water Treatment in Oil Sands Mining

Oil sands extraction can affect water resources by its requirement for large quantities of water during the separation of the oil and sand. The majority of oily wastewater is produced in oil sands mining.

Heavy metals such as vanadium, nickel, lead, cobalt, mercury, chromium, cadmium, arsenic, selenium, copper, manganese, iron, and zinc are naturally present in oil sands and may be concentrated by the extraction process. The technologies reported in Section 5.3 of this book could be considered for oil sands mining wastewater treatment after careful evaluation.

There are currently two methods for recovering bitumen. The first is surface mining of oil sands ores with subsequent separation of bitumen from the other constituents by an alkaline hot water extraction process.

The second method is an in situ thermal process for extracting bitumen from oil sands that are too deep to mine. In surface mining operations, alkaline hot water extraction produces large volumes of surface-mining water that must be stored in on-site tailings ponds to prevent environmental pollution.

The in situ oil sands production method called steam-assisted gravity drainage (SAGD) reuses process wastewater following treatment. Figure 5.6 shows a water treatment system in the SAGD process operated by Japan Canada Oil Sands Ltd. (JACOS).

Asterisks in Figure 5.6 indicate water sample collection points in the SAGD facility. Water sampling points are: #1, groundwater; #2, produced water; #3, de-oiled water; #4, softened water; #5, boiler blow-down; #6, recycled water. Below is the SAGD process water for each sample point.

The treatment and reuse processes that concentrate contaminants in the process water and determine the concentration and dynamics of inorganic and organic contaminants, make-up

Table 5.4 Water treatment processes – problems and potential.

Processes	Problems Associated with Treatment of Produced Water	Significant Technological Advances	Target Chemicals in Oil Sands Process Water
Adsorption	Incomplete pollutant removal; fouling from oil; cleaning and regeneration costs; low adsorption capacity	Organic modifications to clay adsorbents; natural and synthetic polymers with improved adsorption and regeneration properties	Napthenic acids, bitumen, aromatic hydrocarbons, trace metals
Microfiltration and ultrafiltration	Fouling from oil and solids; membrane durability; disposal of retentate	Surface chemistry modifications to reduce fouling and permeate flux decline; chemical additives, aeration and ultrasound to reduce fouling	Bitumen, suspended solids
Nanofiltration and reverse osmosis	Fouling from oil, dissolved organics, and algal growth; membrane replacement costs; brine disposal	Membrane modifications to reduce fouling from organics; ultra low pressure membranes; lower energy consumption	Napthenic acids, hardness, TDS, aromatic hydrocarbons
Biological treatment	Feed water toxicity; incomplete pollutant removal; sludge disposal	Membrane bioreactors facilitate oxidation of recalcitrant compounds and protect biofilm from nfluent toxicity	Napthenic acids, ammonium, bitumen, aromatic hydrocarbons
Advanced oxidation	Incomplete pollutant removal; high energy costs; radical scavengers; oxidation byproducts	Solar photocatalytic systems; photo electrocatalytic process to reduce the effect of radical scavengers	Napthenic acids, ammonium, aromatic hydrocarbons
Constructed wetlands	Flow capacity;Feed water toxicity; removal of salinity; bioaccumulation of toxicants by wetland biota; cold water operation	Subsurface designs for operation in cold climates; implementation of large-scale wetlands to treat oil contaminated water; improved understanding of degradation pathways	Napthenic acids, ammonium, bitumen, aromatic hydrocarbons

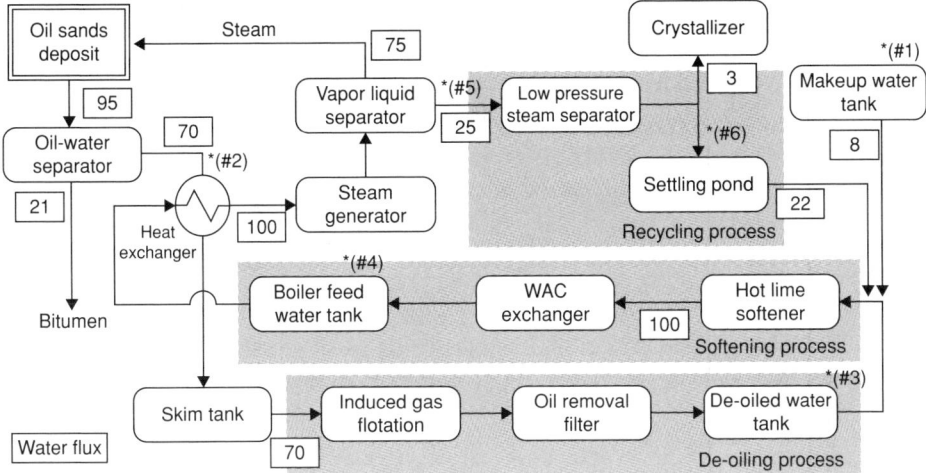

Figure 5.6 Water treatment system in the SAGD process operated by Japan Canada Oil Sands Ltd. (JACOS) (Reproduced with permission from [20] © Elsevier, 2012).

water, and process water from six process steps are shown in Figure 5.7 for the facility shown in Figure 5.6 employing the SAGD process in Alberta, Canada.

The results of the analysis study for samples in Figure 5.6 revealed the chemical abundance and dynamics of organic contaminants in the recycling of process water by the SAGD method.

The contaminants were mainly derived from the oil sands deposits, although traces of saturated fatty acids were contained in the groundwater used as make-up water, which became concentrated in the water recycling system.

Apparently, polar organics such as organic acids, phenols, and ketones were selectively concentrated as the water treatment progressed, whereas non-polar or low-boiling components such as PAHs and BTEX were removed. These results suggest that the removal

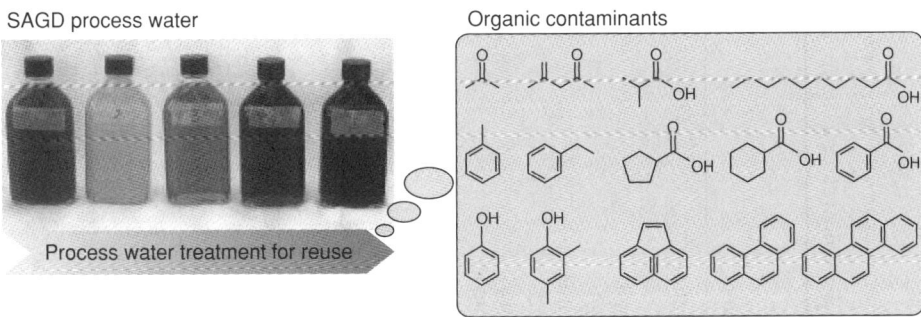

Figure 5.7 Samples of water takes from sample points in Figure 5.6 (Reproduced with permission from [20] © Elsevier, 2012).

Table 5.5 Oily water qualitative characteristic ranges and the treatment capabilities of each piece of equipment.

Parameter	Untreated Range	API Separator		
		Inlet Range	Reduction %	Outlet Range
Temperature, °C	30–60	30–60	–	30–60
pH	7–8	7–8	–	7–8
TDS, mg/L	150–5000	150–5000	–	150–5000
TSS, mg/L	300–800	300–800	67–75	100–200
Oil & Grease, mg/L	3000–5000	3000–5000	90	200–500
BOD, mg/L	300–500	900–1400	50	450–700
COD, mg/L	300–1200	1700–3400	50	850–1700
Cl⁻	50–2000	50–2000	0	50–2000
NH$_3$	20–50	50–100	0	50–100
P	–	–	0	–
Cyanides	1–3	1–3	0	1–3
Phenols	5–20	5–20	0	5–20
H$_2$S	5–10	5–10	0	5–10

Parameter	IGF/DGF			Conventional Activated-Sludge		
	Inlet Range	Reduction %	Outlet Range	Inlet Range	Reduction %	Outlet Range
Temperature, °C	30–60	–	30–60	30–40	–	30–40
pH	7–8	–	7–8	7–8	–	7–8
TDS, mg/L	150–5000	–	150–5000	150–5000	–	150–5000
TSS, mg/L	100–200	75–80	20–50	20–50	75–80	5–10
Oil & Grease, mg/L	200–500	90	10–30	10–30	95–90	2–5

BOD, mg/L	450–700	30	300–500	300–500	90	20–30
COD, mg/L	850–1700	30	600–1200	600–1200	73–80	80–100
Cl⁻	50–2000	0	50–2000	50–2000	0	50–2000
NH₃	50–100	0	50–100	50–100	85–94	<3
P	–	0	–	–	0	<0.5
Cyanides	1–3	0	1–3	1–3	95–98	<0.05
Phenols	5–20	0	5–20	5–20	80–95	<0.5
H₂S	5–10	0	5–10	5–10	>99	<0.05

Activated Sludge with UF + RO

Parameter	Inlet Range	Reduction %	Outlet Range
Temperature, °C	30–40	–	30–40
pH	7–8	–	7–8
TDS, mg/L	150–5000	95	15–300
TSS, mg/L	5–10	>95	<1
Oil & Grease, mg/L	2–5	>97	<1
BOD, mg/L	20–30	90	<15
COD, mg/L	80–100	73–84	<80
Cl⁻	50–2000	90	5–200
NH₃	50–100	85–94	<3
P	<0.5	80	<0.1
Cyanides	1–3	95–98	<0.05
Phenols	5–20	80–95	<0.5
H₂S	5–10	>99	<0.05

of such polar components is important for improving process water recycling efficiency and further work may elucidate other factors that obstruct process water recycling.

To identify and handle the potential risks of these contaminants derived from oil sands deposits to the environment, analytical methods with a higher resolution should be further improved for widespread use and adopted as environmental guidelines for protecting aquatic biota and human society, as well as for sustaining the emerging oil sands development with alternative approaches to minimize fresh water use.

In general, treatment of oily wastewater is a standard process common across the industry. Oily water is first sent to an API gravity separator, followed by Dissolved Air (or Gas) or Induced Air (or Gas) Flotation (DAF/AIF or DGF/IGF) which generates tiny air (or gas) bubbles. The bubbles adhere to small oil particulates and float to the top of the tank where they are skimmed off. Following flotation, oily wastewater can be combined with other waste streams and treated biologically. Table 5.5 lists oily water qualitative characteristic ranges, and the treatment capabilities of each piece of equipment.

6

Wastewater Sewer Systems

Effluents in the form of wastewater are a combination of the liquid and water-borne wastes from buildings and industrial plants, plus groundwater, surface water, or stormwater. Wastewater may be grouped into the following classes:

- **Class 1:**
  Effluents that are non-toxic and not directly polluting but liable to disturb the physical nature of the receiving water. They can be improved by physical means.
- **Class 2:**
  Effluents that are non-toxic but polluting because they have an organic content with high oxygen demand. They can be treated for removal of objectionable characteristics by biological methods.
- **Class 3:**
  Effluents that contain poisonous materials and therefore are often toxic. They can be treated by chemical methods.
- **Class 4:**
  Effluents that are polluting because of organic content with high oxygen demand and, in addition, are toxic. Their treatment requires a combination of chemical, physical, and biological processes.

In general, the aim of any drainage/effluent disposal system should be to segregate uncontaminated water from contaminated water or effluents and to segregate different types of effluents in order to reduce the size, complexity, and costs of any treatment units that may be required for handling the contaminated water and effluents before they are discharged from the "Company's" property.

All wastewater effluents from industries that are discharged to public and/or natural water sources or directed to recycling purposes inside the industry, and may contain a wide variety of matters in solution or suspension, should be controlled according to the requirements imposed by the final destination. However, in any case elimination of the waste or the hazard potential of the waste should be ultimate goal in the management of hazardous wastes.

6.1 Stormwater Sewer System

This system should consist of pipes and open ditches collecting clean and/or oily stormwaters, fire and washing waters from the non-polluted areas. The stormwater should be disposed of to the oily stormwater basin located in the wastewater treatment area through the stormwater network. It should mainly collect the non-polluted area clean waters from:

- Diked and undiked tank areas.
- Unpaved areas.
- Process and utilities non-polluted paved areas (excluding concrete paved areas).
- Roads, yards, and roofs.

The collected stormwater, after oil removal in the oily stormwater basin, should be stored in the clean stormwater basin. The final disposal of such clean waters will be to a:

- Wastewater treating unit at the API separator(s) effluent for further oil removal.
- Evaporation pond(s).
- Ocean/river if it complies with the local conditions of effluent waste streams.

6.2 Oily Water Sewer System

This sewer should collect:

- Process spillages and drainages.
- Drains of all hydrocarbon equipment.
- Pumps and compressor cooling water.
- Oily condensate.
- Cooling water drains that have a chance of becoming polluted with oil.
- Waters coming mainly from all hydrocarbon pollutable paved areas including the following:
  - Process units.
  - Utilities.
  - Non-volatile products truck loading stations.
  - Workshops.
  - Transport and mobile plant garages.
  - Pump stations.
- Pipe trench drains.
- Sample point drains.
- Drainage from level gauges, cocks, and similar equipment.
- Drains of the following fluids are excluded because of flowing in pit for truck disposal:
  - Heavy viscous fluids such as asphalt.
  - Any fluid containing hazardous materials with concentrations more than the allowable figures set out by the environmental and/or biological treatment restrictions.

The system shall consist of drains, funnels, underground piping, clean-outs, catch basins, manholes, sealed manholes, and vent pipes. The final main of the oily water sewer shall flow into API separator(s) in the wastewater treating area through a dedicated underground gravity flow network.

Open ditches should be avoided. Leakages of manifolds can be collected in a suitable collecting basin, located underneath the manifold.

The basin should drain into a sump located outside the manifold/piping area. The sump can be emptied intermittently (e.g., by vacuum truck) or it can be connected to the continuously oil contaminated drain system.

6.3 Non-Oily Water Sewer System

This system should collect special oil free waters containing high total dissolved solids such as:

- Boiler blow-down.
- Desalination unit blow-down.
- Brine drainage.
- Neutralized effluents from all neutralization sumps through the plant.
- Storm, fire, and washing run-off waters from the sulfur solidification and crushing area after removal of sulfur particles by a sedimentation sump.
- Tempered water system drains.
- Cooling water (circulated) blow-down and drains, provided that there is no possibility of oil contamination.

The system shall consist of drains, funnels, underground piping, clean-outs, catch basins, manholes, sealed manholes, and vent pipes. The non-oily water sewer system can be directed to the following disposals where required:

1. Wastewater treatment plant effluent, if it is intended to reuse the treated wastewaters as cooling tower make-up. In this case, the recycled treated water should meet the cooling tower make-up minimum requirements. If needed, total hardness removal facilities should be provided on the non-oily water system before disposal (see also item 2 below).
2. Wastewater treatment plant influents (at API separator(s)' outlet), if non-oily waters need further physical and/or biological treatment in order to meet the final disposal requirements. In this case the following conditions should be met.
   - Non-oily waters should be treated for total hardness removal before any disposal to the API separator(s)' effluent (if required).
   - All materials that will suffer from the biological treatment activities should be removed from the non-oily waters.
3. Evaporation pond(s).
4. Public waters (if non-oily waters comply with the environmental regulations).

However, the non-oily water sewer system should be investigated for an appropriate disposal while considering the following aspects:

- Environmental regulations.
- Availability of refinery/plant raw water.
- Economical aspects.
- Operability of equipment furnished.

6.4 Chemical Sewer System(s)

In this book all sewer systems containing acids, alkalides, chemicals, and all other special organic materials such as Furfural, MEK, etc. are designated as "chemical sewers." The number and route of chemical sewer systems in a plant should be studied based on the geographical location of various units and the feasibility of the gathering and disposal systems.

Chemical sewer streams should include but not be limited to:

- Polluted drains from chemical additives dosing pumps (excluding tetraethyl lead).
- Laboratory building drains (excluding oily drains).
- Drainage and stormwater polluted by acid and/or other chemicals.
- Caustic drains (caustic dissolving unit drains are excluded and should have a closed circuit network inside the unit).
- All waters contaminated with acids and/or chemicals.

6.4.1 Disposal of Chemical Sewers

In general, disposal of any chemical sewer before neutralization/treatment to the environment should be avoided. Segregated chemical sewer network(s) should be provided depending on the variety of chemicals to be disposed of. Neutralization pond effluent should be connected to the non-oily sewer.

The volume of the neutralization pond should be at a minimum equal to the highest batch volume among the streams being disposed of. Chemical effluents from the laboratory building should flow into a dedicated neutralization pit near the laboratory itself.

6.4.2 Neutralization Systems

Adequate facilities for acid and caustic injection systems, agitators, pumps, eductors (if required), steam coils (if required for winterization), etc., should be provided for each neutralization pond.

The necessary instrumentation, such as acid and caustic flow indicator, pH indicator, pH low and high alarm in the control room, temperature controller, etc., should be foreseen for each pond.

The type of operation (manual or automatic) for the neutralization ponds will be instructed by the Company. All acid and caustic handling facilities, such as pipes, tanks, pumps, etc., should be traced for proper temperature maintenance.

6.4.3 Type of Chemical Wastes

Chemical wastes should include all wastes contaminated with acids, alkalides, chemicals, and additives and waters containing hazardous liquids. Disposal of any stream containing hazardous materials to the oily sewer and/or non-oily sewer systems should be avoided before implementation of the necessary treatment processes for removal of the hazards.

The method and extent of treatment will be instructed based on the authorities' approval. The final treated water specifications should comply with the environmental pollution requirements set out by the authorities concerned.

6.5 Sanitary Sewer System

This sewer should collect non-polluted raw sanitary from sanitary facilities of all buildings as required.

The final output should flow into sanitary sewage treatment units. The sanitary sewage treatment plant effluent, if it complies with the required effluent characteristics, can be routed to the wastewater treating plant (at the biological treatment outlet) for recycling purposes.

6.6 Special Sewer Systems

Special sewer systems should be provided where required. In general all fluids containing poisonous/hazardous materials and/or fluids subject to recovery should be segregated and handled away from all the other sewer systems mentioned in this book. The systems should include but not be limited to the following streams:

- Caustic drains inside the caustic dissolving unit.
- Amine drains.
- Solvent drainage.
- Motor gasoline drainage contaminated with Toluene or MTBE.
- Hydrocarbon drains containing benzene in concentrations that are more than those allowed by the environmental regulations.
- All drains contaminated with toxic components such as cyanide, phenol, lead, etc.
- Aluminium chloride drainage.
- Hydrofluoric acid drainage.
- Spent catalysts.
- Others.

6.7 Effluent Sources and Disposals

The majority of effluent source streams can be broadly characterized as one or more of the following types:

- high or low dissolved solids content;
- oily or non-oily;
- high or low in phenols and/or sulfides;
- chemical or non-chemical;
- high or low in suspended solids.

Using these broad characteristics, the effluent source streams should be investigated for suitable disposal.

1. **Unpaved areas**
   The effluent from unpaved, non-process, and non-tank areas will be clean stormwater. The word "clean" is defined as meaning non-oily.
2. **Undiked tank areas**
   The effluents from undiked tank areas will be oily stormwater.
3. **Diked tank areas**
   The normal effluent from diked hydrocarbon tank areas will be oily stormwater. If a tank should rupture, the residual oil after clean-up will probably be washed down and routed to oily stormwater sewer. Diked area drains should be valved at the outlet of dike, so that any accumulation of oily water or oil can be impounded and released under controlled conditions.
4. **Tank bottom draws**
   The water periodically drained from hydrocarbon tanks will be "oily foul water" and these drains should be valved.
   These waters may contain salt and other dissolved solids if the tanks are storing crude oil, sulfides, and/or phenols, or if the tanks are storing untreated intermediate products, and free or emulsified oils.
   These waters normally are discharged to the oily water sewer except in cases where the drains are rich in hazardous materials such as phenols or lead that should be segregated from the ordinary sewer systems. Drains of slops and any other tanks which are rich in H S should be routed to the sour water stripper unit for treatment.
5. **Concrete paved process and utility areas drains**
   Effluent from concrete paved process and utility areas which is contaminated from various sources of oily drips and drains should be connected to the oily water sewer system unless otherwise specified in this book.
6. **Process and utility paved areas**
   Drains from all clean areas adjacent to the streets around process and utilities areas, and/or other clean paved areas which are not subject to any oil spillage, should be routed to the stormwater sewer system.
7. **Pump and compressor cooling**
   Some amount of water cooling will usually be used for hot pump pedestals and glands and compressor jackets. Additionally, some water and/or oil may be used in pump and compressor seals. The drips and drains from these systems should constitute another source of the oily sewer system. Should the utility area include a pump or compressor, then the drain of such areas that will be subject to oil contamination should be connected to the oily water sewer system.
8. **Boiler blow-down and water treating rinses**
   These waters will be free of oil but high in dissolved solids. Hence, they should be discharged to the non-oily sewer system.
9. **Cooling water drains**
   If the tubes in the water-cooled heat exchangers in once-through cooling water and/or circulating cooling water develop leaks, then these waters are liable to contamination with process fluids. Due to the characteristics of the process fluids (light oils or heavy

oils) and also cooling water (once-through or circulated) the following four categories should be taken into consideration. Light oils refers to pentane and lighter.
- Once-through cooling water (light oils): This will be clean water and should be considered as non-oily water.
- Once-through cooling water (heavy oils): This will be oily water to acknowledge the possibility of exchanger tube leaks of non-volatile oil.
- Circulating cooling water blow down (light oils): This will be high solids clean water and should be considered as non-oily water.
- Circulating cooling water blow-down (heavy oils): This will be high solids oily water to acknowledge the possibility of exchanger tube leaks of non-volatile oil. These waters can be routed to the oily water sewer if the quality of the final waste treated water will not suffer from the receipt of such high solids content.

10. **Process drums drains**

    Water that has been withdrawn from process drums containing H_2S or H_2S and NH_3 will be designated as "sour water" and if the water is principally condensed steam, it will be called "sour condensate." The destination of such drains should be the sour water stripper unit through a pump if required. For non-oily, gravity, and low flow cases the destination can be the non-oily sewer system and/or the oily water sewer system under controlled conditions and not in such amounts that they will impact the biological treatment operation.

11. **Laboratory wastewater**
    - **Oily wastes**

      Wastes from laboratory oily sinks that are not contaminated with chemicals should be routed to the oily water sewer system.
    - **Chemical contaminated wastes**

      All chemical contaminated wastes should be directed to the non-oily sewer system after neutralization in the dedicated neutralization pit adjacent to the laboratory building.

12. **Flare seal drum blow-down**

    Continuous drain from the main flare and acid flare seal drums located at the bottom of the flare stack(s) that are rich in H_2S should be pumped to the sour water stripper unit for treatment. The make-up water to the flare seal drum/pot should be condensate.

13. **Floating roof drain of tanks**

    The roof drain(s) of the floating roof tanks should be discharged to the oily water sewer system due to the possibility of hydrocarbon leakage from rubber seals.

6.8 Particular Effluents in Refinery and Petrochemical Plants

Caustic used to scrub pentanes and lighter will essentially contain sodium sulfide. Such streams should be neutralized in the chemical neutralization pit, and be sent to the non-oily sewer after neutralization.

6.8.1 Caustic Scrubs (Heavy Oils)

Caustic used to scrub gasoline, kerosene, and distillate oils may contain naphthenic acids, phenols, and cresols in addition to spent caustic. These drains should be segregated from the non-oily and chemical sewer systems and should be neutralized before entering the oily water sewer.

6.8.2 Desalter Wastewater

Desalter wastewater effluent should be piped in a pressure line to the dedicated desalter oil–water separator in the wastewater treatment area. It should also have the possibility of being routed to the main API separator influent under controlled conditions. The effluent water of the desalter water oil–water separator can be directed into the main API separator effluent water pond if the pollutable materials do not exceed allowable figures.

6.8.3 Foul or Sour Waters

In petroleum refining, various processing operations produce wastewater solutions; principally, they are condensates containing sulfides – generally as hydrogen sulfide – ammonia, mercaptans, phenolics, and possibly small amounts of water-soluble organic acids, nitrogen bases, and cyanides. These wastewaters generally are referred to as "foul waters" or "sour waters." The principal sources of foul waters are condensates from accumulators, reflux drums and knockout pots in catalytic reformers, cracking, hydrocracking, coking, and crude distillation units. Foul waters generally are neither highly alkaline nor highly acidic.

Their content of pollutants is relatively low compared to that of spent caustics. However, the high oxygen demand and the odorous or toxic nature of foul waters make it desirable to treat them for the reduction of these objectionable characteristics before they undergo biological treatment or are discharged into the wastewater system.

Foul waters should be stripped in the sour water stripper(s), where removal of a single/multiple contaminants is desirable. Stripped water from the sour water stripper unit should be piped in an underground pressure line into the main oily sewer system terminating in the wastewater treating plant.

The stripped water should also have disposal access to the desalter water as make-up and evaporation pond(s)/public water in case of being congruent with the local regulations. Characteristics of the stripped water, which are normally instructed by the Company, should be properly controlled such that disposal of this water does not impede any operation/environmental pollution in the upset condition. The waste gases from the stripping operations should be routed to the sulfur recovery unit and/or incinerated.

6.8.4 Spent Caustic Solutions

6.8.4.1 Sources and Characteristics

Typically, caustic solutions are used to neutralize and extract:

- Acidic materials that may occur naturally in crude oil and in any of its fractions.
- Acidic reaction products that may be produced by various chemical treatment processes.
- Acidic materials formed during thermal and catalytic cracking, such as hydrogen sulfide, phenolics, and organic acids.

Spent caustic solutions may therefore contain sulfides, mercaptides, sulfates, sulfonates, phenolates, naphthenates, and other similar organic and inorganic compounds.

6.8.4.2 Disposal Methods

An adequate handling system should be provided for the effective disposal of the spent caustic. The system should include tankage and pipelines to segregate, accumulate, and transfer the spent caustic solutions to the disposal site. The following disposal methods should be taken into consideration in each plant.

- Direct disposal methods.
- Chemical methods.
- Chemical-physical methods.
- Biological methods.

The soundness of these methods must be determined by the individual refiner, paying due regard to applicable laws and regulations.

6.8.4.2.1 Direct Disposal Methods.

1. **Dilution**
   Controlled disposal of spent caustic solutions into large bodies of water, particularly brackish or salt water, or into rivers capable of adequate dilution may be considered if the maximum concentration of sulfates (as SO_4) does not exceed the value set by environmental regulations. Disposal into fresh water lakes and streams must be given more critical consideration, particularly if the waters are used as a source of potable water supplies or for recreation purposes.
2. **Disposal ponds**
   Disposal of spent caustic solutions by means of ponds is to be avoided unless otherwise specified for disposal of small quantities of caustic waste. The following factors are to be considered to prevent subsequent air or water pollution where disposal ponds are concerned:
   - **Location**
     Odor nuisances within the vicinity will affect site selection. Furthermore, geologic formations at the site must be such that contamination of potable water supplies by seepage will not occur.
   - **Capacity**
     To prevent pollution of surface water, the pond must have sufficient capacity to hold the maximum amounts of rainfall, as well as the maximum amounts of waste chemicals to be received and stored. Adequacy of size depends not only on the volume of spent waste to be handled, but also on the anticipated evaporation, seepage, and annual rainfall.
   - **Equipment**
     Pressurized pipelines should be provided to transfer the spent caustic solutions from the process units or the central collection system to the pond. Disposal of any spent caustic stream contaminated with oil to the pond should be avoided. If the spent solutions carry entrained oil to the pond, the necessary equipment for oil removal is required for the purpose of safety, as well as prevention of pollution and loss of oil.

6.8.4.2.2 Chemical Methods. The following methods for chemical treatment of spent caustic can be considered:

- Regeneration.
- Air oxidation.
- Neutralization.

1. **Regeneration**
   Caustic used to extract mercaptans from hydrocarbon streams can be regenerated by:
   - Steam stripping the mercaptans from the solution followed by incineration or recovery of the mercaptans.
   - Oxidation of the mercaptans to disulfides that can be separated as an oil phase. Oxidation of mercaptans should be accomplished by electrolysis or air blowing. The latter should be conducted under pressure, or with the use of oxidation catalysts, or both.
2. **Air oxidation**
   Spent caustics that contain sulfides or sulfites can be pre-treated by air oxidation to reduce their high oxygen demand before they are diluted or further processed by biological treatment.
3. **Neutralization**
   Spent caustics containing hydrogen sulfide, phenolics, or naphthenates (acid oils) can be pre-treated by acid or flue gas neutralization followed by separation and recovery or incineration of these constituents.

6.8.4.2.3 Chemical-Physical Methods. Stripping and extraction following neutralization operation can be used to treat spent caustic solutions. After neutralization, stripping removes residual hydrogen sulfide, mercaptans, and possibly some phenolic compounds. In the neutralization of spent caustics with flue gas, stripping of the volatile components from the solution occurs simultaneously with the neutralization.

6.8.4.2.4 Biological Methods. Biological treatment following pre-treatment can be applied, particularly in areas where the refinery/plant effluent is discharged into brackish or salt water. Due to the high biochemical oxygen demand of spent caustic solutions, the solutions will not be amenable to biological treatment unless they are highly diluted or pre-treated by the chemical or physical methods described previously.

6.8.4.2.5 Caustic Lines Maintaining Temperature. The temperature of the aboveground caustic lines should be maintained at least 22 °C above their freezing points.

6.8.5 MTBE or Leaded Contaminated Streams

1. **Sources**
   The sources of MTBE or leaded contaminated streams will be mainly the following areas already allocated to the finished gasoline production facilities:
   - Finished gasoline tanks drainage.
   - Finished gasoline transfer pumps area drainage.
   - Finished gasoline loading area drainage.
   - Tetra-ethyl lead (TEL) or MTBE storage and injection facilities drainage.

2. **Disposal**

"Disposal of all MTBE or leaded contaminated streams shall be routed to MTBE or Tetra-ethyl lead (TEL) contaminated water pond" provided near the spillage area for further transferral outside the plant battery limit by truck for incineration or safe disposal. The pond should be a covered concrete pit equipped with a flame arrester and sample, inspection and pump-out connections.

6.8.6 Benzene Contaminated Streams

1. **Sources**

Drains of all hydrocarbons containing more benzene than allowed, such as straight run naphtha, platformate, etc., will be sources for such waste liquids. Drainage of all other hydrocarbon streams containing allowable amounts of the benzene component can be handled through ordinary wastewater sewer systems in the plant. However, special attention should be made to the reduction of benzene content in the sources concerned to the maximum extent possible.

2. **Disposal**

A segregated drainage system terminating at the dedicated covered pond should be provided to handle all contaminated benzene drains. The recovered waste streams can be routed to the refinery slops tanks.

6.8.7 Spent Sulfuric Acid Products

6.8.7.1 Sources and Characteristics

In general, sulfuric acid is used extensively both as a treating agent and as a catalyst. The principal sources of sulfuric acid sludges are from the treatment of lubricating oils, heating and diesel oils, and gasoline and naphthas.

The principal sources of spent catalyst acids are alkylation and the manufacture of alcohol and similar products. The hydrocarbon content in spent acids and sludges varies from a few percent up to as much as 60%. The acidity in titratable sulfuric acid can vary from 20 to 90%.

6.8.7.2 Disposal Methods

Disposal of sludges or spent catalyst acid in the refinery or plant effluent stream is not permissible and should be avoided. The following disposal methods are listed only as reference applicable methods allowed in principle.

6.8.7.3 Thermal Decomposition

Thermal decomposition is the most important method of recovering sulfuric acid from spent acids and sludges. Thermal decomposition consists of heating or burning the sludge in the presence of a reducing agent. The organic or carbonaceous material in the spent acid or sludges serves as the reducing agent.

6.8.7.4 Manufacture of Ammonium Sulfate

Fertilizer-grade ammonium sulfate is also made from spent sulfuric acid products. In these processes, acid recovered through a hydrolysis operation is reacted with ammonia, and the resulting ammonium sulfate is crystallized from the water solution.

6.8.7.5 Burning

Some sludges are low in sulfur content and contain enough combustible material to make them utilizable as fuel, either alone or mixed with fuel oil. Due to air pollution problems, use of this method of disposal should be avoided.

6.8.7.6 Sale or Reuse

In the case of spent caustic solutions, spent sulfuric acid catalysts and sludges may be sold or reused in other operations in which the remaining acid quality can be utilized.

6.8.7.7 Offshore Dumping into the Ocean

This method is not permitted unless otherwise allowed by the environmental protection regulations.

6.8.7.8 Burying or Dumping on Waste Land

This method can be used only in rare cases and with permission of the local authorities and/or environmental regulations. In this case the sludge must be either a solid or a semisolid, and special precautions must be taken to prevent underground drainage and pollution of surface waters.

6.8.8 Nitrogen Base Components

Nitrogen base materials, such as pyridine and quinoline, may be produced in the thermal processing of high nitrogen content crudes or distillates. They may enter the refinery/plant effluent in wastes from acid treating operations or in wastewaters from cracking operations. Due to the fact that nitrogen bases of high molecular mass are relatively insoluble in water at pH values above 4, therefore, neutralization and "springing" shall be applied to reduce the gross amounts of these compounds. In such cases streams rich in nitrogen base components should be segregated and routed to the non-oily sewer system after neutralization.

6.8.9 Cyanides

Normally, cyanides are not found in refinery/petrochemical wastewaters in significant concentrations. Small amounts may be found in accumulator condensate waters from catalytic cracking units when processing high nitrogen containing oils. In such cases, when the concentration of cyanides will affect biological treatment operation, the streams concerned should be segregated and treated to convert cyanides to less toxic thiocyanates or ammonia and sodium format.

6.8.10 Aluminum Chloride

1. **Sources**

   Usually, sources are sludge from isomerization or treating processes in which aluminium chloride is the catalyst.

2. **Disposal and treatment**

   The sludge should be neutralized quickly and disposed of as silt or by controlled discharge into the non-oily sewer if it does not include oil and to the oily sewer system when it contains oily materials. The aqueous solution from hydrolyzed sludge can be utilized as a flocculating agent.

6.8.11 Polyelectrolyte

Although liquid or dry powder polyelectrolyte is not a hazardous material, its buffered acidic action is, in some instances, irritating when in contact with the eyes, skin, or mucous membranes.

Normal precautions should be employed to prevent the spraying or splashing of liquid polyelectrolyte, particularly if the material is hot. Areas of special concern include the concentrated liquid polyelectrolyte receiving area hose connections, the transfer pumps, and associated valves and piping. All structures, equipment, valves, and piping with surfaces exposed to the polyelectrolyte should be flushed with plant water prior to being handled. Disposal of all drains contaminated with polyelectrolyte should be through the chemical sewer.

6.8.12 Ferric Chloride

Ferric chloride is not toxic for inhalation at ambient temperatures. At high temperature (over 70 °C) hydrochloric acid exhalation can form with irritant properties on the skin, upper respiratory tract, and lung tissue. Disposal of all contaminated waste drainages should be via the chemical sewer system.

6.8.13 Phosphoric Acid

The spent phosphoric acid catalyst wastes should be neutralized by spreading them in pits filled with limestone, lime, oyster shells, or spent caustic waste. The waste land should be well isolated because these catalysts are deliquescent, and absorbed water will leach the acid from the catalyst.

6.8.14 Hydrofluoric Acid

Hydrofluoric acid tars and gaseous hydrogen fluoride are wastes from alkylation processes in which hydrofluoric acid is the catalyst. The tars and the gas should be disposed of by burning/scrubbing.

6.8.15 Other Spent Catalysts

Spent catalysts that contain high value metals such as nickel, cobalt, molybdenum, platinum, etc. should be reprocessed to recover the metal by the catalyst manufacturer or in recovery plants. Discarding this type of spent catalyst to the environment should be avoided.

6.8.16 Chemical Cleaning Wastes

Spent chemicals from the cleaning of equipment may produce emulsions if discharged directly into the sewer systems. Therefore, spent cleaning solutions should be treated separately to remove iron and solids before being discharged to the wastewater sewer systems. In general, such streams should be neutralized and directed to the evaporation pond through the non-oily sewer system.

6.8.17 Sulfur Solidification and Crushing Facilities and Loading Systems Drainage

The sewer system in this area should collect water drainage (including stormwater) and send it to the non-oily sewer system by gravity. The water drainage from the sulfur solidification and crushing facilities and loading systems area should be free of sulfur particles.

6.8.18 Water Containing Solids, Emulsifying Agents, etc.

Streams impairing gravity separation, such as streams containing solids, emulsifying agents, and/or contaminants that tend to flocculate upon dilution, should not be merged with the continuously oil contaminated water.

6.8.19 Heavy Viscous Oils Drainage

Drainage of any heavy viscous oil (e.g., asphalt) which may be congealed and cause clogging along the sewer system should be avoided. Such streams should be routed to dedicated pits near the sources.

6.8.20 Toxic Metal Contaminated Streams

All drains contaminated with toxic metals, such as arsenic, chromium, mercury, cadmium, lead, selenium, etc., which are most often found in refinery/petrochemical plants' wastewater streams in concentrations that are more than allowed, should be segregated. The method of treatment and disposal should be in accordance with the environmental regulations.

6.8.21 Solvent Processes Drainage

The local sewer system in a solvent process such as extractive distillation, liquid extraction, physical absorption, chemical absorption etc. should segregate and reuse if practical the unavoidable and inadvertent solvent losses from pump seals, flange leaks, etc.

The design of a solvent process plant should give very careful consideration to the possible effluent problems. Every effort should be made to limit/exclude the amount of solvents entering the plant effluent system. It should also be abundantly clear that the aqueous effluents to be expected from the solvent processes may contain a very wide range of organic and inorganic chemical pollutants.

Generally, solvents involved in the solvent processes are phenol, mixtures of phenol, glycols, amines, furfural, sulfolane, sulfur dioxide, sulfuric acid, ammoniacal copper acetate, acetonitrile, ketones, urea etc.

6.8.22 Treating Processes Drainage

These are a broad category of processes for upgrading various intermediate and final product streams. In many cases, the treatment process involves the use of a solvent. All the treatment processes are potential sources of highly undesirable aqueous effluents that should be segregated and treated with similar streams through the refinery/plant.

6.9 Petrochemical Plants' Special Effluents

A careful check should be made of the processes proposed or used for the manufacture of petrochemicals to decrease the possibility of water soluble organics entering water supplies. In general, in order to minimize the wastewaters resulting from petrochemical processes, the following methods are to be taken into consideration:

- Recycling and reuse of waste streams.
- Quenching with oil or chemicals other than water, which do not produce water-borne wastes.
- Use of alternative processes that do not produce water-borne wastes.
- Use of air coolers or of cooling towers in place of once-through cooling water.
- Elimination of waste products in the manufacturing operation before they become associated with waste streams.
- Processing of waste streams to reduce the amount of chemicals in wastewaters leaving the plant.

In design of petrochemical plant processing units, special attention should be paid to the following design notes:

- Extensive use of instruments, alarms, and checks by the operators should be provided to prevent loss of chemicals.
- Adequate facilities should be installed to prevent uncontrolled release of chemicals and wastes to sewers or receiving waters.
- A large storage capable of holding several days production of wastewater should be provided to allow the water to be checked before being released to its final destination.

6.9.1 Summary of Disposal/Treatment Methods

The disposal/treatment methods applied to the reduction/removal of the various compounds that may be found in petrochemical operations should be selected based on the following factors:

- Plant production and/or type of pollutants.
- Final destination of the wastewater effluent.
- The processes ordinarily used for waste disposal.
- Economical aspects.
- Environmental pollution regulations.

6.9.1.1   Treatment

The following methods can be evaluated for wastewater recovery in petrochemical plants and refineries.

6.9.1.2   Physical Treatment Methods

1. **Gravity separation**
   Gravity separation is the usual method of physical treatment for separation of oil–water. It has limitations that may make it necessary to resort to other methods.
2. **Stripping**
   Steam or flue gas stripping, by which phenolics, mercaptans, hydrogen sulfide, and other compounds can be removed, is the most widely used as disposal method for petrochemical wastes. Waste gases from stripping should be burned to prevent air pollution.
3. **Adsorption and extraction**
   Many compounds can be removed from water either by adsorption or by extraction, such as extraction applied for removal of phenolics by using isopropyl ether or other suitable solvents.

6.9.1.3   Other Physical Methods

Other physical methods that can be evaluated are:

- Sedimentation.
- Filtration.
- Flotation.
- Evaporation.

6.9.1.4   Ultimate Disposal

1. **Controlled dilution**
   This method can be used only on materials that are relatively small in quantity and non-toxic, and then only under certain special conditions. A large volume of dilution water in the receiving stream is usually necessary. This method, as with others, must have adequate controls to ensure that the aquatic life and the quality of the receiving waters are not being harmed. In some cases, improved dilution control is obtained by releasing the effluent to the stream through diffusers or jets to obtain better dispersion in the body of the stream.
2. **Incineration**
   This method can be applied only where wastes cannot be successfully disposed of by other methods. In such cases, special incinerator designs should be used to handle the various types of waste that may be encountered, with careful consideration to the resulting air pollution problems. Some of the more important variables to be considered in the design and operation of incinerators are time, temperature, percent of excess air, turbulence, and heat release per unit volume of combustion chamber. When disposal by incineration is contemplated, the composition of the wastes must be considered insofar as it might cause the emission of irritating or odorous gases, chlorine and fluorine, visible plumes, or particulate matter.

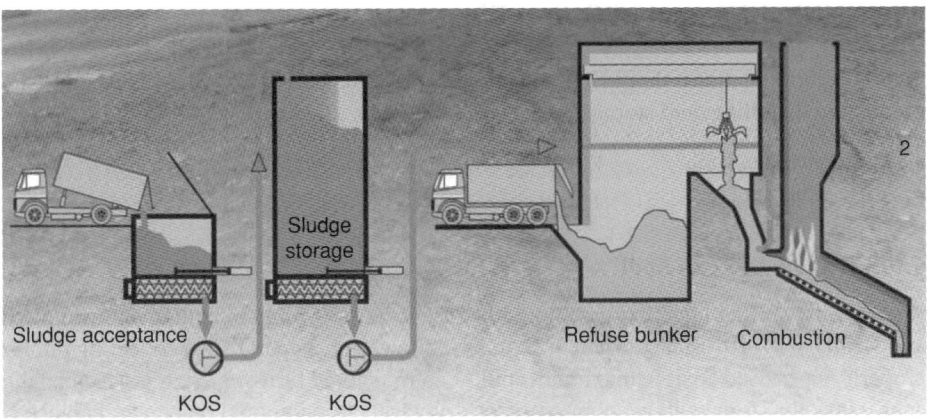

Figure 6.1 A possible design for how sewage sludge incineration can be integrated in a MSW incineration plant (Conradin, K., Kropac, M., Spuhler, D. (Eds.) (2010): The SSWM Toolbox. Basel: seecon international gmbh. URL: http://www.sswm.info).

Incineration of dewatered sludge from wastewater treatment plants reduces the volume of the dry sludge and produces a sterile non-harmful residue that is free from toxic organic chemicals and pathogens. It provides a safe alternative solution when faced with land scarcity for waste disposal and if anaerobic digestion or composting is not an option. Moreover, it also helps to recover some of the energy used in the combustion process, especially in large treatment plants where there is a huge quantity of sludge generation. The disposal of sewage sludge along with the incineration of household rubbish can be integrated both in newly planned plants and retrofitted in older plants (Figure 6.1).

Furthermore, with the implementation of new technologies (e.g., Ash Dec), multi-nutrient fertilizer can be produced out of sewage sludge. The thermo-chemical process in Ash Dec removes toxic heavy metals and makes the sludge harmless to the environment. This allows the closing of the nutrient cycle.

3. **Salvage**
Careful consideration should be given to disposal by salvage, particularly if the waste products are not too dilute. Segregated sewer systems can frequently be used to recover losses with a minimum of dilution, permitting reuse of the material in the manufacturing process.

4. **Deep well injection**
The disposal of waste streams from petrochemical plants into underground strata, which do not contain potable water, may be applied only under certain conditions and if allowed by the local authorities. Very careful pre-treatment or control, or both, of the materials introduced into the strata is necessary to prevent precipitation that would halt the continuation of the injection. It is also important that the disposal well does not contaminate fresh water sources.

5. **Other disposal methods**
Application of any other disposal method, such as disposal at sea, disposal on waste land surfaces, dumping or burial, or spray irrigation, is not allowed unless permitted by the company and environmental regulations.

6.10 NGL, LNG, and LPG Area Effluents

Two specific situations should be considered for effluents from NGL, LNG, and LPG production, handling, and storage areas:

- A liquefied gas spill.
- A fire situation.

6.10.1 Liquefied Gas Spill

Accidentally spilled liquefied gas should be drained away from the equipment as quickly as possible to a safe distance where it should be allowed to evaporate in a collecting pit. For properly supervised areas, a maximum spill of 25 m^3 should be assumed. It is assumed that approximately 15 m^3 will evaporate (under dry weather conditions), leaving a balance of minimum 10 m^3 that should be caught in a collecting pit/liquid gas trap. Spilled liquefied gas and water should be separated as much as possible and the liquefied gas should never be allowed to enter any underground flooded drainage system to avoid blocking due to freezing. It is assumed that only one spill or one fire at a time will occur.

6.11 Gas Treatment Facilities' Effluents

In view of the low risk of encountering heavy pollution (except during line cleaning) the drainage system in a gas treatment plant can be routed to the stormwater sewer system. For line cleaning, the polluted effluent should be discharged into a neutralization pond. The effluent from the neutralization pond should be tested to justify discharge via the non-oily water sewer system for further treatment. Direct discharge from line cleaning into the drainage system is not allowed.

6.12 Effluents from Terminals, Depots, and Product Handling Areas

Effluents from terminals, depots, and product handling areas should be in line with requirements based on local regulations. Due to economic considerations, drainage of terminals, depots, and product handling areas can be integrated with the same type of drainage as the refinery or petrochemical plants installed adjacent to the areas concerned. Special attention should be made to the geographical location and elevation of the areas.

6.13 General Considerations and Conditions for Release of Wastes

With a view to guidelines, standards, or criteria, as well as to regulations, programmes, measures, and discharge permits for the release of refinery/plant wastes, particular attention should be given to the following factors:

1. Regional guidelines, standards, or criteria, as appropriate, for the quality of wastewater used for specific purposes necessary for the protection of human health, living resources, and ecosystems.

2. Regional regulations for the waste discharge and/or degree of treatment for all significant types of sources.
3. Stricter local regulations for waste discharge and/or degree of treatment for specific sources based on local pollution problems and desirable water usage considerations.
4. Polluters should be required to obtain a permit to discharge from the competent local authorities (if needed). Such permits should allow for review and modification of discharge conditions reflecting the periodic updating of regulations.

6.13.1 Characteristics and Composition of Waste

The following requirements should be considered:

1. Type and size of waste source, e.g., industrial process.
2. Type of waste (origin, average composition).
3. Form of waste (solid, liquid, sludge, slurry).
4. Total amount (volume discharged, e.g., per year).
5. Discharge pattern (continuous, intermittent, seasonally variable, etc.).
6. Concentrations with respect to major constituents.
7. Properties:
   - Physical, e.g., solubility and density.
   - Chemical and biochemical, e.g., oxygen demand, nutrients.
   - Biological, e.g., presence of viruses, bacteria, yeast, parasites.
8. Toxicity.
9. Persistence:
   - physical,
   - chemical, and
   - biological.
10. Accumulation and biotransformation in biological materials or sediments.
11. Probability of producing taints or other changes reducing the marketability of resources, e.g., fish, shellfish, etc.
12. Susceptibility to physical, chemical, and biochemical changes and interaction in the aquatic environment with other dissolved organic and inorganic materials.

6.13.2 Characteristics of the Discharge Site and Receiving Environment

The following requirements should be considered:

1. Hydrographic, meteorological, geological, biological, and topographical characteristics of the discharge site.
2. Location and type of discharge (outfall, canal, outlet, etc.) and its relation to other areas, e.g., agricultural areas, spawning, nursery, and fishing areas.
3. Rate of disposal per specific period, e.g., quantity per day, per week, and per month.
4. Initial dilution achieved at the point of discharge into the receiving marine environment, if any.
5. Methods of packaging and containment, if any.
6. Dispersion characteristics such as effects of currents, tides, and wind on horizontal transport and vertical mixing (in case of disposal to a sea or lake).

7. Water characteristics (in case of discharge to a sea/river), e.g., temperature, pH, salinity, stratification, oxygen indices of pollution dissolved oxygen (DO), chemical oxygen demand (COD), biochemical oxygen demand (BOD) – nitrogen present in organic and mineral form including ammonia – suspended matter, other nutrients, and productivity.
8. Existence and effects of other discharges that have been made into the discharge site, e.g., heavy metal background levels and organic carbon content.

6.13.3 Availability of Waste Technologies

The methods of waste reduction and discharge for industrial effluents as well as domestic sewage should be selected taking into account the availability and feasibility of:

- Alternative treatment processes.
- Reuse or elimination methods.
- On-land disposal alternatives.
- Appropriate low-waste technologies.

6.13.3.1 General Considerations

Due consideration should be given to the following:

1. Possible effects on amenities, e.g., presence of floating or stranded materials, turbidity, objectionable odor, discoloration, and foaming.
2. Effects on human health through pollution impact on edible marine organisms, bathing waters, aesthetics, etc.
3. Effects on marine ecosystems, in particular living resources, endangered species, and critical habitats.
4. Possible effects on other uses of the water (in case of discharging to a sea/river), e.g., impairment of water quality for industrial use, underwater corrosion of structure, interference with ship operations from floating materials, interference with fishing or navigation through deposit of waste or solid objects on the sea floor, and protection of areas of special importance for scientific or conservations purposes.

6.14 Effluent Wastewater Characteristics

The quantities and characteristics of wastewater differ considerably for different processes. In general, the major sources of waste contribution in a petroleum refinery are storage tank drain-offs, crude desalting and distillation, and the thermal and catalytic cracking processes, followed by solvent refining, dewaxing, and drying and sweetening.

6.14.1 Flow

Based on total water usage, crude and vacuum distillation units are the largest water users mainly because of the large volumes required by the barometric condensers, desalters, and vacuum ejectors. Catalytic cracking and drying and sweetening are the next largest water users. The extent of water use is significantly affected by the technology level of the processes employed.

6.14.2 Temperature

Crude desalting, especially the electrostatic process, contributes substantial thermal waste loads, as do distillation and cracking. The increased use of cooling towers has played an important role in the reduction in quantities of water discharged and not necessarily by reduction in effluent temperature. Effluent heat loads can have significant adverse effects on the receiving waters since the increased temperature causes decreased oxygen solubility and greater oxygen utilization, both of which reduce the ability of the stream to handle waste loads.

6.14.3 pH

pH indicates the hydrogen ion concentration of a wastewater. However, the extreme values often observed do not truly reflect the buffering capacity of a waste or its ultimate effect upon a receiving water course. Most refinery wastewaters are alkaline, with the cracking (both thermal and catalytic) and crude desalting processes as the principal problem sources; some solvent refining processes also contribute substantial alkalinity. Power house boiler treatment produces alkaline wastewaters and sludges. Hydrotreating also contributes definite alkaline wastes.

Alkylation and polymerization utilize acid processes and have severe acidity problems. In general, petroleum refinery effluents have pH variations, but this is not a major problem from the standpoint of effluent standards. Where a pH range is outside the normal limits, equalization of caustic wastes (and sometimes acid wastes) before bleeding into the sewer system is usually sufficient to maintain pH control. In general, large volumes of cooling and wash waters dilute out strong acid or caustic discharges; thus, pH may become a more significant problem as cooling water volumes decrease.

pH control is also important with regard to the wastewater treatment operation. Very low or very high pH can worsen emulsification of oils already in the sewer. The pH of the wastewater influent to biological treatment processes is an important consideration for effective treatment.

6.14.4 Oxygen Demand

The measurement of the biological and/or chemical oxygen demand an effluent will exert on the oxygen resources of a stream. COD (chemical oxygen demand) and BOD (biochemical oxygen demand) are standard analysis used in this evaluation.

Wastewaters from petroleum refineries or petrochemical plants exert a major, and sometimes severe, oxygen demand.

The primary sources are soluble hydrocarbons and sulfides. The combination of small leaks and inadvertent losses that occur almost continuously throughout a complex can become principal pollution sources.

Crude and product storage and the product finishing operations are the major contributors of COD and BOD, mainly because of the many tanks and vessels used, and the number of times a barrel of oil or product is handled in these operations.

The wastewater discharges from these operations are intermittent. The cracking and solvent refining processes are the major BOD contributors on a continuous basis.

6.14.5 Phenol Content

Catalytic cracking, crude oil fractionation, and product treatment are the major sources of phenolic compounds. Catalytic cracking produces phenols by the decomposition of multi-cyclic aromatics, such as anthracene and phenanthrene. Some solvent refining processes use phenol as a solvent, and although it is salvaged by recovery processes, losses are inevitable. Phenols, particularly when chlorinated, cause taste and odor problems in drinking water.

6.14.6 Sulfide Content

Sulfide waste streams generally originate from the crude desalting, crude distillation, and cracking processes. Sulfides as discussed here are considered to include mercapatns also. Sulfides interfere with subsequent refinery operations and are removed by caustic or diethanolamine scrubbing or appear as sour condensate waters. Hydrotreating processes, which are used to remove sulphides from the feeds or products, naturally produce a rich sulfide waste stream; however, most of the sulfide is removed as H_2S and is usually recovered or burned.

6.14.7 Oil Content

This is a major pollutant characteristic of refinery/plant wastewaters. As free oil, it produces oil slicks and iridescence and coats boats and shorelines if permitted to discharge into the receiving stream. Oil coated solids are particularly troublesome since they are usually of neutral relative density (specific gravity), and are not readily removed by conventional gravity separation techniques. Oil or oil coated solids in the receiving stream also may have a serious detrimental effect on aquatic life.

Oil has limited solubility in water and therefore would be expected to contribute little to effluent BOD or COD. However, crude petroleum and its refined products contain a wide range of soluble hydrocarbons that can ultimately find their way into waste streams through product washes, etc. These product wash streams contribute to effluent BOD and COD.

6.14.8 Light Hydrocarbon Solubility in Water

The solubility of hydrocarbon components in water is of great importance to the environmental sciences. Soluble organics in produced water and refinery effluents are treatment problems for the petroleum industry. Production facilities and refineries have to meet regulatory discharge requirements for dissolved organics. This is expected to become more difficult as environmental regulations become stricter and production from deep water operations increases. However, offshore analysis and remediation of produced water is expensive, and the relatively high polar content of deep-water crude oil also means a higher solubility of organic components in the aqueous phase. In addition, the identities of the water-soluble components are not well known, nor are their concentrations in the produced water brines. Hence, quantitative characterization data are needed as the first step in understanding the dissolution of water-soluble organic compounds in produced water.

Its prediction is usually based on using the pure component solubilities and the mole fraction of the components in the mixture. The solubility of light alkanes (methane and ethane) in pure water has been studied extensively over the past decades. However, due to their extremely low solubilities, it is necessary to predict the aqueous solubility of those components using an accurate method. In this section a simple-to-use correlation (developed by Bahadori et al. 2009) is presented for better prediction of aqueous solubility of light alkanes, equation 6.1 presents the new correlation for predicting the aqueous solubility of light alkanes in which four coefficients are used to correlate the mole fraction of individual components and reduced partial pressure of the component:

$$x_i = 0.001 \left(a + bP_{ri} + cP_{ri}^2 + dP_{ri}^3\right) \tag{6.1}$$

$$a = A_1 + B_1 T_{ri} + C_1 T_{ri}^2 + D_1 T_{ri}^3 \tag{6.2}$$

$$b = A_2 + B_2 T_{ri} + C_2 T_{ri}^2 + D_2 T_{ri}^3 \tag{6.3}$$

$$c = A_3 + B_3 T_{ri} + C_3 T_{ri}^2 + D_3 T_{ri}^3 \tag{6.4}$$

$$d = A_4 + B_4 T_{ri} + C_4 T_{ri}^2 + D_4 T_{ri}^3 \tag{6.5}$$

In above equations, xi is the mole fraction of solute components (i) in water phase; P_r and T_r are the reduced partial pressure and temperature of individual component in KPa and K, respectively. The tuned constants are also given in Table 6.1.

Figures 6.2 and 6.3 illustrate the solubility trends of methane and ethane components in water at different temperatures and pressures, applying the newly developed method. As can be seen from these figures, the aqueous solubility of methane and ethane components is almost independent of temperature at high pressures, while at low pressures the solubility decreases with rising temperature.

Table 6.1 New proposed correlation coefficients for equations 6.2–6.5 to predict aqueous solubility of light alkanes.

Coefficient	Methane	Ethane
A1	261.159102	1.52452624
B1	−257.987470	−0.790117139
C1	0	0
D1	37.2471507	−0.973352139
A2	−294.266511	−198.534348
B2	290.940225	311.386014
C2	0	0
D2	−41.6884515	−109.130286
A3	120.094172	731.199867
B3	−116.988102	−1132.12025
C3	0	0
D3	16.3778867	395.621746
A4	−18.3989831	−905.085194
B4	17.7638515e	1394.71993
C4	0	0
D4	−2.44816293	−486.883551

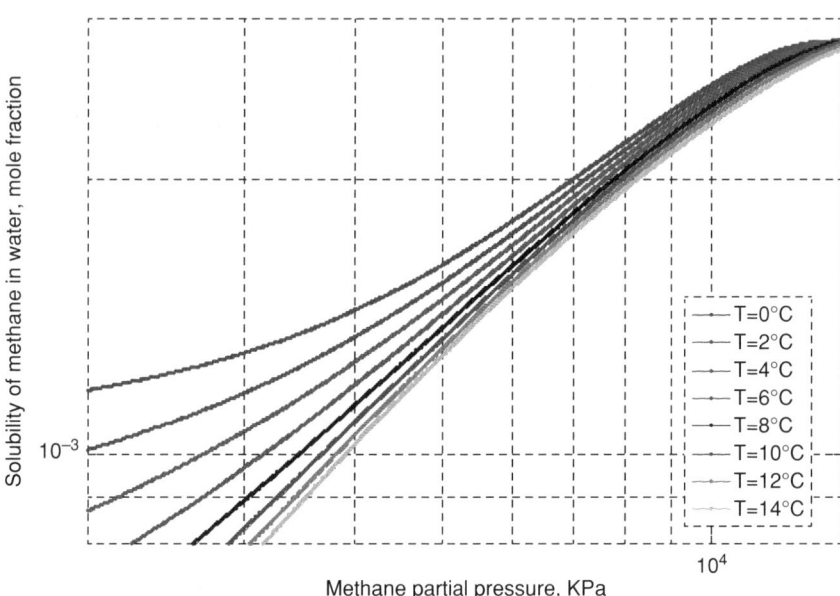

Figure 6.2 Predicting the aqueous solubility of methane (Reproduced with permission from [50] © Taylor and Francis, 2009).

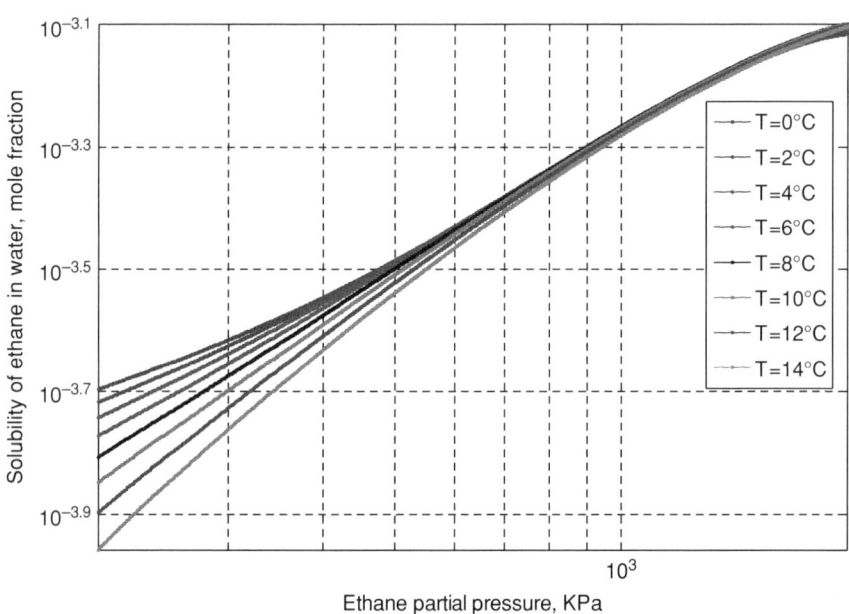

Figure 6.3 Predicting the aqueous solubility of ethane (Reproduced with permission from [50] © Taylor and Francis, 2009).

6.14.9 Predicting Water–Hydrocarbon System Mutual Solubility

Describing the mutual solubilities of hydrocarbons and water is very important in the energy industry. The presence of water in a hydrocarbon mixture can affect the product quality and damage the operation equipment due to corrosion and formation of gas hydrates. Tracing the concentration of hydrocarbons in aqueous media is also important for technical purposes, like preventing oil spills, and for ecological concerns, such as predicting the fate of these organic pollutants in the environment. This section presents a simple-to-use correlation for an excellent prediction of water-hydrocarbon system mutual solubility in a broad range of temperatures between 0 to 120 °C and heavy hydrocarbons between C_3 to C_{10}.

6.14.9.1 Correlation Development

Equation 6.9 presents the new correlation for predicting the water–hydrocarbon mutual solubility in which four coefficients are used to correlate solubility as a function of τ and ξ of the component:

$$\tau = \frac{T}{T_{Ci}} \tag{6.7}$$

$$\xi = \frac{T_{NBPi}}{T_{Ci}} \tag{6.8}$$

$$x_i = \alpha \left(a + b\tau + c\tau^2 + d\tau^3\right) \tag{6.9}$$

$$a = A_1 + B_1\xi + C_1\xi^2 + D_1\xi^3 \tag{6.10}$$

$$b = A_2 + B_2\xi + C_2\xi^2 + D_2\xi^3 \tag{6.11}$$

$$c = A_3 + B_3\xi + C_3\xi^2 + D_3\xi^3 \tag{6.12}$$

$$d = A_4 + B_4\xi + C_4\xi^2 + D_4\xi^3 \tag{6.13}$$

In the above equations, x_i is the mole fraction of solute components (i) in water phase and kg of water dissolved in 100 kg of hydrocarbon. τ and ξ are calculated by Equations 6.7 and 6.8. The evaluated constants are given in Tables 6.2 and 6.3. α is 0.000001 for calculating hydrocarbon solubility in water and it is equal to 1 for water solubility in hydrocarbons. T_{NBPi} and T_{Ci} are the normal boiling point and critical temperature for component "i."

Figures 6.4 and 6.5 illustrate the solubility trends of C_3 to C_{10} components in water at different temperatures. As can be seen, good agreement was observed between the reported data and the obtained results of the new correlation. These graphs also show that light hydrocarbons, such as propane, are more soluble than heavy hydrocarbons, such as decane, in water and that solubility of hydrocarbon in water increases in higher temperatures.

Figures 6.6 and 6.7 illustrate the solubility trends of water in different hydrocarbon components over a wide range of temperatures. These graphs illustrate good agreement between the reported data and the obtained results of the new correlation. These graphs also show that water is more soluble in light hydrocarbons compared with heavy hydrocarbon components.

Table 6.2 Evaluated coefficients for predicting hydrocarbon solubility in water (equations 6.10–6.13).

Coefficient	Propane-Hexane	Hexane-Decane
A1	$1.53061855208904 \times 10^8$	$-7.80318540290318 \times 10^4$
B1	$-6.95425772815835 \times 10^8$	$3.69737747273782 \times 10^5$
C1	$1.05311164420129 \times 10^9$	$-5.77961661971329 \times 10^5$
D1	$-5.31541395374747 \times 10^8$	$2.98529541379832 \times 10^5$
A2	$-4.5084296748633 \times 10^8$	$1.3181984347846 \times 10^6$
B2	$2.04763764916127 \times 10^9$	$-5.78706658568791 \times 10^6$
C2	$-3.09976139197572 \times 10^9$	$8.46135894896114 \times 10^6$
D2	$1.56404762898601 \times 10^9$	$-4.12008439650811 \times 10^6$
A3	$4.47102182743484 \times 10^8$	$-3.85661272900561 \times 10^6$
B3	$-2.03041426889201 \times 10^9$	$1.66978003215558 \times 10^7$
C3	$3.073392747304 \times 10^9$	$-2.40930454837587 \times 10^7$
D3	$-1.5506231578167 \times 10^9$	$1.15845093404961 \times 10^7$
A4	$-1.51100996876207 \times 10^8$	$3.14620688985382 \times 10^6$
B4	$6.86533638859106 \times 10^8$	$-1.35459756919058 \times 10^7$
C4	$-1.03972563665959 \times 10^9$	$1.94391066047826 \times 10^7$
D4	$5.24852883270059 \times 10^8$	$-9.29742057877851 \times 10^6$

Table 6.3 Evaluated coefficients for predicting water solubility in hydrocarbons (equations 6.10–6.13).

Coefficient	Propane-Hexane	Hexane-Decane
A1	$-0.37543995841323 \times 10^5$	$0.36882777088116 \times 10^5$
B1	$1.73693356843739 \times 10^5$	$-1.588017272292665 \times 10^5$
C1	$-2.67896847966559 \times 10^5$	$2.27781165264479 \times 10^5$
D1	$1.37731071844325 \times 10^5$	$-1.0885375042589 \times 10^5$
A2	$0.15898975008232 \times 10^6$	$-0.19720127013985 \times 10^6$
B2	$-0.73780254787291 \times 10^6$	$0.84883989830623 \times 10^6$
C2	$1.14123199781881 \times 10^6$	$-1.21729901998746 \times 10^6$
D2	$-0.58830721436155 \times 10^6$	$0.58164606742025 \times 10^6$
A3	$-0.23187113029754 \times 10^6$	$0.34992894151685 \times 10^6$
B3	$1.078496564000005 \times 10^6$	$-1.50601457844133 \times 10^6$
C3	$-1.6717164862978 \times 10^6$	$2.15954437645887 \times 10^6$
D3	$0.86339618872612 \times 10^6$	$-1.03185031311938 \times 10^6$
A4	$1.15457781637261 \times 10^5$	$-0.20623421783169 \times 10^6$
B4	$-5.37774830265444 \times 10^5$	$0.88757317045455 \times 10^6$
C4	$8.34557228439786 \times 10^5$	$-1.27281419845502 \times 10^6$
D4	$-4.31441338774772 \times 10^5$	$0.60825470939167 \times 10^6$

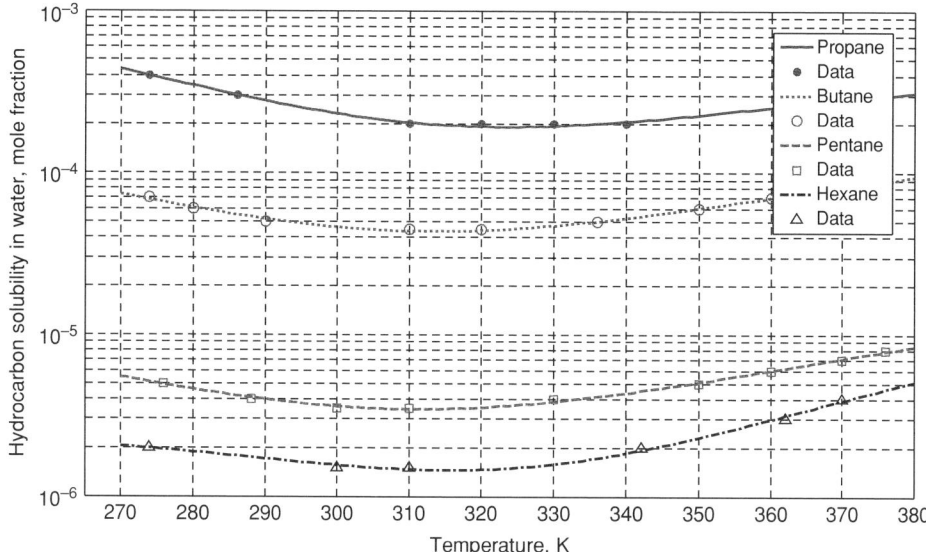

Figure 6.4 Predicting the aqueous solubility of C_3 to C_6 based on the new developed correlation (Reproduced with permission [60] © John Wiley & Son, 2008).

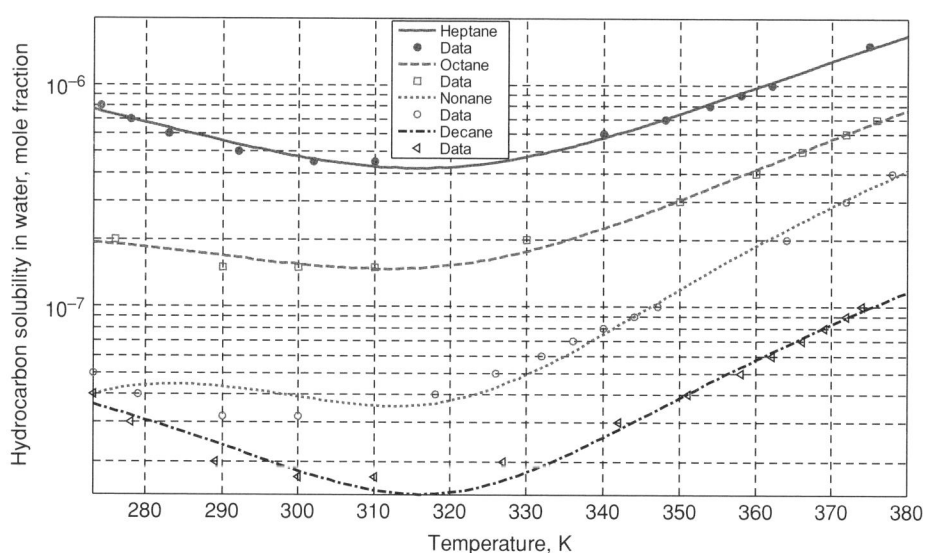

Figure 6.5 Predicting the aqueous solubility of C_6 to C_{10} based on the new developed correlation (Reproduced with permission [60] © John Wiley & Son, 2008).

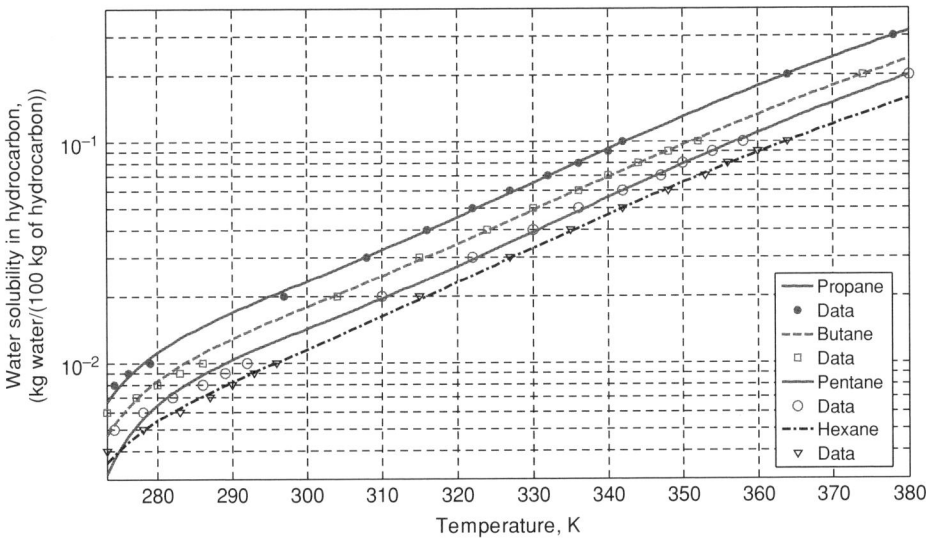

Figure 6.6 Predicting the solubility of water in C_3 to C_6 based on the new developed correlation (Reproduced with permission [60] © John Wiley & Son, 2008).

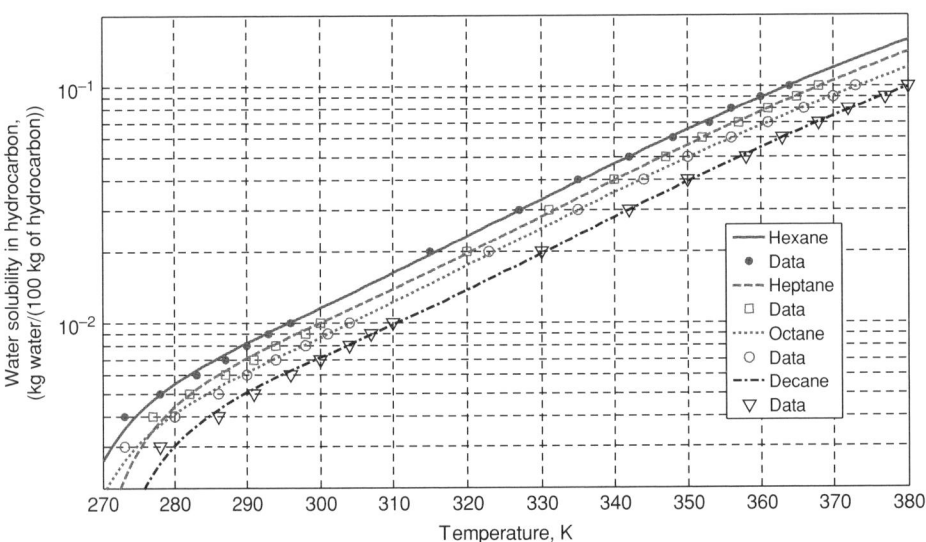

Figure 6.7 Predicting the solubility of water in C_6 to C_{10} based on the new developed correlation (Reproduced with permission [60] © John Wiley & Son, 2008).

6.15 Wastewater Emissions

This section will be divided into two principal parts as follows:

- point source wastewater discharges as released by refinery treatment plants; and
- diffuse wastewater that arises through stormwater, and other miscellaneous run-off from the refinery site that is not captured and treated prior to discharge.

6.15.1 Point Source Discharge

The following tables could be used to provide "default" emission data for refinery effluent discharges that are not classified as transfers (transfers include discharging to sewer). The "dissolved organic carbon" (DOC) content of refinery effluents is usually a known parameter. Hence, the speciation factors for organic compounds in Table 6.4 are based on this parameter. The document from which Table 6.4 was derived indicates a ratio of DOC/COD of 0.267. In the absence of site-specific information regarding DOC, this ratio can be used to determine DOC from measurements of COD.

A similar parameter to DOC was not identified for trace elements and other inorganics in wastewater effluent. Therefore, trace elements and inorganic compound emissions are expressed as default emission factors in Table 6.5.

Tables 6.4 and 6.5 were developed based on experience in industry consultation, and a review of available literature on substance levels in refinery wastewater effluents. It seems reasonable that default values should be derived in the event that no better data is available. The referred methodology usually involves presenting the organic effluent data on a speciated basis, using the dissolved organic content (DOC) of refinery effluents.

The default speciation numbers presented in Tables 6.4 and 6.5 were determined using test data presented in the Ontario Ministry of the Environment (1992) document. This

Table 6.4 Default speciation factors for organics in refinery effluent (Source: © Queen's Printer for Ontario, 2013, pp. A-1).

Substance	Weight Percent of DOC
Toluene	0.00092
Benzene	0.00091
Xylenes	0.0014
Phenol	0.00069
1,2-Dichloroethane	0.00027
Hexachlorobenzene	0.0000044
PAHs	0.0016
Styrene	0.0001
Ethylbenzene	0.00012
1,1,2-trichloroethane	0.000036
Chloroform	0.0025

Table 6.5 Default emission factors for trace elements and inorganics in refinery effluent (Source: © Queen's Printer for Ontario, 2013, pp. A-1).

Substance	Emission Factors (kg/m³ of flow)
Zinc	4.4×10^{-4}
Phosphorous	4.1×10^{-7}
Arsenic	6.7×10^{-6}
Chromium (VI)	7.7×10^{-6}
Selenium	3.1×10^{-6}
Nickel	3.6×10^{-6}
Copper	2.9×10^{-6}
Antimony	5.8×10^{-7}
Cobalt	1.6×10^{-6}
Mercury	1.1×10^{-8}
Cadmium	3.3×10^{-7}
Lead	1.9×10^{-6}
Cyanide	7.6×10^{-9}
Ammonia	1.3×10^{-6}

document provided average effluent concentration levels, as well as the percentage of samples tested, that showed substance concentrations below detection limits.

The averages presented were only based on samples with levels exceeding the detection limit, and so account needed to be taken of the samples that showed non-detectable levels. Therefore, using a conservative approach (where it was assumed that samples with non-detectable levels actually contain the compound at half the detection limit), new "average" effluent concentrations were derived. These average effluent concentrations were then divided by the DOC average presented in the same document to derive speciation fractions.

These speciation factors are applied to this effluent parameter in the following manner:

$$WWE_i = DOC \times (WP_i/100) \times Flow \qquad (6.14)$$

Where:

WWE_i = The wastewater emission of component "i" from the treatment plant (kg/hr);
DOC = The dissolved organic carbon (DOC) content of the treated effluent discharged by the plant (kg/m3);
WP_i = The weight percent of component "i" as provided in Table 6.4 above;
Flow = The wastewater flow-rate discharged to the receiving body of water (m³/hr).

The emission factors in Table 6.5 are applied in the same way as factors are applied to air emissions, with the exception that they are based on the flow of effluent from the treatment plants (i.e., the emission factors are kg per m³ of wastewater flow).

Table 6.6 Typical effluent permissible concentrations.

Contaminants	Discharge to Surface Run-Off, (mg/L)	Discharge to Groundwater, (mg/L)	Irrigation & Agriculture Usage, (mg/L)
Al	5	5	5
Ba	2	1	1
Be	0.1	1	0.1
B	2	1	1
Cd	1	0.01	0.01
Ca	75	–	–
Cr^{6+}	1	1	1
Cr^{3+}	1	1	1
Co	1	1	0.05
Cu	1	1	0.2
Fe	3	0.5	5
Li	2.5	2.5	2.5
Mg	100	100	100
Mn	1	0.5	0.2
Hg	0	0	0
Mo	0.01	0.01	0.01
Ni	1	0.2	0.2
Pb	1	1	1
Se	1	0.01	0.02
Ag	1	0.05	0.01
Zn	2	2	2
Sn	2	2	–
V	0.1	0.1	0.1
AS	0.1	0.1	0.1
Cl^-	Amount of chloride in industrial effluents should not exceed 250 mg/L (ppm) for fresh water.	Amount of chloride in industrial effluents should not exceed 250 mg/L (ppm) for fresh water.	Amount of chloride in industrial effluents should not exceed 250 mg/L (ppm) for fresh water.
F	2.5	2	2
P	1	1	–
CN	0.2	0.02	0.02
C_5H_5OH	1	0	1
CH_2O	1	1	1
NH_4^+	2.5	0.5	–
$NO_2\text{-}NO_2^-$	50	10	–
$NO_3\text{-}NO_3^-$	50	1	–
SO_4^{2-}	300	300	500
SO_3^{2-}	1	1	1
TSS	30	30	100
SS	0	0	0

(Continued)

Table 6.6 (Continued)

Contaminants	Discharge to Surface Run-Off, (mg/L)	Discharge to Groundwater, (mg/L)	Irrigation & Agriculture Usage, (mg/L)
TDS	Total dissolved solids in industrial effluents should not increase the amount of these materials by more than 10% in the underground water/river and any other sources in a distance of 200 m, in which effluent is dumped.	Total dissolved solids in industrial effluents should not increase the amount of these materials by more than 10% in the underground water/river and any other sources in a distance of 200 m, in which effluent is dumped.	Total dissolved solids in industrial effluents should not increase the amount of these materials by more than 10% in the underground water/river and any other sources in a distance of 200 m, in which effluent is dumped.
Oil & Grease	10	10	10
BOD	20	20	100
COD	50	50	200
DO	>2	>2	>2
ABS (detergent)	1.5	0.5	0.5
Turbidity	50	50	50
Color	Color of source water should not exceed more than 16 standard units due to industrial effluent dumped.	75 unit of color	75 unit of color
Temperature	Temperature of industrial effluent should not change the temperature of source water more than ± 3 °C at a distance of 200 m.	—	Temperature of industrial effluent should not change the temperature of source water more than ± 3 °C at a distance of 200 m.
pH	6.5–8.5	5–9	5–9
Radio actives	0	0	0
Digestable Coliform	400/100 mL	400/100 mL	400/100 mL
MPN	1000/100 mL	1000/100 mL	1000/100 mL

6.15.2 Effluent Permissible Concentrations

The maximum permissible concentrations of contaminants in effluents, according to typical environmental regulations, are presented in Table 6.6 as guidance. However, the maximum concentrations of effluent contaminates should be updated during execution of each project in accordance with the latest available information officially issued by the relevant environmental regulation.

The figures indicated in Table 6.6 can be used only for cases not covered in this book.

7
Sewage Treatment

Sewerage is a critical issue, owing to the grave risk that high demand and incapacity place on the environment and the risk to human health from leaks and spills. In addition, aging infrastructure does not hold together well when literally weighed down.

Besides the chemical toxins and wastewater found in sewage; dangerous life-threatening micro-life abound in our flows. These pathogens are deadly parasites, bacteria, and viruses and enter our oceans–beaches and recreational inland water bodies. A release of highly toxic chemical and biological contents into the environment can create health epidemics such as cholera.

1. **Pathogens**
   Some of the lethal health impacts caused by these microorganisms are diarrhoea, stomach cramps, fever, skin fungi, worms, and hepatitis; symptoms may be either acute (short-term) or chronic (long-term).
2. **High Nutrient Content**
   Untreated raw sewage also contains high levels of nutrients (namely from pesticides and chemicals) leading to algal blooms, some of which are toxic to humans and can be passed along the food chain through shellfish consumption or recreational swimming. Symptoms include abdominal pain, vomiting, diarrhoea, and liver failure.
3. **Ecological Risks**
   Besides the human health impacts, raw sewage leaking into the environment decimates other life forms, destroying food sources and places for habitation.
4. **Odors and Gases**
   Not only does sewage reek, it is dangerous in that passing through the system does not only organically produce decomposition gas, but household and industrial wastewater chemicals. Sewer gases may contain a range of gases from ammonia, methane, carbon dioxide, hydrogen sulfide (rotten egg smell). Illegal hazardous discharges of fuels and other chemicals all contribute to the odors, explosion risks, and health risks. Figure 7.1 shows typical images related to the issue of sewage.

196 Waste Management in the Chemical and Petroleum Industries

Figure 7.1 Typical photos related to sewage (Reproduced with permission from © Shutterstock).

This chapter covers the minimum requirements for design of small sanitary sewage treatment plants for oil industries and residential areas. Also the characteristics and chemistry of sewage and its treatment methods are prescribed. Guidance is given for the disposal of the final effluents discharged from sewage treatment works.

7.1 Sewage Effluents

For the protection of the aquatic environment and the maintenance of water quality in lakes, reservoirs, streams, rivers, estuaries, and the sea, sewage-treatment works should be designed to reduce the strength of sewage to a level that may be expected to ensure complete avoidance of nuisance in the circumstances in which the sewage is discharged.

7.1.1 Receiving Water

The setting of receiving water standards comes under the jurisdiction of "environmental regulations." Hence the maximum concentrations of the contaminants for domestic and industrial wastewaters included in the latest issued booklet by local environmental authorities should be respected, with the stipulation that no discharge should create conditions that would violate them.

7.1.2 Final Effluents of Domestic Wastewater Plants

Apart from setting the maximum permissible concentrations with respect to Biochemical Oxygen Demand (BOD_5) and suspended solids, which are the most important factors, it should be obvious that oil, grease, and floating solids should be removed from wastes before discharge to receiving waters.

As regards the BOD_5 and Suspended Solids (SS), the final effluent i.e., the end product of sewage treatment plants is considered satisfactory if it contains not more than 30 parts per million of suspended solids and does not absorb more than 20 PPM of dissolved oxygen in five days.

A final effluent of this strength and turbidity when discharged into rivers assumes that the minimum flow of the river will further dilute the effluent by at least eight volumes of river water to one of effluent. Where this degree of dilution in discharging final effluents of the mentioned effluent standards into watercourses is not available, a higher degree of purification may be advisable. On the other hand, where the dilution is great, as in tidal estuaries, less stringent rules may be permitted.

However, in certain instances of inland discharges, the standard of 30 ppm suspended solids and 20 ppm BOD_5 might be acceptable without further dilution of at least eight times, mentioned above, providing there are no water intakes downstream of the effluent outfall within a safe distance of 50 km.

7.2 Methods of Sewage Treatment: General

The primary aim of sewage treatment is to retain the sewage in circumstances where it is in contact with the air and acted upon by aerobic organisms for sufficient time to oxidize the organic contents to a sufficient degree for the effluent to be safely passed to the natural waters without any fear of causing a nuisance. The contaminants in wastewater are removed by physical, chemical, and biological means. The operations involved are classified as physical unit operations, chemical unit processes, and biological unit processes.

7.2.1 Conventional Methods

The conventional methods of treatment, apart from treatment by dilution, fall under three heads. Treatment on land through septic tanks, treatment in biological "filters," and treatment by the activated-sludge processes.

7.3 Choice of System: General

The type of sewage treatment plant to be adopted in any particular instance depends on:

1. The system, which will involve the minimum of running costs and annual repayments of cost of construction.
2. The extent to which minor nuisance due to flies or sewage work odor matter in the location concerned.
3. The area of land available and the suitability of the land for each of the methods of treatment.

4. The degree of treatment required.
5. The available fall from the incoming sewer at the battery limit of the plant to the point of outfall.

These together form a common sense economic problem to which there should be only one correct answer in any particular instance. In fact, however, decisions as to type of works too often depend on the preferences of individual engineers and on what may be considered fashionable at the time.

7.4 Design of Sewage Treatment Plants: General Guidances

Sewage treatment works are designed on the quantity of sewage to be treated and the strength of the sewage, i.e., the degree of organic pollution. Were sewages generally more consistent than they are, it would be quite possible to design accurately in terms of heads of population to be served, but sewages vary considerably according to the amount of water used per head of population, the degree of infiltration of subsoil water, and the quantity and nature of trade waste. In the case of purely or almost purely domestic sewage from a normal town, where water closets are generally installed, the size of the aeration unit of the sewage works (land, percolating filter, or aeration tank) can be, and sometimes is, determined according to the population to be served. But in most other cases, the aeration unit is sized according to the strength of the sewage as determined by analysis and the average dry-weather flow.

In this engineering standard general guidance is given on adequate design only. Particular requirements should be determined by local conditions. Hence, the given recommendations can be supplemented as required by skilled engineering advice, if any, based on a knowledge of sewage works practice.

Figure 7.2 shows typical sewage treatment facilities.

7.5 Design of Small Sewage Treatment Plants

The design guidance and criteria given under this clause deal with engineering of sewage treatment works suitable for the domestic discharge from domestic and industrial communities ranging from single households up to about 1000 population equivalent and with storage of sewage by means of a cesspool, the contents of which are periodically removed for disposal or treatment.

7.5.1 Collection of Information

For designing small sewage treatment plants, the following main items of basic data should be obtained and considered.

1. Any requirements of the local municipality over and above the permissible effluent standards should be collected from local "Environment Protection Authorities."
2. Minimum and maximum number of persons (resident and non-resident) to be served.

Figure 7.2 Sewage treatment plants (Reproduced with permission from © Shutterstock).

3. Average 24 h water consumption, and any special conditions affecting the composition of sewage and peak rates of flow; data are obtainable from the local water authorities in many instances.
4. Existence of infiltration water.
5. Particulars of site:
   - distance from nearest habitable building;
   - prevailing winds;
   - levels;
   - information as to the nature of the ground including the level and variations of the water table;
   - access for vehicles and plant.
6. Particulars of outfall, e.g., tidal or inland waters, rivers, streams, ditches, or soakage; also the proximity, highest known flood level, and minimum flow of any stream or other watercourse to which discharge of the effluent is possible.
7. Conditions under which the works will normally operate and be maintained.
8. Possibility of the need for future extensions of the works or of their elimination by a comprehensive scheme.
9. Availability of electric power and mains water.
10. Facilities for eventual disposal of sludge and screenings.

Figure 7.3 Grit chamber (Reproduced with permission from © DDM Imagery).

7.6 Preliminary Treatment

Rags and floating debris will inevitably form part of the flow reaching the works and to reduce blockages and fouling of plant, particularly with larger installations, one of the following methods may be adopted:

1. The placing of a small metal screen with 30 to 75 mm clear spacing between the vertical bars in the inlet channel. Provision should be made for overflow or bypass of the screen in the event of blockage. Provision should also be made for the regular and safe disposal of screenings.
2. The provision of a macerator in the inlet channel of pipe to chop up all the debris before it enters the plant.
3. If the sewage has to be pumped at any stage before treatment, a pump incorporating a cutting edge or a separate macerator unit.
4. A grit chamber in which the velocity of waste flow is reduced to a point where the denser sand and other grit will settle out, but the organic solids will remain in suspension (refer Figure 7.3). The settled material is buried or used for fill.

7.7 Primary and Secondary Settlement Tanks

Primary settling tanks (or basins) are usually large tanks in which solids settle out of water by gravity (refer Figure 7.4) where the settleable solids are pumped away (as sludge), while oils float to the top and are skimmed off. It operates by means of the velocity of flow being reduced to about 0.005 m so that the suspended material (organic settle-able solids) can settle out. The usual detention time is 1.5–2.5 hours. Longer periods usually result in depletion of dissolved oxygen and subsequent anaerobic conditions. Removal of suspended solids ranges from 50–65%, and a 30–40% reduction of the five-day biochemical oxygen demand (BOD) can be expected.

Figure 7.4 *Primary settling tank (Reproduced with permission from © DDM Imagery).*

Primary settlement tanks are used to settle out solids prior to biological treatment and thus reduce the BOD load on the units that follow. They should not normally be used for populations of fewer than about 100.

7.7.1 Capacities of Primary Settlement Tanks

The efficiency of a settlement tank is dependent on the velocity of the flow, which is determined by the tank dimensions. In small sewage treatment works in particular, the considerable variations in flow that occur can reduce settlement efficiency.

Settlement tanks may be of the horizontal flow or upward flow type. The latter type is generally more expensive to construct than a horizontal flow tank, but it has distinct advantages.

Facilities should be provided for the regular removal of sludge, which is crucial to the performance of all settlement tanks. In normal operation, tanks should be desludged at least once each week.

Unless otherwise specified, scum retention boards and removal facilities should be provided for settlement tanks, since small sewage treatment works are more likely to receive relatively high proportions of oils, fats, and grease than are large works.

1. **Upward flow tanks**
   The arrangement of an upward flow settlement tank should be such that the nominal upward flow velocity through it is less than the settling velocity of the material to be removed. A figure of 0.9 m/h at maximum flow-rate is recommended. Where the maximum flow-rate is unknown, the surface area of the tank may be calculated from the formula:

$$A = \frac{1}{10} P^{0.85} \tag{7.1}$$

Where:

A is the minimum area (in m^2) of the tank at the top of the hopper; and

P is the design population

This formula allows for increased variability of flow-rates, which occurs as populations decrease. It is based on a dry weather flow of 180 L per head per day but should be adjusted pro rata for other values of the dry weather flow. The dimensions and capacity of the hopper can be determined from knowledge of its volume and surface area.

2. **Primary horizontal flow tanks**

    The calculation of the capacity of a horizontal flow tank should be based on the number of persons to be served and the dry weather flow. The detention period should not exceed 12 h at dry weather flow and the following formula is recommended:

$$C = 180 \, P^{0.85} \tag{7.2}$$

Where:

C is the gross capacity of the tank (in L); and

P is the design population

This formula allows for the increased variability in flow-rates that occurs as populations decrease. It is based on a dry weather flow of 180 L per head per day but should be adjusted pro rata for other values of the dry weather flow.

7.8 Sludge Digesters

The sludge which settles in the sedimentation basin is pumped to the sludge digesters (see Figure 7.5) where a temperature of 30–35 °C is maintained. This is the optimum temperature for the anaerobic bacteria (bacteria that live in an environment that does not contain oxygen). The usual length of digestion is 20–30 days but can be much longer during winter months. Continual addition of raw sludge is necessary and only well-digested sludge should be withdrawn, leaving some ripe sludge in the digester to acclimatize the incoming raw sludge.

7.9 Drying Beds

Digested sludge is placed on drying beds of sand (see Figure 7.6) where the liquid may evaporate or drain into the soil. The dried sludge is a porous humus-like cake which can

Figure 7.5 A sludge digestor (Reproduced with permission from © DDM Imagery).

Figure 7.6 Drying bed (Reproduced with permission from © DDM Imagery).

be used as a fertilizer base. Liquid effluent from the primary settling tank is passed to the secondary part of the system where aerobic decomposition completes the stabilization. For this purpose, a trickling-filter is used. Please refer to Section 4.4 in Chapter 4 for more details about trickling-filters.

7.9.1 Secondary Settlement Tanks

With the majority of the suspended material removed from the sewage, the liquid portion flows over a weir at the surface of the secondary settling tank (see Figure 7.7). Chlorination of the effluent from the secondary settling tank takes place in accordance with state and local laws. Depending on the location, most laws require that a free available chlorine (FAC) residual (usually 0.2 mg/L) be maintained after a 30-minute contact period. This contact period is obtained through the use of chlorine contact chambers which are designed to provide a 30-minute detention time.

From the chlorine contact chamber the treated sewage is normally discharged into a receiving body of water. Secondary settlement tanks, usually known as humus tanks when used in conjunction with biological filters, are essential components of secondary sewage treatment where a 30:20 or better quality effluent is required. They are installed immediately following biological treatment, either as independent units or as integral parts of

Figure 7.7 Secondary settling tank (Reproduced with permission from © DDM Imagery).

packaged systems. For more information about different biological treatments please refer to Chapter 4.

The design principles for secondary settlement tanks are similar to those for primary tanks. For design, constructional, and operational convenience it may be desirable to make secondary settlement tanks of equal size to primary tanks.

7.9.1.1   Capacities of Secondary Settlement Tanks

1. **Upward flow tanks**
   The surface area should be not less than:

$$A = \frac{3}{40} P^{0.85} \tag{7.3}$$

   Where:

   A is the minimum area (in m^2) of the tank at the top of the hopper; and

   P is the design population

   This formula is based on a dry-weather flow of 180 L per head per day and allows for increased variability of flow-rates in small populations. It may be adjusted pro rata for other values of dry-weather flow.

2. **Secondary horizontal flow tanks**
   The calculation of the capacity of a horizontal flow tank should be based on the number of persons to be served and the dry weather flow. The following formula is recommended:

$$C = 135 \, P^{0.85} \tag{7.4}$$

   Where:

   C is the gross capacity of the tank (in L); and

   P is the design population

   This formula is based on a dry-weather flow of 180 L per head per day and allows for increased variability of flow-rates in small populations. It may be adjusted pro rata for other values of dry-weather flow. Use of the formula will give gross detention periods of less than 9 h at dry-weather flow for all values and a population in excess of 100.

7.10   Biological Filters

A biological filtration facility (Figure 7.8) has almost the same structure as a sand filtration facility, so that the suspended solids (SS) get removed through physical filtration. Further, air is passed through the bottom of the filter to create a film of aerobic microorganisms (biofilm) on the surface of the filter. This allows the biodegradable dissolved organics remaining in the raw water to get adsorbed, dissolved, and removed. In this way, cleaner treated water can be obtained.

In a conventional biological filter, the effluent from a septic tank or a primary settlement tank is brought into contact with a suitable medium, the surface of which becomes coated with a biological film. The film assimilates and oxidizes much of the polluting matter

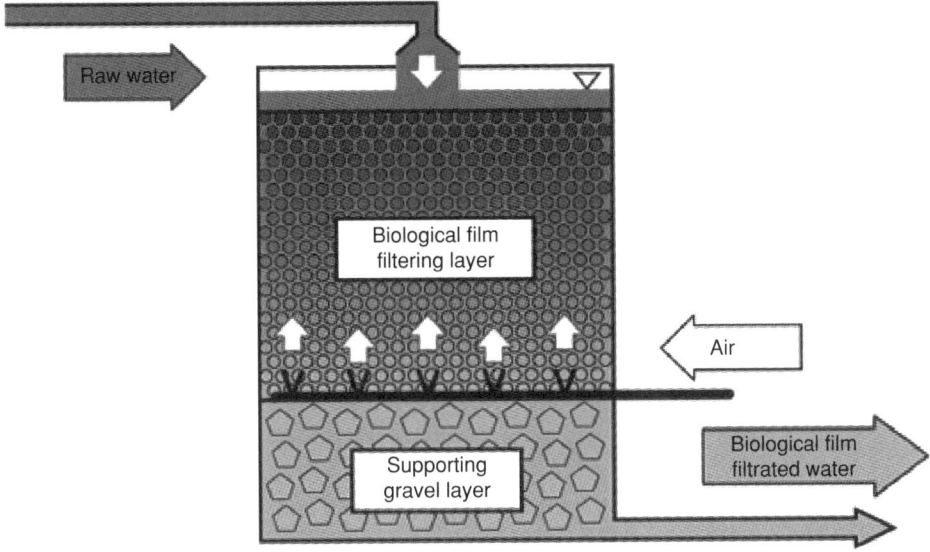

Figure 7.8 Typical biological filter (Reproduced with permission from © Shutterstock).

through the agency of microorganisms. The biological filter requires ample ventilation and an efficient system of under drains leading to an outlet.

7.10.1 Distribution

The effluent should be distributed evenly over the surface of the biological filter, through which it percolates to the floor. Biological filters are usually either rectangular or circular in plan, and various methods of distribution may be used. The filters most suitable for use in small installations are series of fixed channels or a rotating-arm distributor.

7.10.2 Volume of Filter

It is essential that the volume of filter medium provided is sufficient to allow for surge flows, which occur with small installations, such variations being more pronounced the smaller the number of persons served. The volume of mineral medium required can be calculated by the formula:

$$V = 15\, P^{0.83} \tag{7.5}$$

Where:

V is the volume of medium (in m³); and
P is the design population

In Table 7.1, the volumes of medium required for representative numbers of users are given; intermediate values may be interpolated on a linear basis. The volume of medium per user is also given and it can be seen that surge flows are allowed for.

Table 7.1 Filter medium capacity.

P, Design Population	V, Volume of Medium (m³)	V/P
4	4.7	1.18
6	6.66	1.11
8	8.4	1.05
10	10.1	1.01
15	14.2	0.95
20	18	0.90
25	21.7	0.87
30	25.2	0.84
40	32	0.8
50	38.6	0.77
100	69	0.69
200	122	0.61
300	171	0.57
400	217	0.54
500	261	0.52
600	303	0.51
700	345	0.49
800	385	0.48
900	425	0.47
1000	464	0.46

7.10.3 Mineral Filter Media

Mineral filter media should comply with the requirements of BS 1438 and be chosen with regard to the following considerations.

1. It should be strong enough to resist being crushed under its own weight or when walked on.
2. It should be washed and dust-free.
3. It should not contain any toxic substances or other undesirable matter likely to be dissolved into the sewage flow.
4. It should be capable of resisting breakdown due to the flow of the sewage or frost action.
5. The general shape of the individual pieces should be roughly cubical rather than very elongated or flat.
6. The surface of the pieces should preferably be rough and pitted.
7. Consideration should be given to local availability, having regard to suitability.

Several mineral materials are suitable for this purpose, the most usual being hard burnt clinker, blast furnace slag, hard broken stones, and hard crushed gravel. Efficiency is dependent on careful grading; a suitable grading for mineral media is 100 mm to 150 mm at the bottom for a depth of about 150 mm, the remainder being 50 mm nominal maximum size which requires, in accordance with BS 1438, the grading limits given in Table 7.2.

Table 7.2 Grading limits for 50 mm filter medium.

Test Sieves, mm	Proportion by Mass Passing,%
63	100
50	85–100
37.5	0–30
28	0–5

7.11 Activated-Sludge Units

For the purposes of this standard, installations operating on activated-sludge principles are those providing for the aeration of crude unsettled sewage with activated-sludge. An important feature of these installations is that a long period of aeration should be provided at some stage in the process in order to bring about oxidation of sludge, thus reducing the rate of production of surplus sludge and the frequency with which this sludge should be removed. In all activated-sludge systems there is a need regularly to remove quantities of surplus sludge. For location, general requirements, types of installation, and settlement of activated sludge refer to Clause 13 of BS 6297.

7.12 Tertiary Treatment (Polishing) Processes

Conventional biological treatment can produce an effluent of 30:20 standard (SS:BOD), or better, after separation of solids, but for reliable production of higher quality effluents a tertiary or "polishing" stage of treatment is necessary before final disposal. Polishing processes rely mainly on flocculation, sedimentation, or filtration of residual suspended solids directed toward further reduction in ammonia and organic nitrogen, phosphorus, refractory organics, and dissolved solids.

The BOD associated with the solids is removed and some methods also provide further biological purification. Polishing is suitable only for good quality secondary effluents and, in general, will operate efficiently only at works where biological treatment is adequate. If a suitably chosen polishing process is applied to a good quality secondary effluent it should normally be possible to achieve at least 10:10 standard. Several methods are now available. These include slow sand filtration, rapid sand filtration, micro straining, and retention in lagoons. In small sewage treatment works the following methods are more common:

- Treatment over grass plots.
- Upward-flow clarifiers (not normally used with activated-sludge plants).

7.13 Disposal of Final Effluent

After treatment, the disposal of final effluent should be by one of the following methods:

- disposal to inland or tidal water;
- disposal to underground strata;
- disposal on land;
- drying and disposal of sludge.

7.14 Advanced Wastewater Treatment

Many of the substances found in wastewater are not affected, or are little affected by conventional treatment operations and processes. These substances range from relatively simple inorganic ions, such as calcium, potassium, sulfate, nitrate, and phosphate, to an ever-increasing number of highly complex synthetic organic compounds. As the effects of these substances on the environment become more clearly understood, it is anticipated that treatment requirements will become more stringent in terms of the allowable concentration of many of these substances in the effluent from wastewater treatment plants. In turn, this will require advanced wastewater treatment facilities, which nowadays are not used extensively.

7.14.1 Effects of Chemical Constituents in Wastewater

The typical composition of a wastewater is shown in Table 7.3. Most sample wastewaters also contain a wide variety of trace compounds and elements, although these are not measured routinely. If industrial wastewater is discharged to domestic sewers, the distribution of the constituents will vary considerably from that reported in Table 7.3.

Table 7.3 Typical composition of untreated domestic wastewater (all values except settleable solids are expressed in mg/L).

Constituent	Concentration		
	Strong	Medium	Weak
Solids, total	1200	720	350
Dissolved, total	850	500	250
Fixed	525	300	145
Volatile	325	200	105
Suspended, total	350	220	100
Fixed	75	55	20
Volatile	275	165	85
Settleable solids, mL/L	20	10	10
Biochemical oxygen demand, 5-day 200 °C (BOD_5, 20 °C)	400	220	105
Total organic carbon (TOC)	290	160	80
Chemical oxygen Demand (COD)	1000	500	250
Nitrogen (total as N)	85	40	20
Organic	35	15	8
Free ammonia	50	25	17
Nitrites	0	0	0
Nitrates	0	0	0
Phosphorus (total as P)	15	8	4
Organic	5	3	1
Inorganic	10	5	3
Chlorides	100	50	30
Alkalinity (as $CaCO_3$))	200	100	50
Grease	150	100	50

Table 7.4 Typical chemical constituents that may be found in wastewater and their effects.

Constituent	Effect	Critical Concentration mg/L
Inorganic ammonia	Increases chlorine demand; toxic to fish; can be converted to nitrates and, in the process, can deplete oxygen resources; with phosphorus, can lead to the development of undesirable aquatic growths Increases hardness and total dissolved solids	Any amount variable (depends on pH and temperature)
Calcium and magnesium chloride	Imparts salty taste; interferes with agricultural and industrial processes Toxic to humans and aquatic life Stimulates algal and aquatic growth can cause	250 75–200
Mercury	Methemoglobinemia in infants (blue babies)	0.00005
Nitrate	Stimulates algal and aquatic growth Interferes with coagulation Interferes with lime-soda softening	0.3 (For quiescent lakes) 10 (For quiescent lakes)
Phosphate	Cathartic action	0.2-0.4
Sulfate	Toxic to fish and other aquatic life	600–1000
Organic DDT	May be related to the development of cancer May cause taste and odor problems in water	0.001
Hexachloride	–	0.02
Petrochemicals	Cause foaming and may interfere with coagulation	0.005–0.1
Phenolic compounds		0.0005–0.001
Surfactants		1.0–3.0

Some of the substances found in wastewater and the concentrations that may cause problems when discharged to the environment are shown in Table 7.4. This list highlights the fact that a wide variety of substances must be considered and that they will vary with each treatment application.

7.14.2 Advanced Wastewater Treatment Operations and Processes

Unit operations and processes that have been applied to the further treatment of wastewater can be classified as physical, chemical, and biological. A brief description of the physical, chemical, and biological processes involved are given in Table 7.5.

Table 7.5 Advanced wastewater treatment operations and processes.

Description	Type of Wastewater Treated	Principal of Major Use	Waste for Ultimate Disposal
Physical unit operations			
Air stripping of ammonia	Effluent after secondary treatment	Removal of ammonia nitrogen	None
Filtration, multimedia	Effluent after secondary treatment	Removal of suspended solids	Liquid and sludge
Diatornite bed	Effluent after secondary treatment	Removal of suspended solids	Sludge
Micro-strainer	Effluent from biological treatment	Removal of suspended solids	Sludge
Distillation	Effluent after secondary treatment + filtration	Removal of dissolved solids	Liquid
Electrodialysis	Effluent after secondary treatment + filtration + carbon adsorption	Removal of dissolved solids	Liquid
Flotation	Effluent from primary treatment, effluent after secondary treatment	Removal of suspended solids	Sludge
Foam fractionation	Effluent after secondary treatment	Removal of refractors, organics, surfactants, and metals	Liquid
Freezing	Effluent after secondary treatment + filtration	Removal of dissolved solids	Liquid
Gas phase separation	Effluent after secondary treatment	Removal of ammonia nitrogen	None
Land application	Effluent from primary treatment, effluent after secondary treatment	Nitrification, denitrification, removal of ammonia, nitrogen, and phosphorous	None
Reverse osmosis sorption	Effluent after secondary treatment + filtration + effluent from biological treatment	Removal of dissolved solids	Liquid and sludge

Chemical unit processes

Process	Application	Byproduct
Breakpoint chlorination	Effluent after secondary treatment, (filtration)	Removal of ammonia nitrogen	Liquid
Carbon adsorption	Effluent from primary treatment, effluent after secondary treatment (filtration, optional)	Removal of dissolved organics, heavy metals and chlorine	Liquid
Chemical precipitation	Effluent from biological treatment	Phosphorous precipitation, removal of heavy metals, removal of colloidal solids	Sludge
Chemical precipitation in activated sludge	Effluent from primary treatment	Removal of phosphorous	Sludge
Ion exchange	Effluent after secondary treatment + filtration	Removal of ammonia and nitrate nitrogen	Liquid
Electrochemical treatment	Untreated	Removal of dissolved solids	Liquid and sludge
Oxidation	Effluent after secondary treatment	Removal of refractory organics	None

Biological treatment processes

Process	Application	Byproduct
Bacterial assimilation	Effluent from primary treatment	Removal of ammonia nitrogen	sludge
Denitrification	Agricultural return water	Nitrate reduction	None
Harvesting of algae	Effluent from biological treatment	Removal of ammonia nitrogen	Algae
Nitrification	Effluent from primary treatment, effluent from biological treatment	Ammonia oxidation	Sludge
Nitrification-Denitrification	Effluent from primary treatment, effluent from biological treatment	Total nitrogen removal	Sludge

Table 7.6 Potential uses of renovated water.

Use	Direct	Indirect
Municipal	Park or golf course watering; lawn watering with separate distribution system; potential source for municipal water supply	Groundwater recharge to reduce aquifer overdrafts
Industrial	Cooling tower water; boiler feed water; process water	Replenish groundwater supply for industrial use
Agricultural	Irrigation of certain agricultural lands, crops, orchards, pastures, and forests; leaching of soils	Replenish groundwater supply for agricultural overdrafts
Recreational	Forming artificial lakes for boating, swimming, etc.; swimming pools.	Develop fish and waterfowl areas
Other	Groundwater recharge to control salt water intrusion; salt balance control in groundwater; wetting agent–solid waste compaction	Groundwater recharge to control land subsidence problems; soil compaction; oil well repressurizing

7.15 Effluent Disposal and Reuse

A sanitary engineer can design a treatment plant to accomplish as much removal of pollutants as may be required. Ultimate disposal of wastewater effluents will be by dilution in receiving waters; by discharge on land; or, in some desert areas, by evaporation into the atmosphere as well as seepage into the ground.

Disposal by dilution (after secondary treatment) in larger bodies of water, such as lakes, rivers, estuaries, or oceans, is by far the most common method. The proportion of the self-purification capacity, sometimes called the assimilative capacity of the receiving waters, should not be exceeded.

7.15.1 Direct and Indirect Reuse of Wastewater

It is generally impossible to reuse a wastewater completely or indefinitely. The reuse of treated effluent by direct or indirect means is a method of disposal that complements other disposal methods.

The amount of effluent that can be reused is affected by the availability and cost of fresh water, transportation and treatment costs, water-quality standards, and the reclamation potential of the wastewater. Water reuse may be classified according to use as:

- municipal,
- industrial,
- agricultural,
- recreational, or
- groundwater recharge.

Direct and indirect reuse applications for these uses are shown in Table 7.6.

7.15.1.1 Municipal Reuse

Direct reuse of treated wastewater as drinking water, after dilution in natural waters to the maximum possible extent, varies only in degree from the situation existing in many rivers that are used for both water supply and waste disposal.

Advanced methods of wastewater and water treatment, such as demineralization and desalination, are capable of almost complete removal of impurities, and water treated by such methods, after chlorination, is safe to drink. These methods are very expensive and, where they are found to be necessary due to inadequate water supplies, may be economically feasible only if a dual supply system is adopted. In such cases, adequately treated and disinfected wastewater effluents could be reused for flushing toilets, yard watering, and other direct applications.

7.15.1.2 Industrial Reuse

Industry is probably the single greatest user of water in the world, and the largest of the industrial water demands is for process cooling water. However, reuse of the final effluents of wastewater treatment plants within petrochemical plants to replace a portion of the need for industrial water is not recommended.

7.15.1.3 Agricultural Reuse

The types of crops that can be irrigated with reclaimed wastewater depend on the quality of the effluent, the amount of effluent used, and the health regulations concerning the use of treated and untreated wastewater on crops. Health considerations prohibit the use of untreated wastewater.

However, final effluents meeting the secondary treatment requirements can be used for irrigation of certain field crops such as cotton.

7.15.1.4 Recreational Reuse

Golf course and park watering, establishment of ponds for boating and recreation, and maintenance of fish or wildlife ponds are methods for the recreational reuse of water. Today's technology allows the reduction of an excellent effluent that is well suited for the purposes described. The use of treated effluent for park watering has been practiced for many years in western countries.

7.15.1.5 Groundwater Recharge

Groundwater recharge is one of the most common methods for combining water reuse and effluent disposal. Recharge has been used to replenish groundwater supplies in many areas of the United States. Another possible effluent use is in the recharging of oil-bearing strata. Oil companies have conducted much research on flooding techniques to increase the yield of oil-bearing strata.

8
Solid Waste Treatment and Disposal

This chapter covers minimum requirements for the process design and engineering of plant solid waste treatment and disposal facilities in the oil, gas, and chemical industries.

8.1 Basic Considerations

8.1.1 Classification

Solid wastes include those suspended in liquids and are classified in the following categories in order of increasing difficulty of disposal:

1. Inert dry solids, e.g., trash, silt, spent cracking catalyst.
2. Combustible dry solids, e.g., trash, waste paper, scrap lumber.
3. Sludges containing water and solids, e.g., water softener sludges, sanitary sludges.
4. Sludges containing oil, e.g., spent clays.
5. Sludges containing oil, water, and solids, e.g., tank bottoms, oil–water separator bottoms.

8.1.2 Methodology

For proper evaluation and selection of solid waste disposal the following procedures should be performed:

1. Tabulation of sources.
2. Sludge production rates and quantities.
3. Physical characteristics such as pumpability, concentration factor, etc.
4. Analyses for water, oil, and solid contents, volatiles, and ash.
5. Analyses of the water component for pH, sulfide, acidity or alkalinity, lead, and other constituents having potentially significant effects on water pollution.
6. Analyses of the solids component for combustible and non-combustible content and for size distribution of the dry solids.
7. Heating value of sludge (on dry basis).

8.1.3 Sources

8.1.3.1 Solids in the Crude Oil Supply

All crude oils contain some basic sediment and water (BS &W), which is generally composed of a mixture of water, iron rust, iron sulfides, clay, sand, and so forth produced with the crude oil or picked up in transit. Part of the BS &W is charged to the crude oil unit and may settle out in the desalter, entering the oily water sewer system along with the desalter effluent. The balance will settle out in storage tanks, resulting in eventual tank-cleaning problems.

8.1.3.2 Solids from Surface Water

Process waters and all other special drainages throughout the plant/refinery should be isolated from surface run-off. The surface drainage should be collected in a dedicated and separate clean stormwater sewer system. Extensive efforts should be made in the segregation of the surface drainages and avoiding contamination or mix with the oily water sewers.

8.1.3.3 Solids in the Water Supply

Silt may enter the water supply. Depending on the source of the supplied water to the plant/refinery and the characteristics and impurities, provision of sedimentation before the water is used should be investigated. Special attention should be paid to reducing the deposition of solids in cooling tower basins, heat exchangers, and other consumers and also to preventing these solids from entering oily water sewers.

8.1.3.4 Sanitary Solid Wastes

Sanitary wastes should be segregated from all other types of drainage systems.

8.1.3.5 Solids and Sludges in Wastewater Systems

According to the type of plant and the method of plant operation, the sources of solids in a waste water treatment plant can be determined. The principal sources of solids and sludge and the types generated in a conventional wastewater treatment plant are demonstrated in Table 8.1. Solids may also be formed by interaction of waste streams in the sewer. Figure 8.1 shows a typical source of solids and sludge.

Wastewaters contain metal ions, such as iron, aluminium, copper, magnesium etc. from corrosion of the process equipment, chemicals used in treating cooling water, salts in the water intake, and chemicals used in processing. Insoluble metal hydroxide floc may be formed when alkaline wastes are discharged and raise the pH of wastewater above neutral.

Wastes containing considerable concentrations of phenols, sulfides, emulsifying agents, and alkalines should be segregated. In general, discharge of any material to the oily sewer system or other drainage systems should be investigated for the final waste treatment and disposal targets.

8.1.3.6 Catalytic Processes

Catalyst can appear in sewer systems in plants with catalytic process applications. Some means should be provided to minimize catalyst disposal. Hopper trucks or covered portable

Table 8.1 Sources of solids and sludge from a conventional wastewater treatment plant.

Unit Operation or Process	Types of Solids or Sludge	Remarks
Screening	Coarse solids	Coarse solids are removed by mechanical and hand cleaned bar screens. In small plants, screenings are often comminuted for removal in subsequent treatment units.
Grit removal	Grit and scum	Scum removal facilities are often omitted in grit removal facilities.
Preaeration	Grit and scum	In some plants, scum removal facilities are not provided in preaeration tanks. If the preaeration tanks are not preceded by grit removal facilities, grit deposition may occur in preaeration tanks.
Primary sedimentation	Primary sludge and scum	Quantities of sludge and scum depend upon the nature of the collection system and whether industrial wastes are discharged to the system.
Biological treatment	Suspended solids	Suspended solids are produced by the biological conversion of BOD. Some form of thickening may be required to concentrate the waste sludge stream from biological treatment.
Secondary sedimentation	Secondary sludge and scum	Provision for scum removal from secondary settling tanks is a requirement of the US Environmental Protection Agency (EPA).
Sludge-processing facilities	Sludge, compost, and ashes	The characteristics of the end products depend on the characteristics of the sludge being treated and the operations and processes used. Regulations for the disposal of residuals are becoming increasingly stringent.

containers should be provided to prevent catalyst fines from becoming air-borne. In some cases, spent catalyst is slurred with water and pumped directly to ponds where the solids are settled. Spent solid catalysts, which contain vanadium, platinum, or other valuable metals, should be returned to the manufacturer for recovery.

8.1.3.7 Solids from Coking Operations

8.1.3.7.1 Coke Fines. Water used to remove coke from coke chambers in delayed coking units should be recirculated through a settling basin to remove entrained coke fines. The basin can be located near the coke storage pile so that stormwater will drain through the basin to recover coke washed from the storage area. To clean the basin, coke can be transferred directly to the storage pile with appropriate equipment.

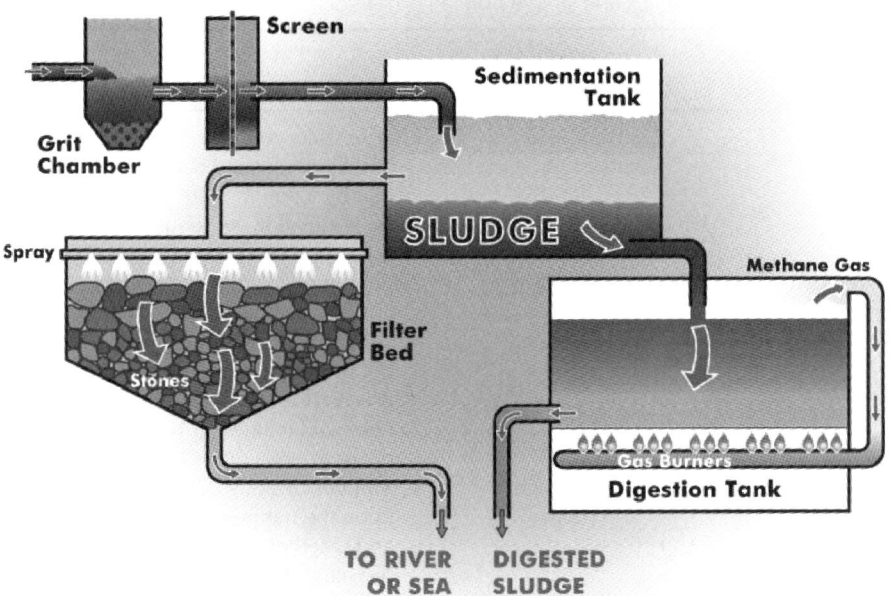

Figure 8.1   A typical source of solids and sludge.

8.1.3.7.2 Wax Tailings. Wax tailings from coking processes present a very difficult disposal problem in the sewer systems. Wax tailings disposal to the oily sewer system should be avoided. Wax deposits may form and clog the oily water sewer or reduce the capacity of the oil–water separators.

In some cases, the wax tailings will rise and be partially absorbed in the oil layer, causing slop oil treatment problems. Investigations should be made to provide facilities to remove wax tailings before discharging to the sewer system. The coker blow-down system may include a scrubber in which a light oil, such as light cycle oil, is recirculated to dissolve wax tailings and remove them from the water. When the API gravity of the light oil has reached a certain point, the oil should be returned to the hot-oil system via the fractionator and be replaced by a new charge.

8.1.3.8   Particulate Matter and Fly Ash

Particulate matter from collectors is sometimes commercially valuable either directly or after further treatment. Therefore, special attention should be paid to collecting the particles for reuse purposes such as:

- addition to concrete in small amounts;
- use as a constituent of clay bricks;
- use as a soil conditioner.

8.1.3.9   Clean-out Wastes

Solids from clean-out operations normally contain considerable amounts of various metals and should not be sent to the oily water system.

Special precautions and procedures are required for the disposal of tank bottoms and other clean-out wastes containing TEL.

8.1.3.10 Other Sources

Other sources of solid wastes such as the following should be taken into consideration:

- storage tank drains;
- process unit drains;
- catalyst contaminated streams;
- others.

8.1.4 Characteristics

Characteristics vary depending on the origin of the solids and sludge, the amount of aging that has taken place, and the type of processing to which they have been subjected. Some of the physical characteristics of sludges are summarized in Tables 8.2, 8.3, and 8.4.

Table 8.2 Characteristics of solids and sludge produced during wastewater treatment.

Solid or Sludge	Description
Screenings	Screenings include all types of organic and inorganic materials large enough to be removed on bar racks. The organic content varies, depending on the nature of the system and the season of the year.
Grit	Grit is usually made of the heavier inorganic solids that settle with relatively high velocities. Depending on the operating conditions, grit may also contain significant amounts of organic matter, especially fats and grease
Scum/grease	Scum consists of the floatable materials skimmed from the surface of primary and secondary settling tanks. Scum may contain grease, vegetable and mineral oils, animal fats, waxes, soaps, food wastes, vegetable and fruit skins, hair, paper and cotton, cigarette tips, plastic materials, condoms, grit particles, and similar materials. The relative density (specific gravity) of scum is less than 1.0 and usually around 0.95.
Primary sludge	Sludge from primary settling tanks is usually gray and slimy and, in most cases, has an extremely offensive odor. Primary sludge can be readily digested under suitable conditions of operation.
Sludge from chemical precipitation	Sludge from chemical precipitation with metal salts is usually dark in color, though its surface may be red if it contains much iron. Lime sludge is grayish brown. The odor of chemical sludge may be objectionable, but is not as bad as primary sludge. While chemical sludge is somewhat slimy, the hydrate of iron or aluminum in it makes it gelatinous.

(Continued)

Table 8.2 (Continued)

Solid or Sludge	Description
	If the sludge is left in the tank, it undergoes decomposition similar to primary sludge, but at a slower rate. Substantial quantities of gas may be given off and the sludge density increased by long residence times in storage.
Activated sludge	Activated sludge generally has a brownish, flocculant appearance. If the color is dark, the sludge may be approaching a septic condition. If the color is lighter than usual, there may have been under aeration with a tendency for the solids to settle slowly. Sludge in good condition has an inoffensive "earthy" odor. The sludge tends to become septic rapidly and then has a disagreeable odor of putrefaction. Activated sludge will digest readily alone or when mixed with primary sludge.
Trickling-filter sludge	Humus sludge from trickling-filters is brownish, flocculant, and relatively inoffensive when fresh. It generally undergoes decomposition more slowly than other undigested sludges. When trickling-filter sludge contains many worms, it may become inoffensive quickly. Trickling-filter sludge digests readily.
Digested sludge (aerobic)	Aerobically digested sludge is brown to dark brown and has a flocculant appearance. The odor of aerobically digested sludge is not offensive; it is often characterized as musty. Well-digested aerobic sludge dewaters easily on drying beds.
Digested sludge (anaerobic)	An aerobically digested sludge is dark brown to black and contains an exceptionally large quantity of gas. When thoroughly digested, it is not offensive, its odor being relatively faint and like that of hot tar, burnt rubber, or sealing wax. When drawn off onto porous beds in thin layers, the solids first are carried to the surface by the entrained gases, leaving a sheet of comparatively clear water. The water drains off rapidly and allows the solids to sink down slowly on to the bed. As the sludge dries, the gases escape, leaving a well-cracked surface with an odor resembling that of garden loam.
Composted sludge	Composted sludge is usually dark brown to black, but the color may vary if bulking agents such as recycled compost or wood chips have been used in the composting process. The odor of well-composted sludge is inoffensive and resembles that of commercial garden-type soil conditioners.
Septage	Sludge from septic tanks is black. Unless the sludge is well digested by long storage, it is offensive because of the hydrogen sulfide and other gases that it gives off. The sludge can be dried on porous beds if spread out in thin layers, but objectionable odors can be expected while it is draining unless it is well-digested.

Table 8.3 Typical chemical composition and properties of untreated and digested sludge.

Item	Untreated Primary Sludge Range	Untreated Primary Sludge Typical	Digested Primary Sludge Range	Digested Primary Sludge Typical	Activated Sludge, Range
Total dry solids (TS),%	2–8	5	6–12	10	0.83–1.16
Volatile solids (% of TS)	60–80	65	30–60	40	59–88
Grease and fats (% of TS) Ether soluble	6–30	–	5–20	18	–
Grease and fats (% of TS) Ether extract	7–35	–	–	–	5–12
Protein (% of TS)	20–30	25	15–20	18	32–41
Nitrogen (N, % of TS)	1.5–4	2.5	1.6–6	3	2.4–5
Phosphorus (P_2O_5, % of TS)	0.8–2.8	1.6	1.5–4	2.5	2.8–11
Potash (K_2O, % of TS)	0–1	0.4	0–3	1	0.5–0.7
Cellulose (% of TS)	8–15	10	8–15	10	–
Iron (not as sulfide)	2–4	2.5	3–8	4	–
Silica (SiO_2, % of TS)	15–20	–	10–20	–	–
pH	5–8	6	6.5–7.5	7	6.5–8
Alkalinity (mg/L as $CaCO_3$)	500–1500	600	2500–3500	3000	580–1100
Organic acids (mg/L as Hac)	200–2000	500	100–600	200	1100–1700
Energy content, Btu/b, (Btu/lb × 2.3241 = kJ/kg)	10 000–12 500	11 000	4000–6000	5000	8000–10 000

Table 8.4 Typical metal content in wastewater sludge.

Metal	Dry Sludge, mg/kg Range	Dry Sludge, mg/kg Median
Arsenic	1.1–230	10
Cadmium	1–3410	10
Chromium	10–99 000	500
Cobalt	11.3–2490	30
Copper	84–17 000	800
Iron	1000–154 000	17 000
Lead	13–26 000	500
Manganese	32–9870	260
Mercury	0.6–56	6
Molybdenum	0.1–214	4
Nickel	2–5300	80
Selenium	1.7–17.2	5
Tin	2.6–329	14
Zinc	101–49 000	1700

Table 8.5 Typical data for the physical characteristics and quantities of sludge produced from various wastewater treatment operations and processes.

Treatment Operation or Process	Relative Density (Specific gravity) of sludge solids	Relative Density (Specific gravity) of sludge solids	Dry solids, kg/10^3m^3	
			Range	Typical
Primary sedimentation	1.4	1.02	108.5–168.7	150.6
Activated-sludge (waste sludge)	1.25	1.005	72.3–96.4	84.3
Trickling-filtration (waste sludge)	1.45	1.025	60.2–96.4	72.3
Extended aeration (waste sludge)	1.30	1.015	84.3–120.5	72.3 (Assuming no primary treatment)
Aerated lagoon (waste sludge)	1.30	1.01	84.3–120.5	72.3 (Assuming no primary treatment)
Filtration	1.20	1.005	12.1–24.2	18.1
Algae removal	1.2	1.005	12.1–24.2	18.1
Chemical addition to primary sedimentation tanks for phosphorus removal Low lime (350–500 mg/L)	1.9	1.04	241–397.6	301.2 (Sludge in addition to that normally removed by primary sedimentation)
Chemical addition to primary sedimentation tanks for phosphorus removal, High lime (800–1600 mg/L)	2.2	1.05	602.4–1325.3	795.2 (Sludge in addition to that normally removed by primary sedimentation)
Suspended-growth nitrification	–	–	–	Negligible
Suspended-growth denitrification	1.20	1.005	12.1–30.1	18.1
Roughing filters	1.28	1.02	–	Included in sludge production from biological secondary treatment processes

8.1.5 Quantities

The quantity of solids entering the wastewater treatment plant fluctuates over a wide range. To ensure capacity capable of handling these variations, the following factors should be taken into consideration:

1. The average and maximum rates of sludge production.
2. The potential storage capacity of the treatment units within the plant.
3. Capabilities to dump short-term peak loads (e.g., by sufficient capacity of equalization basin).

Data on the quantities of sludge produced from various processes and operations and also sludge concentrations are shown in Table 8.5.

8.2 Sludge Handling, Treatment, and Reuse

8.2.1 General

In selecting the appropriate methods of sludge processing, reuse, and disposal, special consideration must be given to the regulations controlling the disposal of sludge from wastewater treatment plants. The following main disposal targets should be investigated:

- application of sludge to agricultural and non-agricultural land;
- distribution and marketing;
- monofilling;
- surface disposal;
- incineration.

Sludge processing and disposal methods are listed in Table 8.6. Thickening (concentration), conditioning, dewatering, and drying are used primarily to remove moisture from sludge; digestion, composting, incineration, wet air oxidation, and vertical tube reactors are used primarily to treat or stabilize the organic material in the sludge.

8.2.2 Sludge and Scum Pumping

Sludge produced in wastewater treatment plants must be conveyed from one plant point to another in conditions ranging from a watery sludge or scum to a thick sludge.

For each type of sludge and pumping application, a different type of pump may be needed. The application and selection of the various types of sludge pumps are summarized below in Table 8.6:

Table 8.6 Sludge processing and disposal methods.

Unit Operation, Unit Process, or Treatment Method	Function
Preliminary operations	
Sludge grinding	Size reduction
Sludge degritting	Grit removal
Sludge blending	Blending
Sludge storage	Storage

(Continued)

Table 8.6 (Continued)

Unit Operation, Unit Process, or Treatment Method	Function
Thickening	
Gravity thickening	Volume reduction
Flotation thickening	Volume reduction
Centrifugation	Volume reduction
Gravity-belt thickening	Volume reduction
Rotary-drum thickening	Volume reduction
Stabilization	
Lime stabilization	Stabilization
Heat treatment	Stabilization
Anaerobic digestion	Stabilization, mass reduction
Aerobic digestion	Stabilization, mass reduction
Composting	Stabilization, product recovery
Conditioning	
Chemical conditioning	Sludge conditioning
Heat treatment	Sludge conditioning
Disinfecting	
Pasteurization	Disinfection
Long-term storage	Disinfection
Dewatering	
Vacuum filter	Volume reduction
Centrifuge	Volume reduction
Belt filter press	Volume reduction
Filter press	Volume reduction
Sludge drying beds	Volume reduction
Lagoons	Storage, volume reduction
Heat Drying	
Flash dryer	Mass and volume reduction
Spray dryer	Mass and volume reduction
Rotary dryer	Mass and volume reduction
Multiple-hearth dryer	Mass and volume reduction
Multiple-effect evaporator	Mass and volume reduction
Thermal Reduction	
Multiple-hearth incineration	Volume reduction, resource recovery
Fluidized-bed incineration	Volume reduction
Co-incineration with solid wastes	Volume reduction
Wet air oxidation	Stabilization, volume reduction
Vertical, deep-well reactor	Stabilization, volume reduction
Ultimate disposal	
Land application	Final disposal
Distribution and marketing	Beneficial use
Chemical fixation	Beneficial use, final disposal
Landfill	Final disposal
Lagooning	Volume reduction, final disposal

8.2.2.1 Types and Selection

8.2.2.1.1 Plunger Pumps. Plunger pumps are frequently used and, if rugged enough for the service, prove quite satisfactory. The advantages of plunger pumps are as follows:

1. The pulsating action of simplex and duplex pumps tends to concentrate the sludge in the hoppers ahead of the pumps and resuspend solids in pipelines when pumping at low velocities.
2. They are suitable for suction lifts up to 3 m and are self-priming.
3. Low pumping rates can be used with large port openings.
4. Positive delivery is provided, unless some object prevents the ball check valves from sealing.
5. They have constant but adjustable capacity, regardless of large variations in pumping heads.
6. High discharge heads may be provided.
7. Heavy solids concentrations may be pumped if the equipment is designed for the load conditions.

The range of plunger pump capacities is from 2.5 to 3.8 L/s per plunger and they are supplied with one, two, or three plungers (called simplex, duplex, or triplex units).

Pump speeds should be between 40 and 50 r/min (rpm). The pumps should be designed for a minimum head of 24 m in small plants and 35 m or more in large plants because accumulations of grease in sludge lines cause a progressive increase in head with use.

8.2.2.1.2 Progressive Cavity Pumps. Progressive cavity pumps are normally used for all types of sludges. The pump is self-priming at suction lifts up to 8.5 m, but it must not be operated dry because this will burn out the rubber stator. They are available in capacities up to 75 L/s and may be operated at discharge heads of 137 m on sludge. For primary sludges, a grinder normally precedes these pumps. The pumps are expensive to maintain because of wear on the rotors and the stators, particularly in primary sludge-pumping applications where grit is present. Advantages of these pumps are:

1. Easily controlled flow-rates.
2. Minimum pulsation.
3. Relatively simple operation.

8.2.2.1.3 Centrifugal Pumps. Centrifugal pumps of non-clog design are commonly used. The selected pumps must have sufficient clearance to pass the solids without clogging and have a small enough capacity to avoid pumping a sludge diluted by large quantities of wastewater overlying the sludge blanket.

Throttling the discharge to reduce the capacity is impractical because of frequent stoppages; hence, it is absolutely essential that these pumps be equipped with variable-speed drives. For pumping primary sludge in large plants, centrifugal pumps of special design-torque flow, that are screw feed and bladeless should be used. Screw feed and bladeless pumps have not been used much in recent applications because of the successful use of torque-flow pumps.

Torque-flow pumps have fully recessed impellers and are very effective in conveying sludge. The size of particle that can be handled is limited only by the diameter of the suction or discharge openings. Pumps used in sludge service should have nickel or chrome abrasion resistant volute and impellers.

The pumps can operate only over a narrow head range at a given speed, so the system operating conditions must be evaluated carefully. Variable speed control should be used where the pumps are expected to operate over a wide range of head conditions. For high pressure applications, multiple pumps may be used and connected together in series. For returning activated sludge to the aeration tanks, slow speed centrifugal, mixed-flow pumps, and screw pumps are commonly used.

8.2.2.1.4 Diaphragm Pumps. Diaphragm pumps are relatively low capacity and low head; the largest available air-diaphragm pump delivers 14 L/s against 15 m of head.

8.2.2.1.5 High Pressure Piston Pumps. High pressure piston pumps are used in high pressure applications, such as pumping sludge long distances, and are very expensive. Several types of piston pump have been developed for high pressure applications and are similar in action to plunger pumps. The advantages of these types of pumps are:

1. They can pump relatively small flow-rates at high pressures, up to 13 800 kPa (ga).
2. Large solids up to the discharge pipe diameter can be passed.
3. A range of solids concentrations can be handled.
4. The pumping can be accomplished in a single stage.

8.2.2.1.6 Rotary-lobe Pumps. Rotary-lobe pumps are positive displacement pumps in which two rotating, synchronous lobes push the fluid through the pump. Rotational speed and shearing stresses are low. For sludge pumping, the lobe shall be made of hard metal or hard rubber.

Their advantage is that lobe replacement is less costly than rotor and stator replacement for progressive cavity pumps. Rotary-lobe pumps, like other positive displacement pumps, must be protected against pipeline obstructions.

8.2.2.2   Application of Pumps to Types of Sludge

Types of sludge that are pumped include primary, chemical, and trickling-filter sludges and activated, thickened, and digested sludges. Scum that accumulates at various points at a treatment plant must also be pumped. The application of pumps to types of sludge is summarized in Table 8.7.

8.2.3   Sludge Piping

1. In treatment plants, conventional sludge piping should not be smaller than DN 150 (6 inch), although smaller diameter glass-lined pipes have been used successfully.
2. Pipe sizes need not be larger than DN 200 (8 inch) unless the velocity exceeds 1.5 to 1.8 m/s, in which case, the pipe should be sized to maintain the velocity. Gravity sludge withdrawal lines should not be less than DN 200 (8 inch) in diameter.

Table 8.7 Application of pumps to types of sludge.

Type of Sludge or Solids	Applicable Pump	Comment
Ground screenings	Pumping screenings should be avoided	Pneumatic ejectors may be used.
Grit	Torque-flow centrifugal	The abrasive character of grit and the presence of rags make grit difficult to handle. Hardened casings and impellers should be used for torque-flow pumps. Pneumatic ejectors may also be used.
Scum	Plunger; progressive cavity; diaphragm; centrifugal	Scum is often pumped by the sludge pumps; valves are manipulated in the scum and sludge limes to permit this. In larger plants, separate scum pumps are used. Scum mixers are often used to ensure homogeneity prior to pumping. Pneumatic ejectors may also be used.
Primary sludge	Plunger; torque-flow centrifugal; diaphragm; progressive cavity; rotary-lobe	In most cases, it is desirable to obtain as concentrated a sludge as practicable from primary sedimentation tanks, usually by collecting the sludge in hoppers and pumping intermittently, allowing the sludge to collect and consolidate between pumping periods. The character of untreated primary sludge will vary considerably, depending on the characteristics of the solids in the wastewater and the types of treatment unit and their efficiency. Where biological treatment follows, the quantity of solids from: 1) waste activated-sludge, 2) humus sludge from settling tanks following trickling- filters, 3) overflow liquors from digestion tanks, and, 4) centrate or filtrate return from dewatering operations will also affect the sludge characteristics. In many cases, the character of the sludge is not suitable for the use of conventional non-clog centrifugal pumps.

Table 8.7 *(Continued)*

Type of Sludge or Solids	Applicable Pump	Comment
Chemical precipitation	Same as for primary sludge	Same as for primary sludge
Digested sludge	Plunger; torque-flow centrifugal; progressive cavity; diaphragm; high-pressure piston; rotary-lobe	Well-digested sludge is homogeneous, containing 5 to 8% solids and a quantity of gas bubbles, but may contain up to 12% solids. Poorly digested sludge may be difficult to handle. If good screening and grit removal is provided, non-clog centrifugal pumps may be considered.
Trickling-filter humus sludge	Non-clog and torque-flow centrifugal; progressive cavity; plunger; diaphragm	Sludge is usually of homogeneous character and can be easily pumped.
Return or waste activated sludge	Non-clog and torque-flow centrifugal; progressive cavity; diaphragm	Sludge is dilute and contains only fine solids so that non-clog pumps may commonly be used. For non-clog pumps, slow speeds are recommended to minimize the breakup of flocculant particles.
Thickened or concentrated sludge	Plunger, progressive cavity; diaphragm; high-pressure piston; rotary-lobe	Positive displacement pumps are most applicable for concentrated sludge because of their ability to generate movement of the sludge mass. Torque-flow pumps may be used but may require the addition of flushing or dilution facilities.

3. A number of clean-outs in the form of plugged tees or crosses instead of elbows should be installed so that the lines can be rodded if necessary.
4. Pump connections should not be smaller than DN 100 (4 inch) in diameter.
5. Pump selection should consider the build-up of head due to grease accumulations inside the piping. In some plants provisions may also be made for melting the grease by circulating hot water, steam, or digester supernatant through the main sludge lines.
6. In the design of long sludge lines, special design features should be considered including:
    o providing two pipes unless a single pipe can be shut down for several days without causing problems;
    o providing for external corrosion and pipe loads;
    o adding facilities for applying dilution water for flushing the line;

- providing means to insert a pipe cleaner at the treatment plant;
- including provisions for steam injections;
- providing air relief and blow-off valves for the high and low points, respectively; and,
- considering the potential effects of waterhammer.

8.2.4 Preliminary Operation Facilities

Sludge grinding, degritting, blending, and storage are necessary to provide a relatively constant, homogeneous feed to sludge processing facilities. Blending and storage can be accomplished either in a single unit designed to do both or separately in other plant components.

8.2.4.1 Sludge Grinding

Some of the processes that must be preceded by sludge grinders for the purpose of preventing clogging are:

- pumping with progressive cavity pumps;
- solid bowl centrifuges;
- belt filter press;
- heat treatment; and,
- chlorine oxidation (for enhancing chlorine contact with sludge particles).

Slow speed, more durable, and more reliable grinders should be applied. The design should include improved bearings and seals, hardened steel cutters, overload sensors, and mechanisms that reverse the cutter rotation to clear obstructions or shut down the unit if the obstruction cannot be cleared.

8.2.4.2 Sludge Degritting

In some plants where separate grit removal facilities are not used ahead of the primary sedimentation tanks, or where the grit removal facilities are not adequate to handle peak flows and peak grit loads, grit removal facilities should be provided before further processing of the sludge. Where further thickening of the primary sludge is desired, a practical consideration is sludge degritting. The most effective method of degritting is through the application of centrifugal forces in a flowing system to achieve separation of grit particles from the organic sludge. Such separation is achieved through the use of cyclone degritters, which have no moving parts. The efficiency of the cyclone degritter is affected by pressure and by the concentration of organic material in the sludge. To obtain effective grit separation, the sludge must be relatively dilute. As the sludge concentration increases, the particle size that can be removed decreases.

8.2.4.3 Sludge Blending

Sludge is blended (if required) to produce a uniform mixture for downstream operations and processes. Uniform mixtures are most important in short detention time systems, such as sludge dewatering, heat treatment, and incineration. Special attention should be given to segregation of oily and non-oily sludges.

Sludge from the various sources can be blended in the several ways.

1. In primary settling tanks:
   The same type (oily or non-oily) of sludge from secondary or advanced wastewater treatment sludges can be returned to the primary settling tanks, where it will settle and mix with the primary sludge.
2. In pipes:
   This procedure requires careful control of sludge sources and feed rates to ensure the proper blend.
3. In sludge processing facilities with long detention times:
   Aerobic and anaerobic digesters (complete-mix type) can blend the feed sludges uniformly.
4. In a separate blending tank:
   This practice provides the best opportunity to control the quality of the blended sludges.

In treatment plants of less than 140 m^3/h capacity, blending is usually accomplished in the primary settling tanks. In large facilities, optimum efficiency is achieved by separately thickening sludges before blending. Blending tanks should be equipped with mechanical mixers and baffles to ensure good mixing.

8.2.4.4   Sludge Storage

Sludge storage must be provided to smooth out fluctuations in the rate of sludge production and to allow sludge to accumulate during periods when subsequent sludge processing facilities are not operating. Sludge storage should be provided ahead of the following processes to establish a uniform feed rate:

- lime stabilization;
- heat treatment;
- mechanical dewatering;
- drying;
- thermal reduction.

Sludge tanks may be sized to retain the sludge for a period of several hours to a few days. If sludge is stored for longer than two or three days, it will deteriorate and will be more difficult to dewater. Sludge is often aerated to prevent septicity and to promote mixing. Means should be provided to reduce and control the odors from sludge storage.

8.2.5   Thickening (Concentration)

Thickening should be used to increase the solids content of sludge by removing a portion of the liquid fraction before the land disposal of the sludge or other applications as required. Representative values of percent total solids from various treatment operations or processes are shown in Table 8.8.

Sludge thickening methods are described in Table. 8.9.

8.2.5.1   Application

In large projects where sludge must be transported a significant distance, such as to a separate plant for processing, a reduction in sludge volume will result in a reduction of

Table 8.8 Expected sludge concentrations from various treatment operations and processes.

Operation or Process Application	Sludge Solids Concentration, % Dry Solids	
	Range	Typical
Primary settling tank		
Primary sludge	4.0–10.0	5.0
Primary sludge to a cyclone	0.5–3.0	1.5
Primary and waste activated-sludge	3.0–8.0	4.0
Primary sludge and trickling-filter humus	4.0–10.0	5.0
Primary sludge with iron addition for phosphorus removal	0.5–3.0	2.0
Primary sludge with low lime addition for phosphorus removal	2.0–8.0	4.0
Primary sludge with high lime addition for phosphorus removal	4.0–16.0	10.0
Scum	3.0–10.0	5.0
Secondary settling tank		
Waste activated-sludge with primary settling	0.5–1.5	0.8
Waste activated-sludge without primary settling	0.8–2.5	1.3
High-purity oxygen activated-sludge with primary settling	1.3–3	2
High-purity oxygen activated-sludge without primary settling	1.4–4	2.5
Trickling-filter humus sludge	1–3	1.5
Rotating biological contractor waste sludge	1–3	1.5
Gravity thickener		
Primary sludge only	5–10	8
Primary and waste activated-sludge	2–8	4
Primary sludge and trickling-filter humus	4–9	5
Dissolved-air flotation thickener (waste activated sludge only)	4–6	5
Dissolved-air flotation thickener (with chemical addition)	3–5	4
Centrifuge thickener (waste activated-sludge only)	4–8	5
Gravity-belt thickener (waste activated-sludge only with chemical addition)	3–6	5
Anaerobic digester		
Primary sludge only	5–10	7
Primary and waste activated-sludge	2.5–7	3.5
Primary sludge and trickling-filter humus	3–8	4
Aerobic digester		
Primary sludge only	2.5–7	3.5
Primary and waste activated-sludge	1.5–4	2.5
Waste activated-sludge only	0.8–2.5	1.3

Table 8.9 Occurrence of thickening methods in sludge processing.

Method	Type of Sludge	Frequency of Use and Relative Success
Gravity	Untreated primary	Commonly used with excellent results. Sometimes used with hydroclone degritting of sludge.
Gravity	Untreated primary and waste activated-sludge	Often used. For small plants, generally satisfactory results with sludge concentrations in the range of 4 to 6%. For large plants, results are marginal.
Gravity	Waste activated-sludge	Seldom used; poor solids concentration (2 to 3%).
Dissolved-air flotation	Untreated primary and waste activated-sludge	Some limited use; results similar to gravity thickeners.
Dissolved-air flotation	Waste activated-sludge	Commonly used; good results (3.5 to 5% solids concentration).
Imperforate basket centrifuge	Waste activated-sludge	Limited use; excellent results (8 to 10% solids concentration).
Solid bowl centrifuge	Waste activated-sludge	Increasing; good results (4 to 6% solids concentration).
Gravity belt thickener	Waste activated-sludge	Increasing; good results (3 to 6% solids concentration).
Rotary drum thickener	Waste activated-sludge	Limited use; good results (5 to 9% solids concentration).

pipe size and pumping costs. Volume reduction is very desirable when liquid sludge is transported by tank trucks for direct application to land as a soil conditioner. In treatment plants with less than 140 m^3/h capacity, separate sludge thickening may not be required. In small plants, gravity thickening is accomplished in the primary settling tank or in the sludge digestion units, or both.

8.2.5.2 Methods of Application

The following thickening methods depending on the type of sludge can be applied (see Table 8.9):

- gravity;
- dissolved air flotation;
- imperforate basket centrifuge;
- solid bowl centrifuge;
- gravity-belt thickener;
- rotary-drum thickener.

For gravity thickening, a circular tank should be used. Provisions for dilution water and occasional chlorine addition should be included to improve process performance.

For flotation types, only dissolved air flotation should be used. Higher loadings can be used with flotation thickeners than are permissible with gravity thickeners because of the rapid separation of solids from the wastewater.

8.2.5.3 Design Considerations

1. In designing thickening facilities, it is important to provide adequate capacity to meet peak demands and prevent septicity, with its attendant odor problems, during the thickening process.
2. To maintain aerobic conditions in gravity thickeners, provisions should be made for adding 24 to 30 $m^3/m^2 d$ of final effluent to the thickening tank.
3. Minimum solids loadings should be used for design of thickeners.
4. The use of polymers as flotation aids is effective in increasing the solids recovery in the floated sludge and in reducing recycle loads.

8.3 Stabilization

Sludges are stabilized to reduce pathogens, eliminate offensive odors, and inhibit, reduce, or eliminate the potential for putrefaction. The means to eliminate nuisance conditions through stabilization, the following facilities should be taken into consideration.

1. The biological reduction of volatile content.
2. The chemical oxidation of volatile matter.
3. The addition of chemicals to the sludge to render it unsuitable for the survival of microorganisms.
4. The application of heat to disinfect or sterilize the sludge.

8.3.1 Design Considerations

When designing a sludge stabilization process, the following factors have to be taken into consideration:

- sludge quantity to be treated;
- the integration of the stabilization process with the other treatment units;
- the objectives of the stabilization process.

The following sludge stabilization processes should be considered and the most suitable technology conforming with the respective environmental regulations should be selected.

1. Lime stabilization.
2. Heat treatment.
3. Anaerobic sludge digestion.
4. Aerobic sludge digestion.
5. Composting.

8.3.2 Lime Stabilization

1. Sufficient lime should be added to the untreated sludge to raise the pH to 12 or higher. The method used should be evaluated for the proper placement of the lime injection.

2. The minimum criteria for lime stabilization in a lime pre-treatment method is to maintain the pH above 12 for about 2 hours so as to ensure pathogen destruction and to provide enough residual alkalinity such that the pH does not drop below 11 for several days.
3. Testing should be performed for specific applications to determine the actual dosage requirements.
4. Because lime stabilization does not destroy the organic material necessary for bacteria growth, the sludge must be treated with an excess of lime or disposed of before the pH drops significantly. An excess dosage of lime may range up to 1.5 times the amount needed to maintain the initial pH of 12.
5. Post-lime stabilization has several significant advantages when compared to pre-lime stabilization, such as:
    - dry lime can be used; therefore, no additional water is added to the dewatered sludge;
    - there are no special requirements for dewatering;
    - scaling problems and associated maintenance problems of lime sludge dewatering equipment are eliminated.

Adequate mixing should be provided for a post-lime stabilization system so as to avoid pockets of putrescible material.

8.3.3 Heat Treatment

Heat treatment, which is also classified as a conditioning process, should be designed for continuous processing and should consider heating the sludge in a pressure vessel to temperatures up to 260 °C at pressures up to 2760 kPa (ga) for short periods of time (approximately 30 minutes). The heat treatment process is most applicable to biological sludges that may be difficult to stabilize or condition by other means.

Advantages for the heat treatment process are as follows:

1. The solids content of the dewatered sludge can range from 30 to 50%, depending on the degree of oxidation achieved.
2. The processed sludge does not normally require chemical conditioning.
3. The process stabilizes sludge and will destroy most pathogenic organisms.
4. The processed sludge will have a heating value of 28 to 30 kJ/g of volatile solids.
5. The process is relatively insensitive to changes in sludge composition.

The major disadvantages are:

1. High capital cost.
2. Close supervision, skilled operators, and a strong preventative maintenance program are required.
3. The process produces sidestreams with high concentrations of organics ammonia nitrogen, and color.
4. Significant odorous gases are produced that require extensive containment, treatment, and/or destruction.
5. Scale formation in the heat exchangers, pipes, and reactor requires acid washing or high pressure water jets.

8.3.4 Anaerobic Sludge Digestion

The following processes must be investigated for proper process selection based on the sludge concentration, rate, and characteristics.

1. Standard-rate digestion.
2. Single stage high-rate digestion.
3. Two stage digestion.
4. Separate sludge digestion.

The design should be based on:

1. The concept of mean cell-residence time.
2. The use of volumetric loading factors.
3. Observed volume reduction.
4. Loading factors based on population.

Anaerobic digester design data should be according to Table 8.10 below:

Table 8.10 Anaerobic digester design data.

Parameter	Value
Number of tanks	Multiple desired
Solids retention time – Mesophilic	20–60 days
Solids retention time – Thermophilic	10–20 days
Temperature – Mesophilic	10–35 °C
Temperature – Thermophilic	38–60 °C
Volume allowance	**0.001–0.01 m^3/capita/day**
Volatile suspended solids loading	0.64–6.4 kg/m^3/day
pH range	6.6–7.6
Digested sludge solids concentration	4–6%

8.3.4.1 Advantages and Disadvantages of Aerobic Digestion

Advantages:

1. Simplicity of operation and maintenance.
2. Lower capital costs.
3. Lower BOD and total phosphorous in supernatant.
4. Few effects from loading, pH, and toxic interference.
5. Insignificant odors.
6. Non-explosive.
7. Reduces grease and hexane solubles.
8. Better sludge fertilizer.
9. Shorter retention periods.
10. Excellent small plant alternative.

Disadvantages:

1. Higher operating costs.

2. Highly sensitive to ambient temperature.
3. No useful byproducts.
4. Variable dewaterability.
5. Lower volatile solids reduction.
6. Questionable economics for larger plants.

Aerobic digester design data should be according to Table 8.11 below:

Table 8.11 Aerobic digester design data.

Parameter	Value
Number of tanks	Multiple desired
Solids retention time at 20 °C – Primary sludge	15–20 days
Solids retention time at 20 °C – Primary and activated sludge or trickling-filter sludge	15–20 days
Solids retention time at 20 °C – Waste activated-sludge only	10–15 days
Solids retention time at 20 °C – Waste activated-sludge without primary settling	12–18 days
Volume allowance	0.8–0.11 m^3/capacity/day
Volatile suspended solids loading	0.32–2.24 kg/m^3/day
Minimum dissolved oxygen	1–2 mg/L
Oxygen requirements– Destroy cell tissue	2:1
Oxygen requirements– Reduce primary sludge	1.6:1–1.9:1
Diffusion aeration – Waste activated-sludge only	20–30 m^3/min/1000 m^3
Diffusion aeration – Mixed or other sludges	Over 60 m^3/min/1000 m^3
Mechanical aeration	20–40 kW/1000 m^3

8.3.5 Composting

The composting process involves the complex destruction of organic material coupled with the production of humic acid to produce a stabilized end product.

The major types of composting systems used are the aerated static pile, window, and in-vessel (enclosed mechanical) systems. The aerated static pile is preferred.

The compost mix should be at least 40% dry solids to ensure adequate composting.

The following important factors should be determined/clarified for the design of composting system:

1. Type of sludge.
2. Amendments and bulking agents.
3. Carbon:nitrogen ratio.
4. Volatile solids.
5. Air requirements.
6. Moisture content.
7. pH.
8. Temperature.
9. Mixing and turning.

10. Heavy metals and trace organics.
11. Site constraints.

8.4 Conditioning

Sludge should be conditioned to improve its dewatering characteristics if required. Based on the economic and technical evaluation of the existing conditions, a suitable conditioning method, such as addition of chemicals, heat treatment, irradiation, or solvent extraction, should be used. The heat treatment process is most applicable to biological sludges that may be difficult to stabilize or condition by other means.

Intimate admixing of sludge and coagulant should be made in the chemical conditioning process for proper conditioning.

The mixing must not break the floc after it has formed, and the detention should be kept to a minimum so that sludge reaches the dewatering unit as soon after conditioning as possible.

8.5 Disinfection

The following methods may be used to achieve pathogen reduction beyond that attained by stabilization:

- pasteurization;
- other thermal processes, such as heat conditioning, heat drying, incineration, pyrolysis, or starved air combustion;
- high pH treatment, typically with lime, at a pH higher than 12 for 3 hours;
- long-term storage of liquid digested sludge;
- complete composting at temperatures above 55 °C and curing in a stockpile for at least 30 days;
- addition of chlorine to stabilize and disinfect sludge;
- disinfection with other chemicals;
- disinfection by high-energy irradiation.

Storage should be taken into consideration in land application systems in order to retain sludge during periods when it cannot be applied because of weather or crop considerations. Because of the potential contamination effects of the stored sludge, special attention must be devoted to the design of these lagoons with respect to limiting percolation and the development of odors.

8.6 Dewatering

Dewatering is a mechanical (physical) unit operation used to reduce the moisture content of sludge for one or more of the following reasons:

1. The costs of trucking sludge to the ultimate disposal site become substantially lower when sludge volume is reduced by dewatering.

2. Dewatered sludge is generally easier to handle than thickened or liquid sludge. In most cases, dewatered sludge may be shoveled, moved about with tractors fitted with buckets and blades, and transported by belt conveyors.
3. Dewatering is normally required prior to the incineration of the sludge to increase the energy content by removal of excess moisture.
4. Dewatering is required before composting to reduce the requirements for supplemental bulking agents or amendments.
5. In some cases, removal of the excess moisture may be required to render the sludge odorless and non-putrescible.
6. Sludge dewatering is required prior to landfilling in monofills to reduce leachate production at the landfill site.

8.6.1 Sludge Dewatering Methods

Dewatering methods and their advantages and disadvantages are presented in Table 8.12.

The dewatering device shall be determined by the type of sludge to be dewatered, characteristics of the dewatered product, and the space available. Some sludges, particularly aerobically digested sludges, are not amenable to mechanical dewatering. Selection of the optimum dewatering device should be made by conducting bench scale or pilot studies.

8.6.2 Vacuum Filtration

The solids content of the feed should be 6 to 8%. Higher solids content makes the sludge difficult to distribute and to condition for dewatering; lower solids content requires the use of larger than necessary vacuum filters.

In addition to the disadvantages listed in Table 8.6 the following factors should also be taken into consideration in selection of this type of dewatering equipment:

- system complexity;
- need for conditioning chemicals;
- high operating and maintenance costs.

8.6.3 Centrifugation

Centrifugation processes are widely used in the industry for separating liquids of different density, thickening slurries, or removing solids.

- Disposal of centrate to wastewater influent should be avoided.
- Before final design decision making, pilot plant tests should be run.
- Special consideration must be given to providing sturdy foundations and sound-proofing because of the vibration and noise that result from centrifuge operation.

8.6.4 Belt Filter Press

Belt filter presses are continuous-feed sludge dewatering devices that involve the application of chemical conditioning, gravity drainage, and mechanically applied pressure to dewater sludge. Belt filter presses can be applied to all types of wastewater sludge.

Sludge blending facilities should be included in the system design where the sludge characteristics are likely to vary widely.

Table 8.12 Comparison of alternative sludge dewatering methods.

Dewatering Method	Advantages	Disadvantages
Vacuum filter	• Skilled personnel not required; • Maintenance requirements are low for continuously operating equipment.	• Highest energy consumer per unit of sludge dewatered; • Continuous operator attention required; • Vacuum pumps are noisy; • Filtrate may have high suspended solids content, depending on filter medium.
Solid bowl centrifuge	• Clean appearance, minimal odor problems; • Fast start up and shut down capabilities; • Easy to install Produces relatively dry sludge cake; • Low capital cost-to-capacity ratio.	• Scroll wear potentially a high maintenance problem; • Requires grit removal and possibly a sludge grinder in the feed stream; • Skilled maintenance personnel required; • Moderately high suspended solids content in centrate.
Imperforate basket centrifuge	• Same machines can be used for both thickening and dewatering; • Chemical conditioning may not be required; • Clean appearance; minimal odor problems; • Fast start up and shut down capabilities; • Very flexible in meeting process requirements; • Not affected by grit; • Excellent results for difficult sludges.	• Limited size capacity; • Except for vacuum filters, consumes more energy per unit of sludge dewatered; • Skimming stream may produce significant recycle load for easily dewatered sludges; • Has highest capital cost-to-capacity ratio.
Belt filter press	• Low energy requirements • Relatively low capital and operating costs; • Less complex mechanically and easier to maintain; • High-pressure machines are capable of producing very dry cake; • Minimal effort required for system shut down.	• Hydraulically limited in throughput • Requires sludge grinder in feed stream; • Very sensitive to incoming sludge feed characteristics; • Short media life as compared to other devices using cloth media; • Automatic operation generally not advised.
Recessed plate filter press	• Highest cake solids concentration; • Low suspended solids in filtrate.	• Batch operation; • High equipment cost high labor cost; • Special support structure requirements;

Table 8.12 (Continued)

Dewatering Method	Advantages	Disadvantages
		• Large floor area required for equipment; • Skilled maintenance personnel required; • Additional solids due to large chemical addition require disposal.
Sludge drying beds	• Lowest capital cost method where land is readily available; • Small amount of operator attention and skill required; • Low energy consumption; • Low to no chemical consumption; • Less sensitive to sludge variability; • Higher solids content than mechanical methods.	• Requires large area of land; • Requires stabilized sludge; • Design requires consideration of climatic effects; • Sludge removal is labor intensive.
Sludge lagoons	• Low energy consumption; • No chemical consumption; • Organic matter is further stabilized; • Low capital cost where land is available; • Least amount of skill required for operation.	• Potential for odor and vector problems; • Potential for groundwater pollution; • More land intensive than mechanical methods; • Appearance may be unsightly; • Design requires consideration of climatic effects.

Safety considerations in design should include adequate ventilation to remove hydrogen sulfide or other gases and equipment guards to prevent loose clothing from being caught between the rollers.

8.6.4.1 Filter Presses

In a filter press, dewatering is achieved by forcing the water from the sludge under high pressure. Advantages of this dewatering equipment include:

- high concentrations of cake solids;
- filterate clarity; and,
- high solids capture.

Disadvantages include:

- mechanical complexity;
- high chemical costs;

- high labor costs; and
- limitations on filter cloth life.

The following filter press types are preferred:

1. Fixed-volume, recessed plate filter press.
2. Variable-volume, recessed plate filter press (diaphragm press).

The following features should be considered in the design of filter press facilities:

1. Adequate ventilation in the dewatering room.
2. High-pressure washing systems.
3. An acid wash circulating system to remove calcium scale when lime is used.
4. A sludge grinder ahead of the conditioning tank.
5. Cake breakers or shredders following the filter press (particularly if the dewatered sludge is incinerated).
6. Equipment to facilitate removal and maintenance of the plates.

8.6.5 Sludge Drying Beds

Sludge drying beds are typically used to dewater digested sludge. After drying, the sludge is removed and either disposed of in a landfill or used as a soil conditioner.

The principal advantages of drying beds are low cost, infrequent attention required, and high solids content in the dried product.

The following types of drying beds may be used. The conventional sand drying bed is preferred:

- conventional sand;
- paved (drainage or decanting);
- artificial media; and
- vacuum assisted.

Open beds may be used where adequate space is available and if the location is sufficiently isolated to avoid complaints caused by occasional odors. Open sludge beds should be located at least 100 m from dwellings to avoid odor nuisance. Covered beds with greenhouse types of enclosure should be used where it is necessary to dewater sludge continuously throughout the year regardless of the weather and where sufficient isolation does not exist for the installation of open beds.

In cold climates, the effects of freezing and thawing should be considered to improve the dewatering characteristics of sludge.

8.6.6 Lagoons

1. Drying lagoons may be used as substitute for drying beds for the dewatering of digested sludge, if permitted by the Company; lagoons should not be used for dewatering untreated sludges, limed sludges, or sludges with a high-strength supernatant because of their odor and nuisance potential.
2. Environmental and groundwater regulations should be considered for application of lagoons; if a groundwater aquifer used for a potable water supply underlies the lagoon site, it will be necessary to line the lagoon.

3. Sludge should be discharged to the lagoon in a manner suitable to accomplish an even distribution of sludge.
4. Facilities for decanting of supernatant should be provided, and the liquid should be recycled to the treatment facility.
5. A minimum of two cells should be provided, even in very small plants, to ensure availability of storage space during cleaning, maintenance, or emergency conditions.

8.7 Heat Drying

1. Sludge drying is a unit operation that involves reducing water content by vaporization of water to the air. Drying is necessary in fertilizer manufacturing so as to permit the grinding of the sludge, to reduce its mass, and to prevent continued biological action. The moisture content of the dried sludge should be less than 10%.
2. The following mechanical processes may be used for drying sludge:
   - flash dryers;
   - spray dryers;
   - rotary dryers;
   - multiple-hearth dryers; and
   - multiple-effect evaporation.

Sludge dryers are normally preceded by dewatering. Flash dryers are preferred in waste-water treatment plants.

- The heat recovered from the dried sludge should be used (if possible) to supply the energy requirements of the process.
- Fly ash and odor control should be considered in heat drying of sludge.

8.8 Thermal Reduction

Thermal reduction of sludge involves the total or partial conversion of organic solids to oxidized end products, primarily carbon dioxide and water, by incineration or wet-air oxidation or the partial oxidation and volatilization of organic solids by pyrolysis or starved-air combustion to end products with energy content. Thermal reduction processes are used by medium to large sized plants with limited ultimate disposal options.

The major advantages and disadvantages are:

Advantages:

1. Maximum volume reduction, thereby lessening the disposal requirements.
2. Destruction of pathogens and toxic compounds.
3. Energy recovery potential.

Disadvantages:

1. High capital and operating cost.
2. Highly skilled operating and maintenance staff are required.
3. The residuals produced (air emissions and ash) may have adverse environmental effects.
4. Disposal of residuals, which may be classified as hazardous wastes, may be uncertain and expensive.

8.8.1 Thermal Reduction Process Applications

8.8.1.1 Multiple-hearth Incineration

Multiple-hearth incineration is used to convert dewatered sludge cake to an inert ash. This process is normally used only in large plants or at the small plants where land for disposal of sludge is limited and at chemical treatment plants for the recalcining of lime sludges.

Feed sludge must contain more than 15% solids because of limitations on the maximum evaporating capacity of the furnace. Auxiliary fuel should be provided.

In addition to dewatering, the required ancillary processes should include ash-handling systems.

8.8.1.2 Fluidized-bed Incineration

The fluidized bed incinerator commonly used for sludge incineration is a vertical, cylindrically shaped refractory. It comprises a lined steel shell that contains a sand bed and fluidized air orifices to produce and sustain combustion. The applications of the fluidized-bed process are similar to multiple-hearth incineration.

8.8.1.3 Co-incineration

The major objective is to reduce the combined costs of incinerating sludge and solid wastes. The process has the advantages of producing the heat energy necessary to evaporate water from sludges, supporting the combustion of solid wastes and sludge, and providing an excess of heat for steam generation, if desired, without the use of auxiliary fossil fuels.

8.8.1.4 Wet-air Oxidation

This process involves wet oxidation untreated sludge at an elevated temperature and pressure. A major disadvantage associated with this process is the high-strength recycle liquor produced as the liquors represent a considerable organic load on the treatment system.

8.8.1.5 Wet Oxidation in a Vertical, Deep-well Reactor

This process consists of discharging liquid sludge in the pressure and temperature controlled environment of a tube-and-shell reactor suspended within a deep well. Advantages cited for this process are:

- low space requirements;
- high removal of suspended solids and organic matter;
- little odor or objectionable air emissions; and,
- low energy requirements because the process is exothermic.

The principal disadvantages are that it does not have a long history of operation and maintenance, and skilled operators are required for process control.

8.9 Land Application of Sludge

The following steps should be evaluated in development of a sludge land application system:

1. Characterization of sludge quantity and quality.
2. Review of pertinent federal, state, and local regulations.

3. Evaluation and selection of site and disposal options.
4. Determination of process design parameters such as loading rates, land area requirements, application methods, and scheduling.

The following factors should be considered in evaluation of the sludge characteristics:

- Organic content (usually measured as volatile solids).
- Nutrients.
- Pathogens.
- Metals.
- Toxic organics.

If regulations require, detailed sampling and analysis of sludge should be carried out to identify and characterize the sludge constituents so as to determine if the sludge is suitable for land application. Maximum annual loading rates should be prescribed as well as permissible cumulative loading rates, depending on whether the land is used for agricultural or non-agricultural purposes.

Sludge applied to the land surface or incorporated into the soil must be treated by a Process to Significantly Reduce Pathogens (PSRP). Sludge applied to land where crops for human consumption are grown must be treated by a Process to Further Reduce Pathogens (PFRP). Examples of PSRP stabilization processes are aerobic digestion, air drying, anaerobic digestion, composting, and lime stabilization. Examples of PFRP stabilization processes are composting, heat drying, and thermophilic aerobic digestion.

For site selection and evaluation the following factors should be taken into consideration:

- land application option(s), such as application to agricultural lands, forest lands, etc.;
- topography;
- soil permeability;
- site drainage;
- depth to groundwater;
- subsurface geology;
- proximity to critical areas;
- accessibility;
- maximum sludge loading rates based on the pollutant limits set forth in regulatory guidelines or on the nutrient loading rates necessary to meet vegetation requirements.

Application of sludge in the liquid state is attractive and dewatering processes are not required in this case. Liquid sludge may be applied to land by vehicle or by irrigation methods.

Liquid sludge can be injected below the soil surface by using tank wagons or tank trucks with injection shanks, or it can be incorporated immediately after the surface application by using plows or discs equipped with sludge distribution manifolds and covering spoons.

Special attention should be made to minimization of potential odors and vector attraction, minimization of ammonia loss due to volatilization, elimination of surface run-off, and minimum visibility leading to better public acceptance.

8.10 Chemical Fixation

The chemical fixation/solidification process is applied to the treatment of industrial sludge and hazardous wastes to immobilize their undesirable constituents. Stabilized sludge may also be disposed of in landfills.

The process consists of mixing untreated or treated liquid or dewatered sludge with stabilizing agents such as cement, sodium silicate, pozzolan (fine-grained silicate), and lime so as to chemically react with or encapsulate the sludge. The process may generate a product with a high pH, which inactivates the pathogenic bacteria and viruses.

8.11 Final Sludge and Solids Conveyance, Storage, and Disposal

The solids removed as sludge from preliminary and biological treatment processes should be concentrated and stabilized by biological and thermal means and reduced in volume in preparation for final disposal if required.

Environmental regulations and subsurface water pollution limitations should be considered in the final selection of disposal method.

8.11.1 Conveyance Methods

Sludge may be transported long distances by pipeline, truck, barge, rail, or a combination of these modes. To minimize the danger of spills, odors, and dissemination of pathogens to the air, liquid sludges should be transported in closed vessels such as tank trucks, railroad tank cars, or covered tank barges.

Stabilized, dewatered sludges may be transferred in open vessels, such as dump trucks, if permitted by the Company and allowed by the regulations. However, if the sludge is hauled long distances, the vessels should be covered.

8.11.1.1 Pipeline

In general, the energy requirements for long-distance transportation of untreated sludges with a solids concentration of more than 6% are likely to be prohibitive.

Adequate flow-rates in large diameter pipes should be maintained to reduce the possibility of grit accumulation, grease build up in unlined pipes, and other problems resulting from low flow conditions.

8.11.1.2 Truck

Trucking is the most flexible and most widely used method for transporting sludge. Either liquid or dewatered sludge may be hauled by trucks to diverse destinations.

Trucking dewatered sludge is usually the most economical method for small to medium sized treatment facilities.

8.11.1.3 Barge

Barge transport is generally economical only for large facilities treating wastewater flows in excess of 15 800 m^3/h or in locations where one barge can serve several plants. Barges can also be used for carrying dewatered sludge in containers.

Barging sludge for ocean disposal should be prohibited.

8.11.1.4 Rail

Rail transportation may be used for sludges of any consistency, but those with high solids content are transported most economically.

The use of rail transportation for small quantities of sludges or for short distances is not justifiable economically.

8.11.2 Environmental Considerations in Sludge Transportation

Environmental features such as air pollutant load, traffic, noise, etc. should be taken into consideration in sludge transportation. On a mass basis, the transportation mode that contributes the lowest pollutant load is piping. Next, in sequence, are barging and unit train rail transportation. the highest pollutant load is from trucking.

8.11.3 Sludge Storage

Sludge that has been digested aerobically before it is disposed of or used beneficially should be stored.

Storage of liquid sludge can be accomplished in sludge storage basins, and storage of dewatered sludge can be done on storage pads.

8.11.3.1 Sludge Storage and Basins

If the basins are not loaded too heavily, provision of an aerobic surface layer through the growth of algae and by atmospheric reaeration should be considered. Alternatively, surface aerators can be used to maintain aerobic conditions in the upper layers.

The number of basins to be used should be sufficient to allow each basin to be out of service for a period of about six months. The depth of the sludge storage basin should be 3 to 5 m.

8.11.3.2 Sludge Storage Pads

1. Where dewatered sludge has to be stored prior to land application, a sufficient storage area should be provided based on the number of consecutive days that sludge hauling could occur without applying sludge to land. Allowances also have to be made for paved access and for space to maneuver the sludge-hauling trucks, loaders, and application vehicles.
2. The storage pads should be constructed of concrete or bituminous concrete and designed to withstand the truck loadings and sludge piles.
3. Provisions for leachate and stormwater collection and disposal should be considered.

8.11.4 Final Disposal

Final disposal for sludge and solids that are not beneficially used usually involves some form of land disposal. Ocean disposal of sludge should be prohibited.

As in the case of land application of sludge, the regulations for other methods of sludge disposal require close attention and review when planning and designing sludge disposal facilities.

8.11.4.1 Landfilling

If a suitable site is convenient, landfill can be used for disposal of sludge, grease, grit, and other solids. Stabilization may be required depending on state or local regulations.

Dewatering of sludge is likely to be required to reduce the volume to be transported and to control the generation of leachate from the landfill.

In a true sanitary landfill, the wastes should be deposited in a designated area, compacted in place with a tractor or roller, and covered with a 30 cm layer of clean soil.

Special attention should be made to minimization of nuisance conditions such as odors and flies.

In selecting a land disposal site, consideration must be given to:

- environmentally sensitive areas such as wetlands, flood plains, recharge zones for aquifers, and habitats for endangered species;
- run-off control to surface water;
- groundwater protection;
- air pollution from dust, particles, and odors;
- disease vectors; and,
- safety as related to toxic materials, fires, and access.

Trucks carrying wet sludge and grit should be able to reach the site without passing through heavily populated areas or business districts.

8.11.4.2 Lagooning

1. Lagooning is economical if the treatment plant is in a remote location.
2. Excess liquid from the lagoon, if there is any, should be returned to the plant for treatment.
3. Lagoons should be located away from roads, highways, and dwellings to minimize possible nuisance conditions and should be fenced to keep out unauthorized persons.
4. Subsurface drainage and percolation should be investigated to determine if the underlying groundwater will be affected.
5. The lagoon shall be lined if excessive percolation is a problem or if regulations require leachate control.

8.11.5 Incineration

Incinerators are generally limited to situations where adequate land for disposal of raw wastes is limited, or where land disposal or other procedures are impractical.

The chief advantage of incineration is reduction of waste containing organic matter to a relatively small volume of inert material. The chief disadvantage is the high initial cost and operating cost.

1. Steam generation should be considered if feasible and economically justified.
2. Air pollution regulations for smoke, fly ash, or hydrocarbon content of stack gases must be considered in the design stage.
3. Incinerator feeds should be limited to solid or semi-solid paste-like materials of low moisture content. Slurry type sludges, such as oil–water separator sludge and tank

bottoms, can be dewatered and deoiled by centrifugation or filtration to produce a solids cake suitable for incineration.
4. Cake can be transported by belt, screw, or ribbon conveyors to a skip hoist which will elevate the cake to a charge hopper. Other solids, including relatively dry spent clay, trash, rubbish, and garbage in trucks can be charged directly to the feed hopper from a truck ramp or indirectly using clam shell equipment. Liquid wastes can be burned with appropriate burners in the firebox.
5. Special attention should be made to provide water scrubbers or cyclone separators if required by air pollution regulations.
6. A smoke monitor and alarm should be provided.
7. Stacks should be sufficiently elevated to disperse the gases and to avoid concentrations of contaminants, such as sulfur dioxide, at high ground levels.
8. The incinerators can be equipped with multiple or single hearths.

8.11.6 Ash Handling and Disposal

The ash may or may not be wetted by a water spray as it is removed. The ash can be dropped into a pit and removed by an elevating device to a hopper which is emptied into a truck when required. Ash may be used for road, fire bank, and tank foundation construction.

8.12 Disposal of Solid Waste

This section gives general descriptions and procedures for solid waste disposal and covers many aspects such as: types, hazard, non-hazard, siting, sources, segregation, reduction, resource recovery, treatment sludge concentration, and sampling equipment. It does not provide all of the detailed information that is needed for design and operation of solid waste system.

8.12.1 Types of Waste

1. Hazardous wastes.
2. Non-hazardous wastes.

8.12.1.1 Hazardous Wastes

Hazardous wastes cover the following units and complexes:

- Refinery waste.
- Petrochemical industry waste.
- Fertilizer industry waste.
- Chemical industry waste.
- Desalting plant and production unit waste.

8.12.1.1.1 Refinery Waste. The following general refinery wastes are considered hazardous:

- Leaded tank bottoms.
- Neutralization hydrofluoric alkylation sludge.
- Dissolved air flotation unit sludge.
- Kerosine filter cake.

Table 8.13 Composition of BPG.

Constituents	Percentage Dry Weight
$CaSO_4, 2H_2O$	90–95
Water of crystallization	19–20
Total phosphate (PO_4)	0.8–1.6
Total fluoride (F)	0.5–1.9
Sodium + potassium ($Na_2O + K_2O$)	0.5–1.0
Silica (SiO_2) and other insolubles	2–3
Organic matter	0.4–0.8
Toxic constituents, mg/kg dry weight except radioactivity	
Cadmium (Cd)	1.25–3.22
Chromium (Cr)	1.47–1.72
Copper (Cu)	4.00–4.76
Lead (Pb)	6.37–6.47
Manganese (Mn)	5.48–10.39
Radioactivity	29 p Ci/gm dry weight

- Lube oil filtration clays.
- Slop oil emulsion solids.
- Exchanger bundle cleaning solvent.
- API separator sludge.

8.12.1.1.2 Petrochemical Industry Waste.   Sources of solid wastes in petrochemical industries are as follows:

- Plastic, rubber.
- Waste catalysts.
- Activated carbon, Japanese acidclay, zeolite.
- Alkali waste (soda, amine) and waste acid.
- Vinyl chloride tar.
- Oil sludge.
- Tar, resin, pitch, sludge.

8.12.1.1.3 Fertilizer Industry Waste. Phosphatic fertilizer industry solid waste By-Product Phospho-Gypsum (BPG) is generated in the fertilizer industry. Only 20% of total BPG is used for producing ammonium sulfate, the remainder being released to landfill or lagooning in slurry form. The general characteristic of BPG and toxic metal constituents are given in Table 8.13.

8.12.1.1.4 Solid Waste from Chemical Industry Pesticides.   DDT and BHC (Benzen Hexa Chloride) are the most commonly used pesticides. Organo phosphorous and carbonate are other possible pesticides. In the process, wastewater releases highly corrosive substances (pH 1–1.7), containing DDT, sulfuric acid, and suspended solids. The wastewaters are neutralized with lime but the settled effluent still contains DDT.

1. In production of Benzen Hexa Chloride (BHC) about 30 m^3 wastewater per ton of BHC containing 0.4 BHC is released, which is highly corrosive and toxic (pH 2.6–3.0).

2. Lime neutralization and sludge drying in sand beds can be utilized. About 90% of BHC in wastewater goes into sludge.
3. Concentrated, unused pesticides should be stored, returned, or disposed off as toxic waste.

8.12.2 Siting of Hazardous Waste Facilities

8.12.2.1 Permit Requirements

Every facility that treats, stores, or disposes of hazardous wastes should obtain a permit from the environmental protection organization to approve the site selected for operation of the facilities mentioned.

Details of all aspects of design, operation, and maintenance must be described in the permit and show compliance with the applicable requirements.

8.12.2.2 Environmental Impacts

Environmental impacts include potential effects on ecological processes, on human health, and on the esthetics of the natural and cultural landscape. For any given facility, these impacts will depend on the volumes and types of wastes to be managed and on the design and operation of the facility.

Table 8.14 lists examples of potential environmental impacts that should be considered, taking into account both expected and possible accidental effects from each element of the facility's operation (e.g., transport, storage, treatment, incineration). Environmental impacts in the absence of a facility should also be considered if the facility will in fact help to remedy disposal practices currently in use that are more hazardous. The point should be made and factually supported. Safeguards to prevent or mitigate such impacts should also be kept in mind.

Table 8.14 Examples of possible environmental impacts.

Groundwater contamination
Surface water contamination
Air pollution
Leaks, spills, accidents
Destruction of wildlife habitat, natural areas, wetlands
Loss of any unique site features (e.g., archaeological)
Permanent contamination of site
Contamination of nearby crops, fisheries
Traffic congestion
Odors
Noise
Visual ugliness
Effects on character of community: magnet for other heavy industry, image as dumping ground

8.12.3 Non-Hazardous Waste

8.12.3.1 Drilling Fluid (Mud)

Functions of drilling fluids are:

- remove drilled solids (cuttings) from the bottom of the hole and carry them to the surface where they are removed;
- lubricate and cool the drill bit and string;
- control downhole pressure;
- deposit an impermeable wall cake on the well-bore wall to seal the formations being drilled so contaminants do not enter the mud and the fluid phase of the mud does not enter the formation.

8.12.3.2 Nature of Drilling Discharge

The solid concentration discharge ranges from 60–90% by weight when drilling a very shallow onshore well. The total volume of mud discharged may be 119 m^3 (1000 barrels) or less. Very deep onshore wells drilled over long period of time may generate up to 3570 m^3 (30 000 barrels) of mud.

For offshore wells the volume of mud discharged ranges from 119 m^3 to 595 m^3 (1000 to 5000 barrels) for a typical well.

8.12.3.3 Environment Effect

Environment effects of drilling discharge onshore include:

1. Excess soluble salts and high exchangeable sodium percentages are the major inhibiting effects of drilling muds to plants and soils.
2. Most drilling muds cause soil dispersion resulting in surface crusting. Proper treatment can minimize or eliminate these effects.

8.12.4 Sources, Segregation, Quantities, and Characteristics of Solid Waste in Refineries

There are numerous sources and types of refinery solid waste. Most wastes can be segregated at source for more efficient handling before reclamation or disposal.

Because of varied refinery configurations, each refinery will have problems specific to its location, process, and local regulations. Even considering the individuality of refineries, there are basic waste materials produced that are common to most refining processes and, therefore, can be handled similarly for resource recovery or disposal. Besides these basic wastes, there are other wastes with characteristics and in quantities that are not easily defined. A list of waste sources and classification of wastes is presented in Table 8.15. Obviously such a list cannot be all inclusive.

8.12.4.1 Sources and Classification

Some of the more easily defined waste materials found are listed by refinery unit source and classified into combustibles, non-combustibles, and biodegradables.

Table 8.15 Sources and characteristics of refinery solid wastes.

Source	Combustible	Classification Non-combustible*	Biodegradable
Refining Processes			
Crude oil storage	Wax bottoms	Sand, rust, slit	–
Product storage		TEL sludge, sand rust, silt	
Crude processing		Sand, rust, silt, salt, slop oil emulsions	
Thermal cracking	Separator coke		
Catalytic cracking		Spent catalyst	
Catalytic reforming		Spent catalyst	
Polymerization		Spent catalyst	
Alkylation		Corrosion products (sludge, tar)	
HF alkylation		Calcium fluoride sludge	
Asphalt manufacture	Asphalt drips	Asphalt emulsions	Emulsions, light solvent
Cooling	Coke fines, wax tailings		
Product treating	Acid sludge	Lead sludge	
	Adsorbents	Adsorbents	
Lubes and grease	Soaps	Clay	
Wax	Slops, drips	–	
Non-refining Operations			
Utilities: Steam generation	–	Boiler blow-down sludge	
Utilities: Feedwater treatment	–	Lime sludge	
Utilities: Cooling towers	–	Suspended solids	
Wastewater Pollution Control			
API separators	Oily sludges, Heavy hydrocarbon	Sand, silt	Oily coated inert solids
Air flotation	Oily froth		
Clarifiers	Biological flocculation	Flocs	Biological flocculation
Flocculation	Floc	Clay	
Filters		Filter cake	
Biological oxidation	Biological flocculation		Biological sludge
Air Pollution Control			
Bag filters		Catalyst dust	
Electrostatic precipitators		Catalyst dust	

*In many cases non-combustibles are associated with combustibles and the whole mass must be processed to remove the combustible before disposal of the inert residue.

8.12.4.2 Segregation of Wastes

By studying Table 8.15 and relating it to a particular operation, prudent segregation practices that can reduce the cost and effort of handling become apparent. The following categories for segregation are to be considered:

- oily process sludges;
- non-oily wastes;
- biological sludges;
- miscellaneous;
- sanitary wastes.

8.12.4.2.1 Oily Process Sludges. Oily sludge is the most difficult to handle. Oil sludges are obtained as sediments from storage tanks, crude desalters, sewer cleaning, vessel cleaning, oil–water separators, dissolved air flotation, lube oil processing, and alkylation. A very thorough study is required to determine which wastes are compatible for mixing prior to recovery.

8.12.4.2.2 Non-Oily Wastes. Non-oily wastes are much easier to handle. These wastes come from storm sewer cleanings, grit chambers, tank cleaning, cooling tower cleanings, water treating, and catalyst replacement. Used catalyst is the most likely material for resource recovery.

8.12.4.2.3 Biological Sludges. Biological sludges are those obtained from biological wastewater treatment plants. These sludges are concentrated by various methods, and when the quality allows, may be used as soil conditioners.

8.12.4.2.4 Miscellaneous Waste. Miscellaneous solid waste materials not related to oil processing can generally be disposed of in municipal landfills.

8.12.4.2.5 Sanitary Wastes. The sanitary sewage collection and disposal system should be segregated from the process wastewater system to eliminate the requirement to chlorinate the treated water effluent. Where refineries do operate sanitary waste treatment systems, adequate information is available in the literature to answer solids disposal questions.

8.12.4.3 Typical Quantities and Characteristics

Major sources of metallic constituents in refineries are API separator bottoms, dissolved air flotation units, waste biological sludge, storm slit, and waste fluid catalytic cracking (FCC) catalyst. Metallic content of wastes varies widely between refineries. The metals in solid waste are, zinc, vanadium, selenium nickel, mercury, lead, copper, chromium, arsenic, and cadmium.

8.12.5 Source Reduction Methods

For source reduction, modification of process and operating procedures should be considered. This is done to reduce the quantity of waste solids and to alter its characteristics. Some source reduction techniques are as follows.

8.12.5.1 Tank Cleaning

Variable angle mixers installed in storage tanks can be used in conjunction with selected solvents (such as, crude, light cycle oil, and water) to reduce the time, manpower, and cost of removing residual solids from storage tanks.

The selected solvent is added to the tank to be cleaned and the mixers are operated for 5 to 15 days in various positions typically ranging 30° from either side of the center line. This sweeps all parts of the tank floor and lifts the solids so that solvent and oily solids are in intimate contact. The resulting deoiling of the solids recovers valuable hydrocarbon for reuse. It also reduces the oil content of the residual solids making them less difficult to dispose of ultimately and reduces the quantity of residual solids by removing the oil and wax content of the solids. An additional advantage of this technique is the greater safety afforded by conducting the major portion of the cleaning activity from outside the tank.

8.12.5.2 Biosludge from Water Treatment

Several techniques are available to reduce the quantity of waste biosludge from wastewater treatment.

8.12.5.2.1 Sludge Age. The sludge age in the biological treatment system can be increased. To accomplish this without creating sludge settling problems usually requires careful pre-treatment to remove colloidal material in the wastewater entering the bio-unit. Removal may be accomplished by use of a well-operated air flotation or sand filter system. The sludge age can usually be increased sufficiently to reduce sludge wastage to very low levels.

8.12.5.2.2 Aerobic Digestion. Aerobic digestion of the waste biosludge will result in a significant reduction in the quantity of sludge. It is also a necessary pre-treatment step for landfarming of biological sludge to prevent odor problems.

8.12.5.2.3 Hydrolysis. Other methods of reducing waste biological sludge include chemical treatment, such as acid treatment, to break down the biological cell wall. Thus, organics can be oxidized and the resultant biological sludge is more easily dewatered for ultimate disposal.

8.12.5.2.4 Air Flotation Float Recycle. Air flotation units are usually operated with chemical additives to improve their efficiency. Alum, ferric chloride, lime, and polyelectrolytes are typical chemical agents used. If possible, the chemical usage should be limited to polyelectrolytes since alum, lime, and ferric chloride type treatment creates a significant volume of oily solids for disposal. The use of polyelectrolytes alone opens the possibility of recycling the float, which is now primarily oil, to the front of the system thus reducing the quantity of solid material for disposal.

8.12.5.2.5 Beneficial Combination. Some wastes may be combined to advantage. For instance, lime from boiler feed water treatment can be used for pH control on landfarm operations.

8.12.5.2.6 Shut Down Planning. The environmental department should be involved in the planning for all types of shut downs so that the quantities and characteristics of wastes to be generated by the shut down are anticipated. The shut down planning should ensure that all possible steps are taken to keep quantities to a minimum, to control the characteristics of the waste to simplify its disposal, and to plan the disposal techniques to be used.

8.12.6 Resource Recovery and Waste Minimization

8.12.6.1 Resource Recovery

The main items that are recoverable and reusable include oil, catalyst, acids, caustics, digested biological sludges, filter clays, and, in some cases, chromate inhibitors used in cooling tower treatment. Almost every refinery can identify recoverable items from waste generated by processes that are individual to that refinery.

8.12.6.1.1 Oil Reclamation. All refineries practice the recovery of waste oils to some degree. The recovered oils may be reintroduced into various process feed streams or sold directly for fuel or other uses. The techniques for recovery include simple gravity separation, emulsion breaking with chemicals and heat, and the use of lighter oils and solvents to aid in emulsion breaking and thinning of heavier oil fractions. After the oil is recovered, the residual waste is amenable to landfarming as the ultimate method of disposal. If a sludge is to be landfarmed, oil should be recovered until the oil content is no more than 15%. Since oil concentration is a limiting factor for the quantity of waste to be applied to a plot of land, the less oil a waste contains the smaller the land area required for landfarming.

8.12.6.1.2 Catalyst Reuse. The metal content of many catalysts is frequently of sufficient quantity to allow the catalysts to be reprocessed for resale or metal recovery.

Most spent catalysts are stable enough that ultimate disposal can be accomplished by landfilling. It is good practice not to mix spent catalyst with other wastes in a landfill. In many cases, landfills become acceptable means for disposal if they are dedicated completely to a particular waste material.

8.12.6.1.3 Acids and Caustics. Waste materials should be neutralized prior to disposal. Reacting waste chemicals in this manner reduces the amount of fresh chemical required for neutralization and can be defined as a reuse process.

Neutralized acid and caustic solutions can be discharged into the wastewater treatment system whenever the total dissolved solids limit is not exceeded. Spent alkylation acid can be returned to sulfuric acid manufacturers for reprocessing.

8.12.6.1.4 Biological Sludges. Biological sludge should not be used for growing food products but waste sludge may be spread in tank farm areas.

8.12.6.1.5 Filter Clays. Filter clays used for oil and wax purification can be regenerated in multiple-hearth furnaces for reuse. A particular clay will withstand several regeneration cycles before particle size and reaction surface deteriorate. Waste filter clay is non-reactive and is amenable to landfill as an ultimate disposal method.

8.12.6.1.6 Chromate Inhibitor Reuse. Chromate continues to be one of the most effective corrosion inhibitors for cooling systems. Because of restrictions, the reduction of chromium content in wastewater has become imperative. Chromate can be recovered from cooling tower blow-down for reuse in the make-up water. This is accomplished by the ion exchange process. Stringent wastewater regulation has made this reuse process more attractive.

The refinery engineer must always consider reuse of waste materials in the design and development of refining processes. Maximum resource recovery might be the difference between success or failure for a particular project.

8.12.6.2 Minimization of Waste Generation

Waste minimization is very often economically beneficial for an industry and also results in improved environmental quality. The following requirements should be considered:

8.12.6.2.1 Design and Operating Features for Waste Minimization

1. **Integrated units**
   Separated components of crude can pass from one unit to the next with little or no intermediate tankage. This reduces the amount of sludge generated from intermediate storage capacity.
2. **In-line blending**
   95% of all prime fuels can be blended into a pipeline as opposed to blending into tankage. Sludge generated from finished product tankage will be minimal.
3. **Crude tank mixers**
   The crude oil tanks at the refinery should be equipped with mixers. The use of mixers prevents the deposit of settleable solids and thereby reduces the frequency for cleaning.
4. **Air cooler maximization**
   Over 70% of the cooling in the refinery should be through air fans as opposed to water-cooled heat exchangers. This reduces potential contamination of cooling water with process fluid. It reduces the volume of cooling water make-up and cooling water in circulation. In addition, this also minimizes the generation of heat exchanger tube bundle cleaning sludge by substituting air coolers for heat exchangers.
5. **Use of demineralized clarified river water**
   This technique reduces the cooling blow-down stream, thereby minimizing the generation of cooling tower treatment sludge.
6. **Closed cooling water system**
   This concept uses a cooling tower system to recirculate the same volume of water for cooling purposes. This further minimizes the volume of water that could be contaminated with process fluids, and the blow-down of significant quantities of cooling water treatment chemicals, e.g., Cr and Zn.
7. **Amine degradation prevention**
   In a refinery that uses DEA (Di-Ethanol Amine) to absorb H_2S and CO_2 acid gases from sour gas streams, DEA forms certain non-regenerable compounds requiring occasional disposal of amine and consequent replacement with fresh amine. To reduce amine disposal, the refinery should use continuous slip-stream filtration in addition to carbon filtration to remove degradation products. In addition to filtration, additives should also be injected into the amine to inhibit deterioration.
8. **Separate sewer systems**
   In refinery sewer design, segregation of all sources is recommended. All non-oily aqueous wastes can be transported via separate sewer systems. The oily aqueous wastes should be transported to the API separator. The segregated sewer design reduces the load on the API separator, thereby increasing its efficiency and minimizing potential formation of API separator sludge from solids of non-hazardous origin.

9. **Forebay skimming of the API separator**
   A slot skimmer should be installed at the head of the API separator to recover as much oil as possible from the separator influent for direct recycling to the refinery. This technique is an oil recovery/recycling process that reduces the oil loading on the separator and decreases the amount of oily sludge generation.
10. **Pressurized air operation of DAF unit**
    DAF (Dissolved Air Floatation) unit sludge generation is minimized by using the pressurized air method of operation, which concentrates the DAF sludge and minimizes the quantity generated and handled.
11. **Independent parallel bay separator design**
    The refinery's API separator should be in a three-bay parallel design with flexibility and capacity so that individual bays can be taken out of service for maintenance without affecting separator efficiency.
12. **Coke fines recovery/reuse**
    The delayed coking unit produces an anode-grade coke as a product. Coke fines are generated by the process on an intermittent basis. Instead of segregating the fines as a waste product, the refinery recovers this material and recycles it for sale along with the marketable product. This minimizes waste stream generation. In addition, the coke fine handling equipment design prevents coke solid introduction into the oily water sewer system and its eventual deposition at the bottom of the API separator as a sludge.
13. **Process catalyst reuse, reclamation, recovery**
    The refinery utilizes catalyst wherever possible that can be reclaimed, recovered, or reused in the various processing operations.
14. **Spent naphthenic caustic reuse**
    A unit should be installed to process spent naphthenic caustics into naphthenic acids for sale as a byproduct.
15. **Slop oil recovery/reuse**
    All slop oils generated should be collected and directed to the crude unit and coker for recovery and production of fuels and petroleum products. This minimizes the sewering of oily materials and production of API separator sludge and slop oil emulsion solids.

Paving of process areas to minimize solids (mostly dirt) entering the refinery oily water sewer system should be undertaken. These solids would otherwise be removed from the API separator as hazardous waste.

Use should be made of the contract filtration services to dewater/deoil wastewater treatment sludges prior to land treatment. By reducing the volume and oil content of these sludges, further land treatment operation can be conducted in a smaller area and at lower oil loading rates and the oil is recovered/reused in the refinery.

An air operated skimmer should be at the head of the API separator to complement existing mechanical skimmer and further enhance oil recovery from the separator inflow. This results in further reducing the oil loading on the separator and subsequent formation and disposition of oily sludges that would be removed as hazardous waste.

The refinery should conduct an assessment of the various additives that can be added to heat exchangers to reduce the current deposit rates. These deposits would eventually be removed as hazardous waste.

258   Waste Management in the Chemical and Petroleum Industries

Sludge storage impoundments should be replaced with above-ground tankage. The tank system should include a mixing tank plus two storage tanks. The system will use heat and chemical addition to remove oil and water from the sludges as an initial step in reducing the volume and oil content of the wastes. These tanks will also eliminate water accumulation in the sludges from rainfall and volatile hydrocarbon air emissions from the sludges.

A filtration service dewaters biological solids generated by the secondary treatment system. The waste biosludge has a very high water content and is dewatered prior to disposal on the land treatment area. Instead of acting as a hydraulic gradient to promote contaminant migration through the treatment zone, the dry biosludge serves to augment the existing land farm microorganisms in waste assimilation and fixation.

The refinery should try to sell spent FCC catalyst for use in the manufacture of Portland cement, in lieu of landfill disposal.

An effort should be made to keep surfactant materials out of the sewers to minimize formation of oil-in-water emulsions that create oily sludges (e.g., API separator bottoms, slop oil emulsion cuff, and DAF float).

Secondary containment should be provided for all intermediate and final product transfer tankage pumps. This measure prevents potential soil contamination in case of a leak or spillage.

A dredge barge in the biological clarifiers improves biosludge recycling and minimizes sludge accumulation in the final polishing pond.

An oil skimming system should be installed in the oily stormwater retention pond. This pond serves as a surge pond for the API separator. The skimming system performs a recovery/recycling function by skimming oil for reprocessing. In addition, the system also reduces the oil loading on the separator and subsequent generation of API separator sludge (hazardous waste) and slop oil emulsion solids (hazardous waste).

A vacuum truck should collect catalyst fines which deposit around the process units during loading and unloading operations. This minimizes the amount of solids introduced to the sewer system. This in turn reduces the overall amount of hazardous (oily) wastewater solids generated at the API separator.

Desalter water should be rerouted from the API separator to intermediate tankage and then back to separator. This allows for oil skimming at the tank which minimizes emulsion carry-over to the separator, decreasing the generation of slop oil emulsion solids (hazardous waste).

A solvent still should be used to reclaim spent paint solvents (hazardous waste). Periodic clean-out of the oily water sewer system should be undertaken to reduce overall generation of hazardous waste at the API separator.

Covers for all oily water sewer openings in the process area should be placed to minimize the amount of solids introduced into the sewer system. This reduces the overall amount of hazardous (oily) wastewater solids generated at the API separator.

8.12.7   Hazardous Waste Reduction

The framework of reduction for control of hazardous waste covers is through:

- generators;
- transporters;

- treaters;
- stores;
- disposers.

8.12.7.1 Generator

The waste generator or handler must determine whether the waste is hazardous. Waste classified as solid hazardous waste must meet two criteria. The first criterion applied is whether the material is a solid waste and the second step is determining whether the solid waste is hazardous. Four characteristics of hazardous waste are shown in Table 8.16.

Generator owners are responsible for establishing standard operating procedures (SOPs), that when followed, satisfy regulation. Furthermore, they must ensure that employees follow the (SOPs) as a condition of employment.

These four characteristics are defined in 40 CFR 261 Resource Conservation and Recovery Act, U.S. Regulary Act (RCRA) 1974, and (HSW 4) Hazardous and Solid Waste 1984. Reference can be made to these sources.

For each hazardous waste, a multi-copy form the called manifest must be filled in, which accompanies waste upon leaving the facilities and continues with it to its ultimate disposal.

The tracking system for the manifest includes the name and address of transporter, treater, storer, or disposer; and a description of the wastes, including type, quantities, and official classifications. The manifest should contain certification that the generator is minimizing hazardous waste as much as economically practical and that the chosen method of disposal minimizes risk to human health and the environment.

8.12.7.2 Transporter Responsibility

Transport of hazardous waste should take every precaution that spills will not occur and report any spill as soon as possible.

8.12.7.3 Treatment, Storage and Disposal Facilities (TSDF)

8.12.7.3.1 Treatment. Any method, technique, or process, including neutralization designed to change the physical, chemical, or biological character composition of any hazardous waste so as to neutralize it or render it non-hazardous or less hazardous, or to recover it, make it safer to transport, store, or dispose of, or amenable for recovery, storage, or volume reduction, are steps that will reduce the quantity of wastes.

8.12.7.3.2 Storage. This is the holding of hazardous waste for a temporary period, and at the expiry time the hazardous waste is treated, disposed of, or stored elsewhere.

Table 8.16 Characteristics of hazardous waste.

Ignitability	RCRA	40 CFR	261.21
Corrosivity	RCRA	40 CFR	261.22
Reactivity	RCRA	40 CFR	261.23
Toxicity	RCRA	40 CFR	261.24

8.12.7.3.3 Disposal. The discharge, deposit, injection, dumping, spilling, leaking, or placing of any solid/waste or hazardous waste into any land or water so that any constituent thereof may enter the environment or be emitted into the air or discharged into any water, including groundwater, are all factors to be considered for prevention and minimizing of environmental pollution.

The procedures can be divided into two categories: standard of performance and permit requirements. Performance standards shall be per RCRA ACT 40 CFR part 264/265 subparts A through E. The permit requirement shall be per RCRA ACT 40 CFR 265.

The general provisions that apply to all treatment, storage, and disposal facilities are as follows:

1. Waste analysis to ensure accurate information concerning properties of the waste for adequate treatment, storage, or disposal.
2. Security measures to prevent unauthorized entry with resulting injury or environmental damage. Inspection to assess the facility and its potential problems.
3. Training of facility personnel to reduce mistakes that could threaten human health, safety, and the environment.
4. Compliance with government standards that include flood plain, earthquake, and hydrological considerations.

Land disposal facility design should be such as to prevent groundwater pollution. New landfills should include double liners and leakage-collection systems. Groundwater quality should be monitored to ensure the protective systems are effective, and corrective action must be taken when the systems fail.

8.12.8 Treatment Prior to Ultimate Disposal

8.12.8.1 Concentration

The rationale for concentrating solids is easily understood when considering that the doubling of solids concentration results in halving the volume for ultimate disposal. Increasing the concentration from 1 to 2, or from 2 to 4%, is especially important because of the large reduction in volume.

Because of new effluent regulations, wastewater treatment sludges have become the single largest source of sludge. These sludges are difficult to process because they are often gelatinous or oily. Furthermore, the physical properties of wastewater sludge may change frequently because of turnarounds or changes in refinery operations.

Selection of the concentration process depends on the characteristics of the sludge and the ultimate disposal methods available at the specific location.

8.12.8.2 Sludge Concentration by Gravity Differential Method

8.12.8.2.1 Air Floating Unit. After removal of the majority of free oil in an API separator, the remaining 50–100 ppm free oil, together with colloidal emulsions and suspended solids, may be further reduced in a Dissolved or Induced Air Flotation unit (DAF or IAF). These units typically concentrate sludge to 4–5 weight percent.

The basic principle of either DAF or IAF is that air bubbles attach to the suspended oil or solids causing the particles to float to the surface where they can be skimmed. For air floating unit specification refer to the API "Manual on Disposal of Refinery Waste."

8.12.8.2.2 Chemical Flocculation. Chemical flocculation is another pre-treatment for removing oil and solids. For more specification refer to API "Manual on Disposal of Refinery Waste."

8.12.8.3 Sludge Concentration by Mechanical Dewatering

8.12.8.3.1 Centrifugation. Three types of centrifuge are used for processing high solids content sludges: scroll, imperforate basket, and disc. Scroll centrifuges are frequently called solid bowl centrifuges; this is confusing since imperforate basket centrifuges also have a solid bowl.

Oily wastes containing a wide range of solids, oil, and water can be processed by centrifugation. Units have also been designed to process emulsions and oily sludges with high solids content, such as separator and tank bottoms. For three types of centrifuges specification refer to the API "Manual on Disposal of Refinery Waste."

8.12.8.3.2 Filtration. The various types of mechanical filters that are generally used to concentrate sludges that have been settled or floated by gravity are:

1. **Pressure filtration**
   A pressure filter should consist of metal plates covered by a fabric filtering medium. The covered plates should hang in a frame equipped with both a fixed and a movable head. The plates should be forced together with a chamber left between the cloth surfaces. Sludge will be pumped through a central opening in the plates to the cloth-lined chamber. Sludge is retained on the fabric and liquid is forced through the fabric to the plate surface where it drains away. At the end of the filtration cycle, the plates are separated, and the sludge cake is discharged from the unit. Feed pressures ranging between 5.5 and 15.5 bar (80 and 225 psi) gauge are common.

   Pressure filters of the plate and frame type should not be used for filtration of oily sludges or oily slurries because the filter media blocks rapidly. This severely limits the filtration rate. If spent inert solids, such as lime sludge, are available, they can be mixed with the waste to provide porosity and improve the ease of removal of the cake from the filter media.
2. **Vacuum filtration**
   A typical filter installation consists of a horizontal compartmented drum, which supports the filter media on its outer surface. The drum is rotated, partially submerged, in a tank containing the waste sludge. As each section of the drum passes through the tank, vacuum pulls the liquid inward and the solids are retained as a thin cake in the filter media covering the outer drum surface. For more detail refer to the API "Manual on Disposal of Refinery Waste," 1980.

8.12.8.4 Sludge Concentration by Incineration

Incineration is the ultimate in volume reduction. It results in an ash which must be landfilled. Gases are passed out of the stack where particulates and acid vapors (CO_2, SO_2, and HCl), if present, are removed. Hydrocarbons present in the sludge reduce the amount of auxiliary feed required. The chief disadvantages are the high capital and operating costs.

These costs, per ton of waste incinerated, are especially high for small incinerators and for large incinerators if not used to their full capacity. Therefore, in evaluating the need and economics for incineration, three basic steps are required. First, make an accurate estimate of the amount and characteristics of the waste to be incinerated. Second, evaluate alternative disposal methods and types of incinerators. Finally, make an economic and environmental comparison of the alternatives.

1. **Major types of incinerator**
   For brief description of major types of incinerator and item application refer to the API "Manual on Disposal of Refinery Waste," Sections 6.4.1 and 6.4.8.
2. **Incinerator design considerations**
   Incinerator design requires close cooperation with the incinerator vendor after the characteristics and amount of sludge are known. Specific design considerations depend on the type of incinerator being considered. General factors include:
   ○ The quality of the feed, including variations in composition, production rate, and combustion characteristics. For example, will the feed to a fluid bed incinerator form a low melting eutectic which will bog a fluid bed or will corrosive fumes attack metal rabble arms.
   ○ Control of incineration temperature to ensure complete and safe combustion with minimum maintenance and operating costs. The need for auxiliary fuel, the control of excess air, turbulence, or mixing heat transfer and residence time in the combustion zone must be considered.
   ○ Control of air pollution due to incomplete combustion.
3. **Pyrolysis**
   Pyrolysis is heating of the waste in the presence of less air than would be required for complete combustion. By doing this some of the hydrocarbon is utilized and some may be partially oxidized.

   These materials along with complete products of combustion are burned in an afterburner for heat recovery, or some material may be collected by condensation as a liquid for subsequent use as a heat or hydrocarbon source. Regeneration of activated carbon is also a pyrolysis process with limited oxygen being used to prevent complete destruction of the carbon.

8.12.8.5   Land Treatment of Hazardous Waste Description

Land treatment is a waste treatment and disposal process whereby waste is mixed with or incorporate into the surface soil and degraded, transformed, or immobilized.

Land-treatment technology, site, and waste evaluation, design and operation criteria provide the potential environmental effects of land treatment.

8.12.8.5.1 Technology

1. **Waste characteristic**
   Biodegradable wastes are suitable for land treatment. The chemical structure, molecular weight, water solubility, and vapor pressure are characteristics that determine the ease of biodegradation.

2. **Soil characteristics**

   The principle soil characteristic affecting land treatment processes are pH, salinity, cation-exchange capacity, redox behavior, texture, aeration, moisture holding capacity internal drainage, and soil temperature.

3. **Waste degradation**

   The factors affecting waste degradation may be adjusted in the design and operation of a land treatment facility, as described here:
   - The optimum pH for bacterial growth is near 7. This pH is generally maintained by liming to promote microbial activity and to immobilize heavy metals in the waste.
   - The soil moisture content should be maintained above a certain minimum usually between 30 and 90% of the water holding capacity of the soil.
   - The soil temperature should be maintained above 10 °C.
   - Nutrients and additional nitrogen must be supplied when highly carbonaceous waste is applied.

8.12.8.5.2 Site Characteristics. Considerations for land treatment sites should include regional and site geology, hydrology, topography, soil, climate, and land use.

Land treatment facilities should not be located in an aquifer recharge zone or within 60 m of a fault which has had displacement since the Holocene Epoch. Current standards require the treatment zone to be less than 137 cm deep and at least 90 cm above the seasonal water table. Although topography can be modified to some extent by grading, the site should not be so flat as to avoid pounding nor so steep as to cause excessive erosion and run-off. The site should not be subject to flooding or washout.

As a treatment medium, the soil should be evaluated in terms of its assimilative capacity (retention and degradation), and the potential for erosion and leaching of hazardous waste constituents. In general, land treatment facilities should not be established on deep, sandy soil because of the potential for groundwater contamination. Silty soils with crusting problems are undesirable because of the potential for excessive run-off. Generally, soils suitable for land treatment are loam, sandy clay loam (ML, SC), silty clay loam, clay loam (CL), and silty clay (CH). Soil properties that should be evaluated include soil depth, texture, drainage, pH, organic matter, soluble salts, cation-exchange capacity, moisture holding capacity, and microbial counts.

Wind, temperature, and rainfall are three variables of climate that are generally considered in site selection. Careful design and good planning can overcome most climatic constraints.

Existing and future land use for the site of the land treatment facility should be evaluated in the planning stages. Zoning restrictions, potential effects on environmentally sensitive areas, proximity to existing or planned developments, and effects on the local economy should all be considered.

8.12.8.5.3 Advantages and Disadvantages of Land Treatment. The advantages of land treatment are summarized as follows:

- The waste is degraded, transformed, or immobilized; thus, the long-term liability is lower than for other land disposal options.
- The treatment area is continually monitored; thus, remedial action can be taken immediately when there are signs of waste migration from the treatment zone.
- The cost for land treatment is lower than for landfills and incineration.

The disadvantage of land treatment is:

- There is potential for adverse environmental impacts if the facility is not properly designed and managed.

8.12.8.5.4 Design Considerations

1. **Area and Facility Layout**
   Land treatment generally requires the commitment of a large parcel of land to accommodate the treatment area, buffer zone, roads, waste storage, equipment storage, retention pond, and on-site structures. For this reason, the area available plays a significant role in the layout and in many design and operating features of the facility.
   A few salient features are shown in Table 8.17.
2. **Potential environmental impacts**
   Soil, a natural acceptor of wastes, has been viewed as a physical, chemical, and biological filter that can effectively deactivate, decompose, or assimilate a wide range of waste materials. Factors affecting this assimilative capacity must be considered in order to develop sound systems for land treatment. Without a detailed analysis of land treatment components and a sound design and management plan, adverse environmental effects can result from land treatment practices.
3. **Water quality**
   Operators of land treatment facilities must adequately protect the surface and subsurface waters from contamination with hazardous waste constituents or byproducts from waste degradation. Waste constituents that are not degraded by microorganisms, transformed, or immobilized in soil may leach into the groundwater. Run-off resulting from excess waste application or heavy rains may carry the constituents and contaminated sediments to nearby streams and lakes.

   Incorporation of wastes into the surface soil may result in soluble constituents moving downward by leaching. Thus, it would be a race between mobility and degradability of organic waste constituents. Under regulations, the seasonal groundwater table beneath a land treatment facility can be as shallow as 2.40 m below grade, or about 90 cm below the treatment zone. In some cases, an impermeable layer with a leachate collection system placed a meter or so below the treatment zone can prevent waste constituents or their metabolites from contaminating the groundwater.

   Surface water contamination resulting from improper water or damaged diversion structures and retention ponds is probably a major environmental concern. In a land

Table 8.17 Design and operational considerations.

Land requirement	pH adjustment and nutrient supply
Facility layout	Frequency and rate of waste application
Access road	Application and mixing methods
Equipment selection	Contingency plan
Run-on and run-off control	Vegetation
Erosion control	Site security and inspection
Odor and air-emission control	Site monitoring
Waste storage	Record keeping
Land preparation	Closure and post-closure care

treatment operation, wastes are concentrated at the soil surface; the concentration of waste constituents in the run-off water may be sufficiently high to have deleterious effects on certain trophic levels in the aquatic ecosystem. Waste constituents may be transported as particulate, dissolved, or bound on eroded soil particles.

4. **Air quality**

   Air quality at a land treatment site can be impaired by odors, dust, volatile substances, and aerosols emitted from waste spreading and incorporating processes. Many industrial hazardous wastes contain heavy metals and organic substances that are highly toxic in particulate or volatile form. The level of volatile compounds present in the air is not necessarily related to the severity of the odor problems. There is insufficient data on adverse effects of land treatment practices on air quality. For air monitoring at waste management facilities and for worker protection refer to ASTM-D4844-88.

8.12.8.5.5 Landfill of Hazardous Waste. Landfill contains isolated wastes that are not recoverable to present long-term environmental protection. Disposal to landfill sites of oily wastes, such as sludges obtained from the bottom of separator systems, needs the selection of a suitable location.

Approval of the relevant authorities is necessary before the selection of a site for oily waste disposal. Oil may leach from the site into groundwater, if the rock strata are porous or fissured this could lead to contamination of the drinking water supplies. In addition, the site has to be selected such that the surface water run-off is either contained or intercepted. The oil must be further stabilized to minimize its mobility. This can be achieved in a number of ways including co-disposal. For co-disposal there should be at least 4 m of mature domestic refuse between the oily waste level and the base of the landfill. Maturing of the waste takes 3–4 years, by which time most of the putrescible and solid organic matter has degraded. In an emergency the oily waste should be dumped on the oldest available refuse.

For high level wastes, e.g., oil or emulsions containing 25% oil and 75% water, the mixing rate should not exceed 5% of the domestic refuse volume. For low level waste, i.e., leached material with an oil content less than 5% by volume, the loading rate should be less than 30% of the underlying domestic waste volume. Ditches or strips for high level wastes should not exceed 300 mm in thickness (preferably 100 mm), and for low level wastes 500 mm is the maximum thickness; 2 m layers of refuse should cover any oily waste layers.

The addition of fly ash or quicklime (CaO) can also be used to render oily sludge inert, and this can be accomplished prior to disposal at a landfill site. Fly ash, the residual ash powder from coal-fired power stations, has a very high surface area and stabilizes the oil by adsorption. The fly ash and sludge may be mixed on site using impellers or screw mixers. The fly ash is added as a polymer-stabilized slurry and the resulting sludge/fly ash mix must be dewatered before dumping. An alternative and less expensive method is to mix the sludge and fly ash at the disposal site using conventional bulldozers and diggers available on site. Quicklime can be used in a similar way to produce a friable, inert material which on compaction can be used as a base infill for low-grade roads. Sludges with oil contents of 5–10% require 15% quicklime addition, while higher oil contents can be accommodated simply by adding more quicklime. Waste oils can be used for spraying on dirt roads to hold down the dust.

8.12.9 Disposal of Waste Generated in Drilling Wells

8.12.9.1 Onshore Drilling

A reserve pit (sump) is used to store drilling mud and cuttings to serve as the means for final disposal. The walls of the reserve pit must be high enough to provide 1 to 1.5 m of native topsoils on top of the mud and cuttings after backfilling. There are three major methods of disposal.

1. Dewatering the pit wastes with subsequent backfilling using the pit walls.
2. Landfarming the wastes into the surrounding soils.
3. Transfer by vacuum truck removal to an approved disposal site.

8.12.9.1.1 Backfilling. Backfilling a reserved pit is a common method employed. Before the backfilling operation begins, it is first necessary to remove the top aqueous layer. After the oil is skimmed, the aqueous layer can be divided by mixing the pit area with organic flocculents such as polyacrylamide, or inorganic ones such as gypsum. The flocculation should be as complete as possible. When this process is over the aqueous layer should be removed. Sometimes it is allowed to evaporate, which can take considerable time. In some cases the wall of the pit dike is cut and the fluid driven out. An analysis should be conducted to ensure that the release of water meets the guidelines of environmental protection and requirements of the system.

The aqueous layer can also be removed by vacuum truck and injected:

1. into the well that was drilled (if it is plugged and abandoned);
2. into the drilled well's annulus (if it is completed as a producing well);
3. or it can be transported to a nearby injection well.

After the aqueous layer has been removed, the actual backfilling at the reserve pit can be performed.

8.12.9.1.2 Landfarming. The second major disposal method utilizes landfarming techniques. Landfarming essentially consists of spreading the contents of the reserve pit evenly over the drilling location with subsequent incorporation into the soil using basic soil tilling equipment. Landfarming is especially useful for wells that produce hydrocarbons.

8.12.9.1.3 Vacuum Truck Removal. In this method both the aqueous and the solid phases of the reserve pit should be pumped to trucks while dirt moving equipment "squeezes" the pit dikes together.

8.12.9.2 Offshore Drilling

In offshore drilling the amount of oil discharged is usually restricted. Different disposal techniques are as follows:

Feasible alternatives:

- Shunting: The drilling mud and cutting through a pipe extending below the surface of the water (usually near to the sea floor) should be released. Shunting minimizes the physical transport of the wastes and environmental control.

Impractical alternatives:

- transport to an ocean dump site;
- transport to a land disposal area;
- pipelining to another area.

8.12.10 General Sampling Considerations

Sampling equipment must be selected that is chemically compatible with the type of waste and type of analyses.

Generally, plastic sampling equipment is not suitable for waste containing, or to be analyzed for, organic parameters. Stainless steel, glass, and plastic are acceptable for most samples to be analyzed for inorganics. It is up to the user to ensure that the equipment will not contaminate or bias the analyses.

The sampling equipment must be capable of extracting a sample from the desired location, depth, or point and at the same time provide protection from cross-contamination during sampling. For instance, one very common problem is extracting a sludge sample from beneath a top layer of wastewater or sludge without contaminating the sample with the overlying wastewater or sludge. This situation, as well as many others, requires special equipment. The collector is therefore required in many instances to fabricate the necessary equipment.

Recommended sampling procedures are for collection of samples from the edge of ponds or lagoons or from piers or catwalks. Sampling from boats is not recommended and should be attempted only if the collector knows the waste poses no real health problem and every possible safety measure has been taken.

Tanks and drums containing unknown substances also pose potential health risks for the collector due to the possibility of fire, explosion, or the release of deadly gases upon opening. In these situations only spark-proof, remote opening devices should be used and only fully trained and experienced personnel should attempt to do this.

Samples that have not been properly mixed or were mixed in such a way that constituents such as volatile organic compounds were destroyed or driven off are not truly representative. Some of the most difficult types of samples to composite are cohesive fine textured solids and sludges. Lime-neutralized metal finishing wastes are an example of these wastes. Because of the sticky cohesive nature of these wastes, mixing is very difficult in the field. It may be best to have the laboratory mix the sample. Depending on the type of analysis required, it may be possible to slurry and mix the sample by the addition of water without affecting the analysis. The laboratory personnel should be consulted before attempting this. If the composite has not been appropriately mixed prior to shipment, this should be clearly indicated on the sample.

The volume of the sample required depends on the type of analyses needed and the testing equipment and procedure to be followed. There are two rules-of-thumb that can be applied.

1. If low chemical concentrations of a parameter to be tested are expected, take a large volume of sample.
2. If high chemical concentrations of a parameter to be tested are expected, smaller volumes will suffice.

When possible, it is recommended that sampling proceed from the least contaminated to the most contaminated areas to reduce the problem of cross-contamination.

8.12.10.1 General Planning of Waste Sampling

The analysis and testing of solid waste requires the collection of adequately sized, representative samples. Wastes are found in various locations and physical states.

Each sampling routine must be tailored to fit the waste and situation. Wastes often occur as non-homogeneous mixtures in stratified layers or are poorly mixed to form a mass. For example, wastes are commonly stored or disposed of in surface impoundments with stratified or layered sludges covered by ponded wastewater. In these situations, the collector may be faced with sampling the wastewater, the sludge, and some depth of soil beneath the sludges. Collecting representative samples in these situations requires a carefully assessed, well-planned, and well-executed sampling routine.

8.12.10.1.1 Significance and Use.   The procedures covered in this guide are general and provide the user with information that is helpful for writing sampling plans, safety plans, labeling and shipping procedures, chain-of-custody procedures, general sampling procedures, general cleaning procedures, and general preservation procedures.

8.12.10.1.2 Sampling Plans.   A sampling plan is a scheme or design to locate sampling points so that suitable representative samples descriptive of the waste body can be obtained. Development of sampling plans requires the following:

- Review of background information about the waste and site.
- Knowledge of the waste location and situation.
- Decisions as to the types of samples needed.
- Decisions as to the sampling design required.
- Background data on the waste is extremely helpful in pre-assessment of the waste's composition, hazards, and extent.
- (Note: background information is needed to determine the necessary safety equipment, safety procedures, sampling equipment and sampling design, and procedures to be used. For more details see ASTM D4687.)
- The waste's location and site conditions greatly influence a sampling plan. The most common waste locations include lagoons, landfills, pipes, point discharges, piles, drums, bins, tanks, and trucks. The site conditions include the physical condition of the waste – is it a solid, liquid, or gas – and describes the condition in which it was disposed.

1. **Considerations**

    The collector must obtain samples based on these considerations.

    The types of samples that may be collected are either composite or single. The sample collector must sort the samples and provide representative samples.

    A composite sample is a well-mixed collection of subsamples of the same waste taken from different points. A composite sample can be used to determine an average measure of a parameter, and is taken when differences in the waste exist.

    A single sample is a well-mixed sample taken from a single point. It is used to measure a particular parameter or parameter set at a given point or within a unique homogeneous layer or throughout the strata at one or several locations.

Sampling plans or schemes should be prepared well in advance of sampling. The most common sampling schemes involve the selection of sampling points, coordinate systems, or grid systems.

2. **Coordinate sampling system**

   This system uses a one or two coordinate system and involves collecting samples at random points from the origin of the coordinates. Random numbers can be generated using the random number tables available in most statistic texts. The origin of the coordinate system is normally placed at some corner of the site and marked off in steps, centimeters, meters, etc., for sampling landfills, waste piles, and lagoons. For storage areas containing barrels, the numbers of barrels from the origin are often used as intervals along the coordinate. For sampling from a flowing stream, the origin may be taken as time-zero (start), and samples are collected at random time intervals over the period of interest.

3. **Grid system**

   This system also involves taking samples at regular intervals, grid points, along an imaginary grid system laid out over the site. The number of sampling points will vary with the size of the grid. Such sampling schemes are used when a statistically sound sampling program is required. They should be used only when the waste body is known to be homogeneous, or when the strata have been defined. If the waste is stratified, a separate grid system may be required for each stratum.

8.12.10.1.3 Sampling Equipment

1. **Composite liquid waste samples (coliwasa)**

   Coliwasa is a device employed to sample free-flowing liquids and slurries contained in drums, shallow tank, pits, and similar containers.

   The coliwasa consists of a glass, plastic, or metal tube and should be equipped with an end closure that can be opened and closed while the tube is submerged in the material to be sampled (Figure 8.2).

2. **Weighted bottle**

   This sampler (Figure 8.3) consists of a glass or plastic bottle, a sinker, a stopper, and a line that is used to lower, raise, and open the bottle. The weighted bottle samples liquids and free-flowing slurries. A weighted bottle with lime should be built to the specifications in ASTM Methods D270 and E300.

3. **Dipper**

   The dipper (Figure 8.4) consists of a glass, metallic, or plastic beaker clamped to the end of a two- or three-piece telescoping aluminum or fiberglass pole that serves as the handle. A dipper samples liquids and free-flowing slurries. Dippers are not available commercially and must be fabricated.

4. **Thief sampler**

   A thief (Figure 8.5) consists of two slotted concentric tubes usually made of stainless steel or brass. The outer tube has a conical pointed tip that permits the sampler to penetrate the material being sampled. The inner tube is rotated to open and close the sampler. A thief is used to sample dry granules or powdered wastes whose particle diameter is less than one-third of the width of the slots. Thief samplers are available at laboratory supply stores.

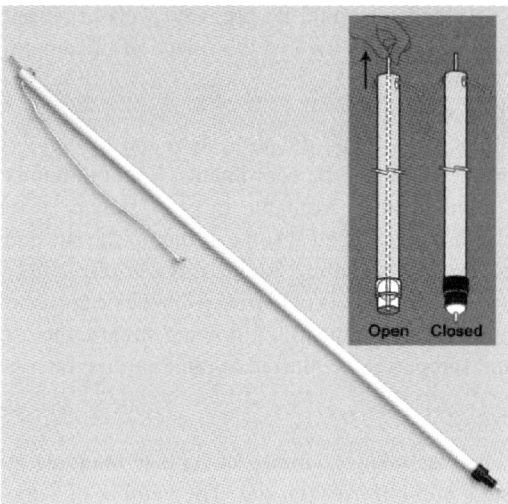

Figure 8.2 *Composite liquid waste sampler (Reproduced with permission © Enasco).*

5. **Trier Sampler**

   A trier (Figure 8.6) consists of a tube cut in half lengthwise with a sharpened tip that allows the sampler to cut into sticky solids and to loosen soil. A trier samples moist or sticky solids with a particle diameter less than one-half the diameter of the trier. Triers 61 to 100 cm long and 1.27 to 2.54 cm in diameter are available at laboratory supply stores. A large trier can be fabricated.

   This sampler in Figure 8.6, made from 304 stainless steel, is tapered for extracting a plug of material from solids (soil, clay) or semi-solids (grease, sludge). Length is 18 cm.

Figure 8.3 *Weighted bottle sampler (Reproduced with permission © Clarkson Lab).*

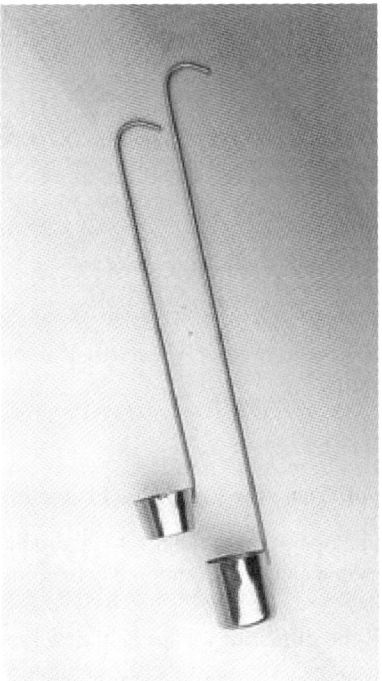

Figure 8.4 A dipper (Reproduced with permission © Caulfield Industrial, Shutterstock).

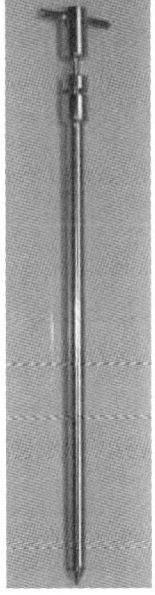

Figure 8.5 A thief sampler (Reproduced with permission © QAQC LAB, Shutterstock).

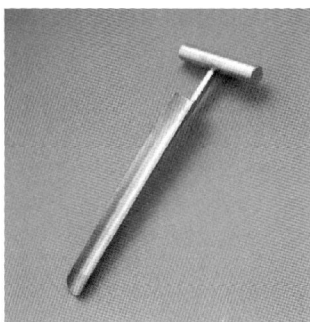

Figure 8.6 *A trier sampler (Reproduced from Wheaton with permission © Krackeler Scientific, Inc, 2013 (www.wheaton.com)).*

8.12.11 Air Monitoring of Waste for Employee Protection

This section applies to routine operations at an active treatment, storage, or disposal site or the extra ordinary conditions that can be encountered in opening and cleaning up a remedial action site.

The user must predict all the difficulties that could develop at the waste facility due to hazardous airborne transmissions. Although the air contaminant measurements obtained may indicate acceptable or tolerable levels of toxic agents are present, care must still be exercised before concluding that all atmospheric contaminants at the site are under control.

8.12.11.1 Significance and Use

The techniques of air monitoring are many and varied. This guide is intended to describe the standard approaches that are used in designing an air monitoring program to protect waste management site workers.

When entering a remedial action site to initiate an investigation or a clean-up operation, operating personnel may be faced with the extreme hazards of fire, explosion, and acute or chronic health hazards. A thorough health and safety program, including a site-specific health and safety plan, must be in place to direct worker activity. Details for such plans can be found in the OSHA Interim Final Rule for Hazardous Waste Operations and Emergency Response. Air monitoring is an integral part of such a program. Clause 12.6.4 describes sampling procedures which can be used to evaluate the airborne hazard potential so as to gain and maintain control over the situation at the site.

Upon obtaining readings at the site, a decision must be made as to whether conditions are under control or not. That decision will depend on the nature of the contaminants (toxicity, reactivity, volatility, etc.), the extent (area affected, number of workers, etc.) of the problem, and the level of worker protection available. Since all such parameters will be site specific, the necessary decision-making is beyond the range of this chapter.

This section does not include the monitoring of sites containing radioactive materials, nor does it cover general safety aspects, such as access to emergency equipment or medical support of emergency needs. These items should be covered in a health and safety plan.

8.12.11.2   Establishing a Test Protocol

Various combinations of equipment and sampling techniques are used in workplace air monitoring. The best monitoring program is one that combines accuracy with a timely response in a cost-effective manner.

The particular test protocol that is selected for an industrial hygiene study depends on the nature of the contaminants and the end purpose of the monitoring effort (that is, routine monitoring, searching for worst case exposure, looking for contaminant leaks in a process).

8.12.11.3   Selecting Specific Methods

The choice of sampling method is most often tied in with the analytical method. There may be no difference in the analytical work whether it is for a 15-min ceiling sample or a 7-h full day sample. If the analytical method has poor sensitivity, however, it may be necessary to increase the pump flow-rate for the short duration sample to make certain that sufficient sample is collected. Such fine adjustments must be worked out between the sampling personnel and the laboratory personnel.

A number of sources of information are available to describe the general methodology. Recommended Practice ASTM D1605 lists some of the classic methods that have been used when sampling for gases or vapors. The American Conference of Governmental Industrial Hygienists offers a publication that provides a review of newer equipment and methodology. The final combination of equipment and procedures is predicted on the precision, accuracy, and sensitivity needed to support the test protocol.

Once the goals and protocol for the sampling program have been set, the specific sampling/analytical method must be selected. Within the Annual Book of ASTM Standards, Volume 11.05 is dedicated to atmospheric analysis and to occupational health and safety issues. Some applicable methods from that reference are listed in Annex A1. Other sources of health and safety support include the NIOSH Manual of Analytical Methods and the OSHA Analytical Methods Manual. The specific equipment and sampling media for a particular set of air-borne contaminants are selected from sources such as these.

8.12.12   Procedures

8.12.12.1   Operating Site

The procedures described in this section apply to air monitoring activities at an operational waste treatment storage or disposal site. At an operating site, controls (work practices, engineering controls, and personal protective equipment) will be in place to minimize the exposure of workers to hazardous conditions. These are defined in the health and safety plan.

8.12.12.1.1 Knowledge of Materials.   Knowledge of the materials arriving at or present in an operating site is critical to the design of a sampling plan. If hazardous wastes are arriving, be sure that they are listed on the manifest. The results of waste sample analyses will also help to identify the contaminants of greatest concern in an incoming shipment. It is also likely that specific users of the disposal site will tend to be consistent in the types of wastes they send to the site based on the generating process and history of shipment. For example, paint manufacturers will most likely send mixtures of solvents, resins, and pigments, whereas plating firms will generally send alkaline sludge of heavy metal waste, and so on. Deviation from established patterns, however, is possible and should not be discounted in sampling plan design.

8.12.12.1.2 Worker Sampling. Of all the different techniques for workplace air monitoring, personal sampling of the worker's breathing zone is paramount. While some workers may be quite sedentary in an operations trailer at a control panel, others may be out covering all areas of the work site. For this reason, the assessment must be capable of following the activity of the worker.

The first order of personal monitoring is long duration Time Weighted Average (TWA) sampling. For an 8-h work shift, be sure that TWA samples are at a minimum of 7-h duration either as a single sample or a series of two or more samples. For any other work hour situation, the procedure is to sample for the duration of the shift less 1 h.

For workers handling organic wastes (for example, vapor degreaser solvent waste) the program will call for charcoal tube sampling with analysis for one or two of the chlorinated solvents most likely to be present in the waste. Such TWA monitoring, as well as the following information, will be repeated periodically to ensure that worker exposure is not increasing.

Another form of personal monitoring that will be carried out is for peak exposures. For example, 15-min ceiling samples might be taken while a set of containers is being opened to inspect or remove the contents. The same type of sampling might be done while pumping the contents of a truck into a holding tank. At these times, personal protective equipment (for example, respiratory protection) is often used to minimize worker exposure to vapors. Ceiling samples will help ensure that workers are using respirators with a high enough protection factor.

Ceiling samples may be the only form of monitoring for certain toxic agents. If a waste acid pickling solution were to come in from a steel mill for neutralization, it might be appropriate to sample for hydrogen chloride. In this instance, only 15-min samples would be of interest, because this is how exposure to HCl is controlled by health/- regulatory agencies.

New equipment has come into use to cover both TWA and peak sampling. Some personal dosimeters, worn by employees, give an overall average exposure and also record the instantaneous exposures of the worker during the day. These units, which are read out on a portable computer, are generally good for only one particular contaminant, though all the different types are read using the same computer. These might be very useful in monitoring a heavy equipment operator for carbon monoxide or a waste treatment plant attendant for sulfur dioxide.

Another concept to be considered in both the monitoring and health and safety plans is the additive effect of certain substances. Once this is done, eventual comparison with permissible exposure limits must be done using a safety factor. This safety factor is intended to take account of the possible effects of other similar compounds that are likely to be present but are not measured routinely.

8.12.12.1.3 Area Monitoring. A good complement to personal monitoring is fixed location area monitoring. This can be done with either sample collecting-type equipment, direct reading instruments, or specialized fixed-parameter monitors such as those described in Section 7.1.3.5. Area monitoring offers the advantage of potentially providing an early warning.

A combustible vapor meter in a solvent storage area can give a warning before an employee walks in to find a leak.

A carbon monoxide monitoring system around a pyrolyzer or incinerator can warn both the operator in the control room and workers in the loading area of a system upset.

An oxygen meter permanently mounted in a below-ground pit can warn an employee of an oxygen deficient atmosphere before he enters the confined space.

Direct reading colorimetric tubes offer a convenient means for obtaining a quick reading. Besides their suitability for qualitative checks, they also provide reasonable quantitative estimates. For more detail please refer to ASTM D4844-88.

8.12.13 Hazards

Proper safety precautions must always be observed when sampling wastes. Persons collecting samples must be aware that the waste can be a strong sensitizer and can be corrosive, flammable, explosive, toxic, and capable of releasing extremely poisonous gases. The background information obtained about the waste should be helpful in deciding the extent of safety precautions to be observed and in choosing protective equipment to be used.

Personnel should wear protective equipment when response activities involve known or suspected atmospheric contamination; when vapors, gases, or air-borne particulates may be generated; or when direct contact with skin-affecting substances may occur. Respirators can protect the lungs, the gastrointestinal tract, and the eyes against air toxicants. Chemical-resistant clothing can protect the skin from contact with skin-destructive and absorbable chemicals. Good personal hygiene limits prevents ingestion of material.

8.12.14 Quality Assurance Consideration

Quality assurance for solid waste sampling should include adherence to the sampling plan and safety plan and, in some cases, the use of quality control samples.

Four types of quality control samples relate to the quality assurance of field sampling:

1. field blanks,
2. split samples,
3. field rinsates, and
4. field spikes.

The selection of the type of quality control samples to be used should be made prior to the sampling event and be included in the sampling plan. The nature of the sampling, the intended uses of the data, and the material being sampled all impact upon the selection of quality control samples to be used in an event.

Definitions and Terminology

Abandonment Discontinuing use of a system component or components by removing them or rendering them inaccessible and inoperable.

Absorption Process of incorporation or assimilation by which one substance is physically taken into and included with another substance; for example, bacteria assimilating nutrients from effluent.

Activated-Sludge A flocculent microbial/mass, produced when sewage is continuously aerated. Sludge that has undergone flocculation forming a bacterial culture, typically carried out in tanks. Can be extended with aeration.

Activated-Sludge Process Wastewater treatment process that uses activated sludge to biologically convert non-settleable (suspended, dissolved, and colloidal) organic materials to a settleable product using aerobic and facultative microorganisms; typically followed by clarification and sludge return.

Acute A stimulus severe enough to rapidly induce an effect; in aquatic toxicity tests, an effect observed in 96 hours or less is typically considered acute. When referring to aquatic toxicology or human health, an acute effect is not always measured in terms of lethality.

Additive Product added to a sewage treatment system marketed to improve performance.

Adsorption Adhesion of a substance to the surface of solid bodies or liquids with which it is in contact.

Advanced Primary Treatment The use of special additives to raw wastewater to cause flocculation or clumping to help settling before the primary treatment, such as screening.

Advanced Wastewater Treatment Any advanced process used above and beyond the de facto typical minimum primary and secondary wastewater treatment.

Aeration The process of bringing about intimate contact between air and a liquid.

Aeration, Active Introduction of air via either mechanical means or diffused aeration.

Aeration, Diffused Process of introducing air bubbles under pressure into a treatment unit using a compressor or blower and a diffuser.

Aeration Liquor Mixed liquor. The contents of the aeration tank, including living organisms and material carried into the tank by either untreated wastewater or primary effluent.

Aeration, Mechanical Process of introducing air into a treatment component by physical agitation using a device such as a paddle, paddle wheel, spray nozzle, or turbine.

Aeration, Passive Process of introducing air into a treatment component without mechanical means

Aeration Tank A chamber for injecting air into water. The tank where raw or settled wastewater is mixed with return sludge and aerated. The same as "aeration bay," "aerator," or "reactor."

Aerobic Wastewater treatment depending on oxygen for bacterial breakdown of waste.

Aerobic Bacteria Bacteria that require free (elementary) oxygen for growth.

Aerobic Treatment Unit (ATU) A mechanical wastewater treatment unit that provides secondary wastewater treatment for a single home, a cluster of homes, or a commercial establishment by mixing air (oxygen) and aerobic and facultative microbes with the wastewater. ATUs typically use a suspended growth process (such as activated-sludge extended aeration and batch reactors), a fixed-film process (similar to a trickling-filter), or a combination of the two treatment processes.

Aeration Chamber Chamber or tank in which wastewater is brought into contact with air to facilitate biological degradation such as in (but not limited to) the activated-sludge process.

Aeration System Piping, diffusers, air sources, vents, and all other necessary devices for an active aeration process.

Aerobic Decomposition Decomposition and decay of organic material in the presence of "free" or dissolved oxygen.

Aerobic Digestion The breakdown of wastes by microorganisms in the presence of dissolved oxygen. This digestion process may be used to treat only waste activated-sludge, or trickling-filter sludge and primary (raw) sludge, or waste sludge from activated-sludge treatment plants designed without primary settling. The sludge to be treated is placed in a large aerated tank where aerobic microorganisms decompose the organic matter in the sludge. This is an extension of the activated-sludge process.

Aerobic Process A waste treatment process conducted under aerobic (in the presence of "free" or dissolved oxygen) conditions.

Aerobic Treatment Unit (ATU) **1.** Treatment component that utilizes oxygen to degrade or decompose wastewater, with or without mechanical means; **2.** Term traditionally used to describe proprietary devices that use direct introduction of air into wastewater by mechanical means to maintain aerobic conditions within the pre-treatment component.

Aerobic Wastewater Treatment Oxygen dependent wastewater treatment requiring the presence of oxygen for aerobic bacterial breakdown of waste.

Agglomeration The grouping, or coming together, of dispersed suspended matter into larger particles, called "iloc," which settle more rapidly.

Aggregate **1.** Primary soil particles that cohere to each other more strongly than other surrounding particles; **2.** Naturally occurring inorganic material (crushed rock or gravel) screened to sizes for various uses.

Air scour The use of air either alone or in combination with backwash to enhance filter cleaning.

Algae Microscopic plants that contain chlorophyll and live floating or suspended in water. They also may be attached to structures, rocks, or other similar substances.

Algae produce oxygen during sunlight hours and use oxygen during the night hours. Their biological activities appreciably affect the pH and the dissolved oxygen of the water.

Alkaline The condition of water or soil that contains a sufficient amount of alkali substances to raise the pH above 7.0.

Alkalinity A measure of a substance's ability to neutralize acid. Water containing carbonates, bicarbonates, hydroxides, and occasionally borates, silicates, and phosphates can be alkaline. Alkaline substances have a pH value over 7.

Ammonia (NH_3) A chemical combination of hydrogen (H) and nitrogen (N) found extensively in nature. An indicator of fresh pollution.

Anaerobic Wastewater treatment in which bacteria break down waste without using oxygen.

Anaerobic Bacteria Bacteria that grow in the absence of free oxygen and derive oxygen from breaking down complex substances.

Anaerobic Wastewater Treatment Wastewater treatment in the absence of oxygen, anaerobic bacteria break down waste.

Anoxic Condition in which all constituents are in their reduced form (no oxidants present); conditions in a septic tank are generally anaerobic, but not anoxic.

An Open Dump An open dump is a facility that does not meet the criteria for a sanitary landfill and is not a facility for disposal of hazardous waste.

Aquifer Geologic formation, group of formations, or part of a formation that is saturated and sufficiently permeable to transmit water.

Areal Fill Above-grade soil treatment area designed and installed such that: the entire infiltrative surface is located above the original ground elevation using suitable imported soil material for fill; it utilizes gravity, pressure-dosed gravity or low-pressure distribution; a final cover of suitable soil stabilizes the completed installation and supports vegetative growth.

Antidegradation Policies that ensure protection of water quality for a particular water body where the water quality exceeds the levels necessary to protect fish and wildlife propagation and recreation on and in the water. This also includes special protection of waters designated as outstanding natural resource waters. Antidegradation plans are adopted by each state to minimize adverse effects on water.

Attached-Growth Process Configuration wherein the microorganisms responsible for treatment colonize a fixed medium.

Average Monthly Discharge Limitations The highest allowable average of daily discharges over a calendar month, calculated as the sum of all daily discharges measured during that month divided by the number of days on which monitoring was performed (except in the case of fecal coliform).

Average Weekly Discharge Limitation The highest allowable average of daily discharges over a calendar week, calculated as the sum of all daily discharges measured during a calendar week divided by the number of daily discharges measured during that week.

Backfill: 1. Material placed in an excavation; **2.** To place material in an excavation; **3.** Portion of an excavation above the haunch zone; for straight-walled tanks or structures, that portion of an excavation above the bedding.

Backfill, Initial Portion of an excavation above the haunch zone or bedding with a depth of 6–12 inches (15 to 30 cm) above the piping, conduit tank, or structure.

Backfill Final Portion of an excavation extending from above the initial backfill to final grade.

Backflow Reverse direction of flow, with liquid returning to the source.
Backflow Prevention Device Any device, method, or configuration used to prevent a reversal of flow.
Backflush To reverse the direction of flow to clean laterals or filter media.
Backwash Cleaning a granular media bed by means of an upward flow of clean water.
Bacteria Single cell microscopic living organisms lacking chlorophyll, which digest many organic and inorganic substances. An essential part of the ecosystem including within human beings.
Bacteria, Aerobic Bacteria that can metabolize only in the presence of molecular oxygen.
Bacteria, Anaerobic Bacteria that is able to metabolize in the absence of molecular oxygen.
Bacteria, Facultative Bacteria that can metabolize with or without molecular oxygen present in the environment.
Bacteria, Mesophilic Bacteria which grow best at temperatures between 20 and 50 °C (68 and 122 °F) with optimum growth between 25 and 40 °C (77 and 104 °F).
Bacteria, Psychrophilic Bacteria which grow best at temperatures between 10 and 30 °C (50 and 86 °F) with optimum growth between 12 and 18 °C (54 and 64 °F).
Bacteria, Thermophilic: Bacteria which grow best at temperatures between 35 and 75 °C (95 and 167 °F) with optimum growth between 55 and 65 °C (131 and 149 °F).
Baffle Physical barrier placed in a component to dissipate energy, direct flow, retain solids and FOG, and/or draw water from a specific depth.
Best Available Technology Economically Achievable (BAT) Technology based standard established by the Clean Water Act (CWA) as the most appropriate means available on a national basis for controlling the direct discharge of toxic and non-conventional pollutants to navigable waters. BAT effluent limitations guidelines, in general, represent the best existing performance of treatment technologies that are economically achievable within an industrial point source category or subcategory.
Bioassay A test used to evaluate the relative potency of a chemical or a mixture of chemicals by comparing its effect on a living organism with the effect of a standard preparation on the same type of organism.
Biochemical Oxygen Demand (BOD) The rate at which organisms use the oxygen in water or wastewater while stabilizing decomposable organic matter under aerobic conditions. In decomposition, organic matter serves as food for the bacteria and energy results from its oxidation. BOD measurements are used as a measure of the organic strength of wastes in water.
Biodegradation The breakdown of organic matter by bacteria to more stable forms that will not create a nuisance or give off foul odors.
Bioengineering A low-tech construction method using living plants as a functioning, self-sustaining part of the system. Examples include control of erosion of stream banks, water quality treatment, and flood control and habitat restoration.
Biofilm Thin coating of microbial growth, organic matter, and microbial secretions on a solid substrate particle.
Biofilter Media filter in which the media used is biological in origin (i.e., peat or coir).
Biological Filter A bed of relatively inert material (such as slay, molded plastics, clinker, etc.) to promote or assist natural aerobic degradation of sewage.

Biological Nutrient Removal (BNR) Use of microbiological activity for removal of nitrogen and phosphorus in a wastewater treatment system.

Biological Oxidation The process by which bacteria and other types of microorganisms consume dissolved oxygen and organic substances in wastewater, using the energy released to convert organic carbon into carbon dioxide and cellular material.

Biomass A mass or clump of organic material consisting of living organisms feeding on the wastes in wastewater, dead organisms, and other debris.

Biosolids Rich organic material left over from aerobic wastewater treatment, essentially dewatered sludge that can be reused.

Blackwater Portion of the wastewater stream that originates from toilet fixtures, dishwashers, and food preparation sinks.

Bleed To drain a liquid or gas, as in bleeding accumulated air from a water line or bleeding (draining) a trap of accumulated water.

BOD_5 Refers to the five-day biochemical oxygen demand. The total amount of oxygen used by microorganisms decomposing organic matter increases each day until the ultimate BOD is reached, usually in 50 to 70 days. BOD usually refers to the five-day BOD or BOD_5.

Branches Branches are collections from various drain funnels, catch basins, and area drains and tie into sublaterals. They are called T, Y, T-Y, double Y, and V branches according to their respective shapes.

Bubble, Coarse Bubble of 3 to 8 mm diameter generated by an air diffuser.

Bubble, Fine Bubble of 0.2 to 3 mm diameter generated by an air diffuser.

Bubble, Micro Bubble of less than 0.2 mm diameter generated by an air diffuser.

Bulking Inability of sludge solids to separate from the liquid under quiescent conditions; under aerobic conditions may be associated with the growth of filamentous organisms, low DO, or high sludge loading rates; under anaerobic conditions, may be associated with attachment of gas bubbles to solids.

Building Sewer Piping that conveys wastewater to the first system component or the sewer main.

Buoyancy Tendency of a body to float in water or other liquid; upward force that a fluid exerts on an object that is less dense than itself.

Bypass The intentional diversion of wastestreams from any portion of a treatment (or pre-treatment) facility.

Catch Basins Catch basins are used to collect surface drainage and process wastes in individual drainage areas and to trap sediment at the point nearest the source.

Catchment Area The catchment area is an area defined by a number of effluent streams which have a common discharge directed into a surface water drainage system, or water course.

Cesspool A covered watertight tank used for receiving and storing sewage from premises that cannot be connected to a public sewer and where ground conditions prevent the use of a small sewage treatment works including a septic tank. A further stage of treating sewage.

Chamber Pre-formed manufactured distribution medium with an open-bottom configuration commonly used in soil treatment areas.

Chemical Oxygen Demand (COD) Chemical oxygen demand (COD) is the equivalent amount of oxygen consumed under specified conditions in the chemical oxidation of the organic and oxidizable inorganic matter contained in wastewater, corrected for the influence of chlorides. In American practice, unless otherwise specified, the chemical oxidizing agent is hot acid dichromate.

Chlorination The application of chlorine to water, sewage, or industrial wastes, generally for the purpose of disinfection, but frequently for accomplishing other biological or chemical wastewater treatment results.

Chlorine Term commonly used to describe a chlorine source such as sodium hypochlorite, a highly reactive chemical used as a disinfectant and oxidizing agent.

Chlorine, Combined Available Chlorine that has combined with ammonia in wastewater to form chloramines; although they are slow-reacting, chloramines also serve as disinfectants.

Chlorine Residual Total amount of chlorine (free and combined available forms) remaining in water, sewage, or industrial wastes at the end of a specified contact period after the chlorination process.

Clarification Any process or combination of processes the main purpose of which is to reduce the concentration of suspended matter in a liquid.

Clarifier A large circular or rectangular tank or basin in which water is held for a period of time during which the heavier suspended solids settle to the bottom. Clarifiers are also called settling basins and sedimentation basins. May also be a tank or basin in which wastewater is held for a period of time during which the heavier solids settle to the bottom and the lighter materials float to the water surface.

Cleanout Device designed to provide access for removal of deposited or accumulated materials, generally from a pipe.

Clear Water Fraction of the wastewater stream including, but not limited to surface water, groundwater, condensate, ice machine drainage, and/or discharge from swimming pools, hot tubs, and water treatment devices.

Coagulant A chemical that causes very fine particles to clump (floc) together into larger particles. This makes it easier to separate the solids from the liquids by settling, skimming, draining, or filtering.

Coagulant Aids Materials added to enhance the action of coagulants, generally by affecting the electrical balance of the particles.

Coagulation The clumping together of very fine particles into larger particles (floc) caused by the use of chemicals (coagulants). The chemicals neutralize the electrical charges of the fine particles, allowing them to come closer and form larger clumps. This clumping together makes it easier to separate the solids from the water by settling, skimming, draining, or filtering.

Cold Climate Limitations Cold temperatures, ice cover, plant dormancy, equipment performance, ice buildup, and reduced microbial action create design challenges for cold weather wastewater treatment.

Colloids Very small solids; particulate or insoluble material in a finely divided form that remain dispersed in a liquid for a long time due to their small size and electrical charge.

Combined Sewer Carries both sanitary sewage and stormwater run-off. Sewer systems in which the stormwater and sanitary waste are combined. A benefit is that non-point pollution flushed from the watershed during moderate rain is treated, but the system can be overwhelmed during severe storms, resulting in untreated waste being flushed into the receiving waters as a combined sewer overflow (CSO).

Combined Wastewater A mixture of storm or surface run-off and other wastewater such as domestic or industrial wastewater.

Comminution A mechanical treatment process that cuts large pieces of wastes into smaller pieces so they won't plug pipes or damage equipment (shredding).

Comminutor A device used to reduce the size of the solid chunks in wastewater by shredding (comminuting). The shredding action is like many scissors cutting or chopping to shreds all the large solids material in the wastewater.

Constructed Wetland A wetland constructed for the purpose of pollution control and waste management. The flow-rate, residence time, and other factors are controlled to enhance the removal of BOD, SS, and N. A waterproof barrier is usually placed below the substrate to isolate the wastewater from the groundwater. Plants such as cattails, bulrushes, and reeds provide a dense cover and an oxygenating substrate for bacteria in the root zone.

Corrosion The gradual decomposition or destruction of a material by chemical action, often due to an electrochemical reaction. Corrosion may be caused by: (1) stray current electrolysis, (2) galvanic corrosion caused by dissimilar metals, or (3) differential-concentration cells. Corrosion starts at the surface of a material and moves inward.

Corrosion Inhibitors Substances that slow the rate of corrosion.

Corrosive A chemical that can cause burns to the skin, eyes, or the respiratory system.

Corrosive Gases In water, dissolved oxygen reacts readily with metals at the anode of a corrosion cell, accelerating the rate of corrosion until a film of oxidation products such as rust forms. At the cathode where hydrogen gas may form a coating on the cathode and slow the corrosion rate, oxygen reacts rapidly with hydrogen gas forming water, and again increases the rate of corrosion.

Conventional Pollutants Pollutants typical of municipal sewage, and for which municipal secondary treatment plants are typically designed; defined by Federal Regulation [40 CFR 401.16] as BOD, TSS, fecal coliform bacteria, oil and grease, and pH.

Conventional Septic System A wastewater treatment system consisting of a septic tank and a typical trench or bed subsurface wastewater infiltration system.

Daily Discharge The discharge of a pollutant measured during any 24-hour period that reasonably represents a calendar day for purposes of sampling. For pollutants with limitations expressed in units of mass, the daily discharge is calculated as the total mass of the pollutant discharged during the day. For pollutants with limitations expressed in other units of measurement (e.g., concentration) the daily discharge is calculated as the average measurement of the pollutant throughout the day.

Dechlorination Removal of the free and combined chlorine residual to reduce the potentially toxic effects of chlorinated effluents.

Degradation The conversion or breakdown of a substance to simpler compounds. For example, the degradation of organic matter to carbon dioxide and water.

Denitrification An anoxic process that occurs when nitrite or nitrate ions are reduced to nitrogen gas and nitrogen bubbles are formed as a result of this process. The bubbles attach to the biological floc in the activated sludge process and float the floc to the surface of the secondary clarifiers. This condition is often the cause of rising sludge observed in secondary clarifiers or gravity thickeners.

Design Population Design population means the minimum and maximum number of persons (resident and non-resident) to be served.

Detention Time (Retention time; residence time) The average period of time wastewater stays in a treatment system. Detention times vary for different types of wastewater treatment systems and can range from hours to weeks.

Dewatered Sludge The sludge after it's been dewatered, also known as sludge cake.

Dewatered Sludge Cake The sludge after dewatering that is cake like, compressed. The lower the water content the better for wastewater treatment purposes.

Dewatering Removing water from sludge or other solids.

Diffuser Part or device that injects air under pressure into wastewater (e.g., submerged porous plate, perforated pipe, or orifice).

Digestion The breaking down of sludge and other waste biologically by microorganisms. Results in byproducts such as methane gas, carbon dioxide, sludge solids, and water. Aerobic digestion requires oxygen, anaerobic digestion the absence of oxygen.

Denitrification Biologically removing nitrate and converting it to nitrogen gas.

Discarded Material Discarded material is used or spent material that is not reused in any way and is committed to final disposition.

Disinfection The use of chemicals to kill any disease causing organisms in the polished wastewater. UV light can also be used.

Dispersion: **1.** Scattering and mixing; **2.** Repellant action of an electric potential on fine particles in suspension in water, as in a stream carrying clay.

Disposal Disposal is the discharge, deposit, injection, dumping, spilling, leaking, or placing of any solid waste or hazardous waste into or onto any land or water so that such solid waste or hazardous waste or any constituent thereof may enter the environment, be emitted into the air, or be discharged into any waters, including groundwaters. Therefore, waste deposited in a landfill, landfarmed wastewater discharged to a basin or surface impoundment, or stormwater run-off diverted to a percolation or settling pond, is disposal.

Disposal Well A disposal well is a deep well used for the disposal of liquid wastes.

Dissolved Oxygen (DO) Dissolved oxygen (DO) is the oxygen dissolved in sewage, water, or other liquid, usually expressed in milligrams per liter or percent of saturation. It is the test used in BOD determination.

Dissolved Solids Solids physically suspended in sewage that cannot be removed by proper laboratory filtering.

Dosing, Demand Configuration in which a specific volume of effluent is delivered to a component based upon patterns of wastewater generation from the source.

Drain, Interceptor Subsurface drain used to intercept and divert laterally moving groundwater or perched water away from the soil treatment area or other system component to an effective outlet.

Drainage Network of natural or artificial groundwater or surface water features including agricultural drain tile, cut banks, and ditches which intercept and divert surface water and/or lower groundwater.

Drains Drains are small sewer connections discharging through a sealed connection to the nearest catch basin from points such as pump bases, equipment drips, low points of floors, funnels, etc.

Drawdown Drop in the liquid level of a tank as a result of some phase of operation.

Drawdown Test Measurement of the drop in liquid level in a dosing tank measured over time to calculate dosing/delivery rate; may be expressed as a pump delivery rate (PDR) or siphon delivery rate.

Drip System   The drip system is a separate drain system for recovery of oil from contaminated fluids.

Ecological Engineering   The design, management, or reconstruction of sustainable ecosystems that serve human needs such as providing clean water and food while requiring low energy inputs. Ecological engineering has enhanced our understanding of environmental problems such as wastewater treatment, wetlands damage and mitigation, the effect of non-point pollution on ecosystems, and ecosystem restoration.

Effluent   Effluent is a liquid that flows out of a containing space, and/or sewage, water or other liquid, partially or completely treated, or in its natural state, as may be flowing out of a reservoir, basin, or treatment plant, or part thereof.

Effluent Limitation   Effluent limitation is any restriction (including schedules of compliance) established by a governmental authority on quantities, rates, and concentrations of chemical, physical, biological, and other constituents which are discharged from point sources into navigable waters, the waters of the contiguous zone, or the ocean.

Effluent Screen   Removable, cleanable (or disposable) device installed on the outlet piping of a septic tank for the purpose of retaining solids larger than a specific size and/or modulating effluent flow-rate.

Effluent Quality   Physical, biological, and chemical characteristics of a liquid flowing from a component or device.

Emerging Contaminants   Newly identified compounds or substances that have the potential to adversely affect public health or the environment and for which there is no currently published health standard.

Emulsification   Suspension of solids as a result of decreased surface tension due to the presence of an emulsifying agent or some substance that alters or prohibits normal microbial activity.

Emulsifying Agent   Agent capable of modifying the surface tension of emulsion droplets to prevent coalescence; examples are soap and other surface-active agents, certain proteins and gums, water-soluble cellulose derivatives, and polyhydric alcohol esters and ethers.

Emulsion   A mixture of two or more liquids that cannot be combined, therefore one liquid is "suspended" in the other.

Endogenous Respiration   Auto-oxidation by organisms in biological processes.

Environmental Sensitivity   Relative susceptibility of the natural environment to adverse impacts from an outside constituent.

Extended Aeration   An aeration system that adds aerobic sludge digestion to the activated sludge process.

Extension   Alteration of a wastewater treatment system resulting in an increase in capacity, lengthening, or expansion of the existing collection, treatment, or dispersal component.

Facultative Ponds   A wastewater treatment pond that includes surface aeration and algal photosynthesis for oxygen replenishment.

Filter   Device that removes constituents through processes such as sieving, stagnation, adsorption, or absorption; a filter has both area and depth with respect to flow.

Filter, Activated Carbon   Device filled with a porous form of carbon that is used to decolorize liquids, recover solvents, and remove toxins and odors from water and air.

Filter, Bottomless Media   Media filter that does not incorporate a liner or other physical barrier between the media and the existing soil on which it has been placed; used as a final treatment and dispersal component.

Filter, Coir   Media filter that uses organic fibric material (coir) from the outer husk of coconut.

Filter, Disc   Device consisting of concentrically grooved discs stacked one upon the other and used for removal of particles larger than a specific size; typically used in drip distribution systems.

Filter, Media   Device that uses materials designed to treat effluent by reducing BOD and/or removing suspended solids in an unsaturated environment; biological treatment is facilitated via microbial growth on the surface of the media.

Filter Medium   The material of which the biological filter is formed and on which a biological film containing bacteria and fungi develops.

Filter, Peat   Media filter that uses appropriate organic fibric material (peat) as the media; typically packaged as pre-fabricated modular units with the media in a container; a type of biofilter.

Filter, Sand   Media filter which uses sand of particular specifications as the media.

Filter, Screen   Filter consisting of a mesh material configured as a cylinder and used to remove particles larger than a specific size in pressurized systems.

Filter, Trickling   Type of media filter which uses a variety of media such as rigid plastics of varying shapes, stone, or tire chips; includes a clarifier in its configuration and may include a recirculation mode.

Filter, Upflow   Media filter through which wastewater flows from a lower to a higher elevation; usually characterized by an anaerobic environment.

Filtration   Removal of suspended materials using processes such as sieving, stagnation, adsorption, absorption, and possibly biochemical degradation.

Filter Run   The time of filter operation between backwashes.

Final Effluent   The effluent discharged from a sewage treatment plant.

Final Treatment and Dispersal   Last treatment component (or combination of components) through which effluent is returned to the hydrologic cycle via a soil treatment area or a discharging outfall.

Floc   The agglomeration of smaller particles in a gelatinous mass that can be more easily removed from the liquid than individual small particles.

Flocculant   Same as flocculating agent, the catalyst substance that causes the chemical reaction with TSS to form flocculent often encapsulating the solids.

Flocculent   The "floc" or wooly mass of clusters that is formed in flocculation. Often used interchangeably with "flocculant" however truly refers to the floc mass and not the catalyst flocculating agent.

Flocculation   The provision of retention time with gentle agitation to allow the floc particles or precipitate, associated with the impurities, to increase in size by agglomeration.

Flocculating Agent   The flocculant or chemical used to cause flocculation.

Flow, Instantaneous   Highest recorded flow occurring within a short, specific period of time (expressed in gallons per minute).

Flow, Peak   Highest flow occurring within a specified time (minutes, hours, days, etc.); may be further expressed as peak hourly flow, peak daily flow, peak monthly flow, peak seasonal flow, etc.

Flow, Surge   Flow of effluent that occurs in a short enough period of time that it upsets the function of one or more components of the treatment train.

Flow Equalization   System configuration that includes sufficient effluent storage capacity to allow for uniform flow to a subsequent component despite variable flow from the source.

Free Water Surface Wetland (FWS)   A lined basin or channel with porous plant substrate and wetland vegetation in which the shallow water is exposed to the air.

Fungi   Small, non-chlorophyll bearing plants, without roots, stems, or leaves, which tend to overpower bacteria at low pH and dissolved oxygen concentrations. They generally have a filamentous type structure and are therefore not welcome in a secondary process clarifier.

Grab Sample   A sample which is taken from a wastestream on a one-time basis without consideration of the flow rate of the wastestream and without consideration of time.

Granular Media   The material used to effect filtration.

Gray Water   Water captured from non-food preparation sinks, showers, baths, spa baths, clothes washing machines, and laundry tubs.

Grease   Fats, soaps, oils, waxes and etc. in wastewater.

Grit   Heavy, inorganic matter, such as sand or pebbles.

Grit Chamber   Usually in municipal wastewater treatment, a chamber or tank in which primary influent is slowed down so heavy, typically inorganic, solids can drop out, such as metals and plastics.

Groundwater   Subsurface water in the saturation zone from which wells and springs are fed. In a strict sense the term applies only to water below the water table. Also called "phreatic water" and "plerotic water."

Hazardous Waste   Hazardous waste is solid waste that poses specified health and environmental hazards.

Headworks   The beginning of the treatment plant where the influent begins treatment.

Heavy Metals   Metals that can be precipitated by hydrogen sulfide in an acid solution, including lead, silver, gold, mercury, bismuth, and copper.

Hydrogen Ion Concentration [H^+]   The weight of hydrogen ion in moles per liter of solution. Commonly expressed as the pH value, which is the logarithm of the reciprocal of the hydrogen ion concentration.

Hydrogen Sulfide Gas (H_2S)   Hydrogen sulfide is a gas with a rotten egg odor. This gas is produced under anaerobic conditions. Hydrogen sulfide gas is particularly dangerous because it dulls the sense of smell so that you don't notice it after you have been around it for a while. In high concentrations, hydrogen sulfide gas is only noticeable for a very short time before it dulls the sense of smell. The gas is very poisonous to the respiratory system, explosive, flammable, colorless, and heavier than air.

Hydrolysis   Conversion of organic nitrogen to ammonia by enzymes secreted by bacteria, plants, and animals in a reaction that adds water.

I&A   Innovative and Alternative. A term defined by the EPA to describe non-conventional technologies. "Alternative systems" are fully proven systems that reclaim or reuse wastewater, productively recycle wastewater components, recover energy, or eliminate the discharge of pollutants. A variety of wastewater treatment systems have been included in the definition, including land treatment, aquaculture, containment ponds, and on-site treatment using small diameter or vacuum sewers.

Immediate Oxygen Demand (IOD) Immediate oxygen demand (IOD) is the amount of oxygen that is utilized by the components of a wastewater within 15 minutes (unless otherwise specified) after being introduced into water that contains dissolved oxygen.

Impermeable Not permitting the passage of fluid through pores; in practical terms, some small level of hydraulic conductivity may occur, but at so low a level (e.g., 1×10^{-7} cm/s) that it is considered to be negligible.

Industrial Ecology Industrial Ecology (IE) focuses on combining perpetually desirable outcomes in the environment, the economy, and technology sustainably. The primary tenet is that all systems mimic nature and are thus closed loop, continuous, circular. In wastewater treatment, industrial ecology would mean that all so called "waste" is re-input into the same or another process. For example, bio-solids as fertilizer can be considered a use of sludge consistent with industrial ecology. Recycling wastewater into the treatment plant, manufacturing, or other process is another example.

Industrial Sources Non-municipal, or industrial sources, often generate wastewater that is discharged to surface waters. The types of wastewater generated at a facility depend on the specific activities undertaken at a particular site, and may include manufacturing or process wastewaters, cooling waters, sanitary wastewater, and stormwater run-off.

Industrial Wastes The solid and liquid wastes originating from industrial processes.

Industrial Wastewater Treatment Wastewater treatment for industries such as manufacturing, food processing, corrugators, printing, and so on. Paper and pulp mills' treatment of wastewater is an example of industrial wastewater treatment. Municipal wastewater treatment would be an example that is not considered to be industrial.

Infiltration The seepage of groundwater into a sewer system, including service connections. Seepage frequently occurs through defective or cracked pipes, pipe joints and connections, interceptor access risers and covers, or manhole walls.

Infiltration/Inflow The total quantity of water from both infiltration and inflow without distinguishing the source.

Inflow Water discharged into a sewer system and service connections from such sources as, but not limited to, roof leaders, cellars, yard and area drains, foundation drains, cooling water discharges, drains from springs and swampy areas, around manhole covers or through holes in the covers, cross-connections from storm and combined sewer systems, catch basins, stormwaters, surface run-off, street wash waters, or drainage. Inflow differs from infiltration in that it is a direct discharge into the sewer rather than a leak in the sewer itself.

Influent Water, wastewater, or other liquid – raw (untreated) or partially treated – flowing into an interceptor, reservoir, basin, treatment process, or treatment plant. The untreated wastewater or raw sewage coming into a wastewater treatment plant.

Influent Screens Screens used to remove large inorganic solids from the waste stream.

Inhibitory Substances Materials that kill or restrict the ability of organisms to treat wastes.

Injection Well Well by which effluent is transmitted to an underground formation; in most cases these are regulated and require a permit from a regulatory authority.

Inorganic Material such as sand, salt, iron, calcium salts, and other mineral materials. Inorganic substances are of mineral origin, whereas organic substances are usually of animal or plant origin. Also see "organic."

Inorganic Waste Waste material such as sand, salt, iron, calcium, and other mineral materials which are only slightly affected by the action of organisms. Inorganic wastes are chemical substances of mineral origin; whereas organic wastes are chemical substances of an animal or plant origin.

Inorganic Material Material that will not respond to biological action (sand, cinders, stone). Non-volatile fraction of solids.

Lagoon Constructed basin lined with either soils with very low permeability or a synthetic material, surrounded with berms and which contains at least three feet of wastewater which utilizes sunlight, wind or mechanical aeration, and natural bacteria to break down waste via physical, chemical, and biological processes.

Lagoon, Evaporation Lagoon where wastewater is stored and the water is allowed to evaporate over time.

Lagoon, Storage Lagoon where some form of wastewater is stored before it is either conveyed to another component for further processing or is reused.

Laterals Laterals are sewers collecting the effluent from two or more sublaterals discharging to the "Mains."

Lagoon Sludge Lagoon sludge is a relatively shallow basin, or natural depression, used for the storage or digestion of sludge, sometimes for its ultimate detention or dewatering.

Lime Any of several compounds consisting of calcium hydroxide ($Ca(OH)_2$) or calciumoxide (CaO).

Liquefaction Liquefaction as applied to sludge digestion means the transformation of large solid particles of sludge into either a soluble or a finely dispersed state.

Liquid Solids Separation The process of separating the liquids and solids in a given wastewater. Liquid/solids separation comes in one of three processes: 1. If the solids sink (specific gravity greater than 1) use a clarifier; 2. If the solids float (specific gravity less than 1) use a floattion unit (DAF); 3. If they neither sink nor float (specific gravity is 1) try using a screen (rotary or parabolic).

Loading Quantity of material applied to a device at one time.

Loading Rate, Areal Quantity of effluent applied to the footprint of the soil treatment area (or the absorption area of an above-grade soil treatment area) expressed as volume per area per unit time, e.g., gallons per day per square foot (gpd/sq.ft.).

Loading Rate, Biochemical Quantity of BOD_5 delivered to a treatment component expressed as mass per time (e.g., pounds of BOD_5 per day).

Loading Rate, Biological Quantity of organic matter delivered to a treatment component expressed mass per time (e.g., pounds per day).

Loading Rate, Mass Sum of organic and inorganic effluent constituents delivered to a treatment component in a time interval, expressed as mass per time.

Loading Rate, Nutrient Sum of organic and inorganic nutrients (primarily nitrogen and phosphorus) delivered to a treatment component in a specified time interval expressed as mass per time.

Loading Rate, Organic Biodegradable fraction of chemical oxygen demand (biochemical oxygen demand, biodegradable FOG, and volatile solids) delivered to a treatment component in a specified time interval expressed as mass per time or area; e.g., pounds per day or pounds per cubic foot per day (pre-treatment); pounds per square foot per day (infiltrative surface or pretreatment); typical residential system designs assume biochemical loading equals organic loading;

Local Limits   Conditional discharge limits imposed by municipalities upon industrial or commercial facilities that discharge to the municipal sewage treatment system.

Main Sewer   A sewer line that receives wastewater from many tributary branches and sewer lines and serves as an outlet for a large territory or is used to feed an intercepting sewer.

Makeup Water   Water supplied to replenish the water of a system.

Malfunction   Condition in which a component is not performing as designed/installed.

Manholes   Manholes are used in sewer mains as junction points and sediment traps, and to provide access for maintenance and inspection.

Media   The material in a trickling-filter on which slime accumulates and organisms grow. As settled wastewater trickles over the media, organisms in the slime remove certain types of wastes thereby partially treating the wastewater. Also the material in a rotating biological contactor or in a gravity or pressure filter.

Microorganisms   Very small organisms that can be seen only through a microscope. Some microorganisms use the wastes in wastewater for food and thus remove or alter much of the undesirable matter.

Mixed Liquor   The combination of primary effluent and active biological solids (return sludge) in the activated sludge process that is fed into the aeration tank.

Mixed Liquor Suspended Solids (MLSS)   Suspended solids in the mixed liquor of an aeration tank.

Mixed Liquor Volatile Suspended Solids (MLVSS)   The organic or volatile suspended solids in the mixed liquor of an aeration tank. The volatile portion is used as a measure or indication of the microorganisms present.

Mixing Zone   An area where an effluent discharge undergoes initial dilution and is extended to cover the secondary mixing in the ambient water body. A mixing zone is an allocated impact zone where water quality criteria can be exceeded as long as acutely toxic conditions are prevented.

Mixed Media Gravity Filter   A filter using more than one filtering media (such as coal and sand).

Natural Systems   Ecologically based biological wastewater treatment systems such as constructed wetlands having minimal dependence on mechanical elements.

N: Nitrogen   This nutrient is present in various forms in wastewater, principally ammonia and nitrate.

Neutralization   Addition of an acid or alkali (base) to a liquid to cause the pH of the liquid to move toward a neutral pH of 7.0.

Nitrification   An aerobic process in which bacteria change the ammonia and organic nitrogen in wastewater into oxidized nitrogen (usually nitrate). The second-stage BOD is sometimes referred to as the "nitrification stage" (first-stage BOD is called the "carbonaceous stage").

Nitrification Stage   A stage of decomposition that occurs in biological treatment processes when aerobic bacteria, using dissolved oxygen, change nitrogen compounds (ammonia and organic nitrogen) into oxidized nitrogen (usually nitrate). The second-stage BOD is sometimes referred to as the "nitrification stage" (first-stage BOD is called the "carbonaceous stage").

Nitrifying Bacteria   Bacteria that change the ammonia and organic nitrogen in wastewater into oxidized nitrogen (usually nitrate).

Nitrobacteria Principal genera of autotrophic bacteria responsible for the second step of biological nitrification: conversion (oxidation) of nitrite to nitrate.

Nitrogen Fixation The conversion of nitrogen gas to organic nitrogen, ammonia or nitrate. Nitrogen fixation can occur biologically (i.e., conversion of nitrogen gas to organic nitrogen by certain photosynthetic blue-green algae), by natural physical processes (i.e., conversion of nitrogen gas to nitrate by lightning), or by industrial processes (manufacture of fertilizers and explosives).

Nitrogenous A term used to describe chemical compounds (usually organic) containing nitrogen in combined forms. Proteins and nitrates are nitrogenous compounds.

Nutrient Any substance that is assimilated (taken in) by organisms and promotes growth. Nitrogen and phosphorus are nutrients which promote the growth of algae.

There are other essential and trace elements which are also considered nutrients.

Nutrient Cycle The transformation or change of a nutrient from one form to another until the nutrient has returned to the original form, thus completing the cycle.

Odor Quality of gases, liquids, or particulates that stimulates the olfactory organ.

Oil Interceptor The oil interceptor is a device designed to remove small oil globules by gravity from the water by limiting the flow velocity and the overflow rate.

On-site Local wastewater treatment for a single house or small community.

Organic Substances that come from animal or plant sources. Organic substances always contain carbon. (Inorganic materials are chemical substances of mineral origin.)

Organic Loading The pounds of BOD per day applied to a unit process.

Organic Waste Waste material which comes mainly from animal or plant sources. Organic wastes generally can be consumed by bacteria and other microscopic organisms. Inorganic wastes are chemical substances of mineral origin.

Organics 1. A term used to refer to chemical compounds made from carbon molecules. These compounds may be natural materials (such as animal or plant sources) or man-made materials (such as synthetic organics). Also see "organic." 2. Any form of animal or plant life.

Overland Flow Land Treatment Partially treated wastewater is applied to relatively impermeable soils at the top of a grass-covered gradient. The waste is cleaned by the vegetation and microbial action, and excess water is captured at the bottom of the slope.

Oxidation The conversion of organic material to a more stable form using bacteria, chemicals, or oxygen.

Oxidation Ponds Oxidation ponds are basins in which wastewater undergoes a biological oxidation treatment by action of algae and bacteria.

Oxidation Direct Oxidation direct is oxidation of substances in sewage without the benefit of living organisms, by the direct application of air or oxidizing agents such as chlorine.

Oxidation Sewage Oxidation sewage is the process whereby, through the agency of living organisms in the presence of oxygen, the organic matter that is contained in sewage is converted into a more stable or a mineral form.

Oxygen Consumed Oxygen consumed is the quantity of oxygen taken up from potassium permanganate in solution by a liquid containing organic matter. Commonly regarded as an index of the carbonaceous matter present. Time and temperature must be specified.

P (Phosphorus) This nutrient, which is present in wastewater, acts as a fertilizer for algae in surface waters.

Parts Per Million (ppm) Parts per million (ppm) is parts by mass in sewage analysis, ppm by mass is equal to milligrams per liter divided by the relative density (specific gravity). In water analysis ppm is always understood to imply mass/mass ratio (mg/kg), even though in practice a volume may be measured instead of a mass.

Pathogenic Disease causing or harmful to man.

Pathogenic Organisms Organisms, including bacteria, viruses, or cysts, capable of causing diseases (giardiasis, cryptosporidiosis, typhoid, cholera, dysentery) in a host (such as a person). There are many types of organisms that do not cause disease. These organisms are called non-pathogenic.

Peak Demand The maximum momentary load placed on a water treatment plant, pumping station, or distribution system. This demand is usually the maximum average load in one hour or less, but may be specified as the instantaneous load or the load during some other short time period.

Peaking Factor Ratio of a maximum flow to the average flow, such as maximum hourly flow or maximum daily flow to the average daily flow.

Percent Saturation The amount of a substance that is dissolved in a solution compared with the amount that could be dissolved in solution, expressed as a percent.

Percolation The flow of liquid through a filtering medium.

Permeability Ability of a porous medium such as soil to transmit fluids (liquids or gases).

pH A measure of acidity or alkalinity of water, or any given substance. The scale is 1 to 14 with 7 being neutral. Over 7 is alkaline or caustic, under 7 is acid or base.

Phosphorus (P) Chemical element and nutrient essential for all life forms, occurring as orthophosphate, pyrophosphate ($P_2O_7^{-4}$), tripolyphosphate ($P_3O_{105}^-$), and organic phosphate forms; each of these forms, as well as their sum (total phosphorus), is expressed in terms of milligrams per liter (mg/L) elemental phosphorus; occurs in natural waters and wastewater almost solely as phosphates; excess levels of phosphorous in fresh surface waters may contribute to eutrophication.

Phosphorus, Inorganic Forms of phosphorus from mineral sources, such as orthophosphate, pyrophosphate ($P_2O_7^{-4}$), and tripolyphosphate ($P_3O_{105}^-$).

Phosphorus, Organic Phosphorus formed primarily by biological processes; sources of organic phosphorus in sewage include bodily wastes, food residues, and the conversion of orthophosphates in biological treatment processes.

Physical Waste Treatment Process Physical waste treatment processes include the use of racks, screens, comminutors, clarifiers (sedimentation and flotation), and filtration. Chemical or biological reactions are important treatment processes, but not part of a physical treatment process.

Pollutant Any substance that causes impairment (reduction) of water quality to a degree that has an adverse effect on any beneficial use of the water.

Pollutant, Conservative Pollutants that do not readily degrade in the environment, and which are mitigated primarily by natural stream dilution after entering receiving bodies of waters. Included are pollutants such as metals.

Pollutant, Non-Conservative Pollutants that are mitigated by natural biodegradation or other environmental decay or removal processes in the receiving stream after in-stream mixing and dilution have occurred.

Pollution The impairment (reduction) of water quality by agricultural, domestic, or industrial wastes (including thermal and radioactive wastes) to such a degree that the natural water quality is changed to hinder any beneficial use of the water or render it offensive to the senses of sight, taste, or smell or when sufficient amounts of wastes create or pose a potential threat to human health or the environment.

Population Equivalent   The equivalent, in terms of a fixed population, of a varying or transient population, of a hospital or restaurant, based upon a figure of 0.06.kg BOD per head per 120L per head per day.

Porosity   1. Open space or interstices in rock, other Earth materials, or synthetic media; 2. Ratio of the open space to the total volume often described as a percentage.

Post-Denitrification   Biological wastewater treatment process for nitrogen removal that utilizes an anoxic zone located at the effluent end of an aeration tank. Due to lack of organic carbon, methanol addition is typically required.

Post-chlorination   The addition of chlorine to the plant effluent, following plant treatment, for disinfection purposes.

Potable Water   Water that does not contain objectionable pollution, contamination, minerals, or infective agents and is considered satisfactory for drinking.

Pre-Aeration   The addition of air at the initial stages of treatment to freshen the wastewater, remove gases, add oxygen, promote flotation of grease, and aid coagulation.

Pre-Chlorination (Wastewater):   The addition of chlorine in the collection system serving the plant or at the headworks of the plant prior to other treatment processes mainly for odor and corrosion control. Also applied to aid disinfection, to reduce plant BOD load, to aid in settling, to control foaming in Imhoff units, and to help remove oil.

Precipitation   When a substance dissolved in a liquid passes out of solution and into solid form.

Precursor, THM   Natural organic compounds found in all surface and groundwater. These compounds may react with halogens (such as chlorine) to form trihalomethanes (THMs); they must be present in order for THMs to form.

Pre-Denitrification   Biological wastewater treatment process for nitrogen removal that utilizes an anoxic zone located at the influent end of an aeration tank. Organic matter present in the wastewater serves as a carbon source for denitrifying bacteria.

Preliminary Treatment   The removal of metal, rocks, rags, sand, eggshells, and similar materials which may hinder the operation of a wastewater treatment plant. Preliminary treatment is accomplished by using equipment such as racks, bar screens, comminutors, and grit removal systems.

Pre-treatment   Any component or combination of components that provides treatment of wastewater prior to conveyance to a final treatment and dispersal component or reuse; often, this treatment is designed to meet primary, secondary, tertiary, and/or disinfection treatment standards.

Pre-treatment Facility   Industrial wastewater treatment plant consisting of one or more treatment devices designed to remove sufficient pollutants from wastewaters to allow an industry to comply with effluent limits established by the US EPA General and Categorical Pre-treatment Regulations or locally derived prohibited discharge requirements and local effluent limits. Compliance with effluent limits allows for a legal discharge to a POTW.

Preventive Maintenance   Regularly scheduled servicing of machinery or other equipment using appropriate tools, tests, and lubricants. This type of maintenance can prolong the useful life of equipment and machinery and increase its efficiency by detecting and correcting problems before they cause a breakdown of the equipment.

Primary Clarifier A wastewater treatment device that consists of a rectangular or circular tank that allows those substances in wastewater that readily settle or float to be separated from the wastewater being treated.

Primary Settlement Tank A tank in which the majority of settleable solids are removed from the crude sewage flowing through it.

Primary Treatment Primary treatment is water purification based on the difference in density of the polluting substance and the medium, the former being removed either by rising or settling. This process can include screening, grit removal, sedimentation, sludge dig. A wastewater treatment process that takes place in a rectangular or circular tank and allows those substances in wastewater that readily settle or float to be separated from the water being treated.

Primary Wastewater Treatment Removal of sand, grit, and larger solids from wastewater by screens, settling tanks and/or skimming devices, and sludge disposal.

Priority Pollutants The EPA has proposed a list of 126 priority toxic pollutants. These substances are an environmental hazard and may be present in water. Because of the known or suspected hazards of these pollutants, industrial users of the substances are subject to regulation. The toxicity to humans may be substantiated by human epidemiological studies or based on effects on laboratory animals related to carcinogenicity, mutagenicity, teratogenicity, or reproduction. Toxicity to fish and wildlife may be related to either acute or chronic effects on the organisms themselves or to humans by bio-accumulation in food fish. Persistence (including mobility and degradability) and treatability are also important factors.

Process Variable A physical or chemical quantity that is usually measured and controlled in the operation of a wastewater treatment plant or an industrial plant.

Propeller Fan Fan in which the air enters and leaves the impeller in a direction substantially parallel to its axis.

Protozoa A group of motile microscopic animals (usually single-celled and aerobic) that sometimes cluster in colonies and often consume bacteria as an energy source.

Prussian Blue A paste or liquid used to show a contact area.

Psychrophilic Bacteria A group of bacteria that grow and thrive in temperatures below 68 °F (20 °C).

Primary Waste Treatment Mechanical separation of solids, grease, and scum from wastewater. With the aid of flocculating agents, primary treatment can eliminate 50 to 65% of the suspended solids. Solids removed by primary treatment may comprise as much as 30 to 40% of the original BOD of the water.

Purification The removal, by natural or other methods, of pollution from a given medium.

Putrefaction Biological decomposition of organic matter resulting in the production of foul-smelling products associated with anaerobic conditions.

Raw Sewage Untreated sanitary wastewater.

Raw Wastewater Plant influent or wastewater before any treatment.

Reagent A substance that takes part in a chemical reaction and is used to detect and measure another substance.

Recalcination A lime-recovery process in which the calcium carbonate in sludge is converted to lime by heating to 1800 °F (980 °C).

Recarbonation A process in which carbon dioxide is bubbled into the water being treated to lower the pH.

Receiving Body A stream, lake, or other waterway into which treated or untreated waste is discharged.

Receiving Water A stream, river, lake, ocean, or other surface or groundwater into which treated or untreated wastewater is discharged. This means rivers, lakes, or other water sources that receive treated or untreated wastewaters.

Reclaimed Water Reusable wastewater from wastewater treatment such as tertiary treatment of wastewater in biological and other systems.

Reducing Agent Any substance, such as base metal (iron) or the sulfide ion, that will readily donate (give up) electrons. The opposite is an oxidizing agent.

Reduction Reduction is the addition of hydrogen, removal of oxygen, or the addition of electrons to an element or compound. Under anaerobic conditions (no dissolved oxygen present), sulfur compounds are reduced to odor producing hydrogen sulfide (H_2S) and other compounds.

Refractory A material having the ability to retain its shape and chemical composition when subjected to high temperatures, or the area of an incinerator or similar equipment that contains the high temperatures.

Remediation Act or process of correcting a fault or deficiency without changing system structure or form.

Residuals Solids generated and retained in wastewater treatment components during the treatment of sewage, including sludge, scum, and pumpings from grease interceptors, septic tanks, aerobic treatment units, or other components.

Residual Chlorine The amount of free and/or available chlorine remaining after a given contact time under specified conditions.

Residuals The solids generated and/or retained during the treatment of wastewater. They include trash, rags, grit, sediment, sludge, biosolids, septage, scum, grease, as well as those portions of treatment systems that have served their useful life and require disposal, such as the sand or peat from a filter. Because of their different characteristics, management requirements can differ as stipulated by the appropriate regulations.

Respiration The process in which an organism uses oxygen for its life processes and gives off carbon dioxide.

Respirator A device designed to protect the wearer from a hazardous atmosphere.

Retention Time The amount of time that water, sludge, or solids are retained or held in a clarifier or sedimentation tank.

Return Air Air entering a space from an air-conditioning, heating, or ventilating apparatus.

Return Sludge The recycled sludge in a (Publicly Owned Treatment Works) POTW that is pumped from a secondary clarifier sludge hopper to the aeration tank.

Return Sludge Ratio (R/Q) The ratio of the return sludge flow to the wastewater flow.

Reuse The use of water or wastewater after it has been discharged and then withdrawn by another user.

Rising Sludge Rising sludge occurs in the secondary clarifiers of activated sludge plants when the sludge settles to the bottom of the clarifier, is compacted, and then starts to rise to the surface, usually as a result of denitrification.

Run-off Run-off is that part of rainfall that flows off the surface to reach a sewer or river.

Run-off Reduction The process whereby practices are implemented to minimize the quantity of stormwater run-off generated, and/or attenuate run-off near its source using storage, infiltration, and/or uptake by vegetation.

Run-off Volume Amount of precipitation (and/or irrigation) minus surface storage, infiltration, evapotranspiration, and interception, that exits a defined area.

Sand Soil particle between 0.05 and 2.0 millimeters in diameter.

Sanitary Collection System The pipe system for collecting and carrying liquid and liquid-carried wastes from domestic sources to a wastewater treatment plant.

Sanitary Sewer A pipe or conduit (sewer) intended to carry wastewater or waterborne wastes from homes, businesses, and industries to the POTW (Publicly Owned Treatment Works). Stormwater run-off or unpolluted water should be collected and transported in a separate system of pipes or conduits (storm sewers) to natural watercourses.

Sanitary Wastewater (domestic) Wastewater, including toilet, sink, shower, and kitchen flows, originating from human domestic activities.

Scale A combination of mineral salts and bacterial accumulation that sticks to the inside of a collection pipe under certain conditions. Scale, in extreme growth circumstances, creates additional friction loss to the flow of water. Scale may also accumulate on surfaces other than pipes.

Screen 1. Porous material or mesh configured as a plate or cylinder that allows the passage of particles smaller than particular size, (e.g., an effluent screen) according to a specific flow/pressure relationship; a screen has area but no depth with respect to flow; 2. Use of a porous material or mesh in order to separate particles by size;

Scum A layer or film of foreign matter (such as grease, oil) that has risen to the surface of water or wastewater; a residue deposited on the ledge of a sewer, channel, or wet well at the water surface; a mass of solid matter that floats on the surface. Usually fatty material in wastewater that floats.

Seals (Hydraulic Seals) Seals (Hydraulic Seals) are used to isolate various parts of a sewer system, preventing vapor travel and spread of fire or explosion.

Secondary Clarifier A wastewater treatment device that consists of a rectangular or circular tank that allows those substances not removed by previous treatment processes that settle or float to be separated from the wastewater being treated.

Secondary Settlement Tank A tank in which settleable solids or humus are separated from the effluent flowing through it from biological filters or an activated sludge plant.

Secondary Treatment A wastewater treatment process used to convert dissolved or suspended materials into a form more readily separated from the water being treated. Usually the process follows primary treatment by sedimentation. The process commonly is a type of biological treatment process followed by secondary clarifiers that allow the solids to settle out from the water being treated.

Secondary Waste Treatment Processing by various types of systems that employ aeration and biological oxidation stages to decompose dissolved and colloidal organic contaminants (inorganic plant nutrients may also be partially removed).

Secondary Wastewater Treatment Biological removal or organics and solids from wastewater. Secondary wastewater effluent limits are generally 30 mg/L BOD_5 and 30 mg/L of TSS.

Sediment Soil that has washed or eroded from a land surface.

Sedimentation   Settling of solid material out of a liquid, typically accomplished by reducing the velocity of the liquid below the point at which it can transport the suspended material; may be enhanced by coagulation and flocculation.

Sedimentation (Wastewater)   The process of settling and depositing of suspended matter carried by wastewater.

Sedimentation usually occurs by gravity when the velocity of the wastewater is reduced below the point at which it can transport the suspended material.

Sedimentation Basin   (Clarifier, Settling Tank.) A tank or basin in which wastewater is held for a period of time during which the heavier solids settle to the bottom and the lighter materials float to the water surface.

Sedimentation Tanks   Provide a period of quiescence during which suspended waste material settles to the bottom of the tank and is scraped into a hopper and pumped out for disposal. During this period, floatable solids (fats, oils) rise to the surface of the tank and are skimmed off into scum pipes for disposal.

Seed Sludge   In wastewater treatment, seed, seed culture or seed sludge refers to a mass of sludge which contains populations of microorganisms. When a seed sludge is mixed with wastewater or sludge being treated, the process of biological decomposition takes place more rapidly.

Septic (Wastewater)   A condition produced by anaerobic bacteria. If severe, the wastewater produces hydrogen sulfide, turns black, gives off foul odors, contains little or no dissolved oxygen, and the wastewater has a high oxygen demand.

Septic Tank   A system sometimes used where wastewater collection systems and treatment plants are not available. The system is a settling tank in which settled sludge and floatable scum are in intimate contact with the wastewater flowing through the tank and the organic solids are decomposed by anaerobic bacterial action.

Used to treat wastewater and produce an effluent that flows into a subsurface leaching (filtering and disposal) system where additional treatment takes place. Also referred to as an "interceptor;" however, the preferred term is "septic tank."

Septic Tank Effluent Pump (STEP) System   A facility where effluent is pumped from a septic tank into a pressurized collection system, which may flow into a gravity sewer, treatment plant, or subsurface leaching system.

Septicity   The condition in which organic matter decomposes to form foul-smelling products associated with the absence of free oxygen. If severe, the wastewater produces hydrogen sulfide, turns black, gives off foul odors, contains little or no dissolved oxygen, and the wastewater has a high oxygen demand.

Series Operation   Wastewater being treated flows through one treatment unit and then flows through another similar treatment unit.

Settling   Process of subsidence and deposition of suspended matter carried by a liquid; typically accomplished by reducing the velocity of the liquid below the point at which it can transport the suspended material.

Settling Time   Time during which suspended, aggregated, precipitated, or colloidal substances settle by gravity.

Sewage   Sewage is the fluid discharged from medical, domestic, and industrial sanitary appliances. The water-borne wastes of a community. The used water and added waste of a community which is carried away by drains and sewers. The used household water and water-carried solids that flow in sewers to a wastewater treatment plant. The preferred term is "wastewater."

Sewage System Sewage system is any of several drainage systems for carrying surface water and sewage for disposal.

Sewer A sewer is an underground pipe or open channel in a sewage system for carrying water or sewage to a disposal area. A pipe or conduit that carries wastewater or drainage water. The term "collection line" is often used also.

Sewerage Sewerage is a system of sewers and ancillary works to convey sewage from its point of origin to a treatment works or other place of disposal.

Sludge The accumulated suspended solids of sewage deposited in tanks or basins. A mixture of solids and water produced during the treatment of wastewater. Biosolids remaining after secondary or tertiary treatment. Sludge may be applied to agricultural fields as a soil amendment, composted, or palletized. Or the settleable solids separated from liquids during processing or the deposits of foreign materials on the bottoms of streams or other bodies of water.

Sludge Dewatering Removing the remaining water from sludge for reuse and to lighten the sludge for reuse or disposal.

Sludge Digestion The process of changing organic matter in sludge into a gas or a liquid or a more stable solid form. These changes take place as microorganisms feed on sludge in anaerobic (more common) or aerobic digesters.

Sludge Gasification A process in which soluble and particulate organic matter are converted into gas by anaerobic decomposition. The resulting gas bubbles can become attached to the settled sludge and cause large clumps of sludge to rise and float on the water surface.

Sludge Return Process that sends the material (sludge) settled in a clarifier back to a septic or processing tank for further treatment or to maintain adequate microbial populations for treatment.

Sludge Volume Index (SVI) This is a calculation used to indicate the tendency of activated sludge solids (aerated solids) in the secondary clarifier to thicken or to become concentrated during the sedimentation/thickening process.

To determine SVI, allow a mixed liquor sample from the aeration basin to settle for 30 minutes. Also determine the suspended solids concentration for a sample of the same mixed liquor. Calculate SVI by dividing the measured (or observed) wet volume (mL/L) of the settled sludge by the dry weight concentration of MLSS in grams/L. When mixed liquor has an SVI well above 100 mL/gram of solids, it tends to form a thin slurry or billowing sludge blanket or to form bulky sludge.

Soil 1. Unconsolidated mineral and/or organic material on the immediate surface of the earth that serves as a medium for the growth of plants; 2. Unconsolidated mineral or organic matter on the surface of the earth that has been subjected to and shows effects of pedogenic and environmental factors of climate (including water and temperature effects), and macro- and microorganisms, conditioned by relief, acting on parent material over a period of time.

Solids, Settleable Suspended solids that will settle out of suspension within a specified period of time, expressed in milliliters per liter (mL/L).

Solids, Suspended That portion of total solids that is retained on a filter of 2.0 μm (or smaller) nominal pore size under specified conditions.

Solids, Total (TS) Material residue left in a vessel after evaporation of a sample subsequent to drying to a constant weight in an oven at 217 to 221 °F (103 to 105 °C); includes total suspended solids (TSS) and total dissolved solids (TDS); typically expressed in mg/L.

Solids, Total Dissolved (TDS) Material that passes through a filter of 2.0 μm (or smaller) nominal pore size, evaporated to dryness in a weighed dish and subsequently dried to constant weight at 180 °C; typically expressed in mg/L.

Solids, Total Suspended (TSS) Measure of all suspended solids in a liquid, typically expressed in mg/L; to measure, a well-mixed sample is filtered through a standard glass fiber filter and the residue retained on the filter is dried to a constant weight at 217 to 221 °F (103 to 105 °C); the increase in the weight of the filter represents the amount of total suspended solids.

Solid Waste Solid waste is any garbage, refuse, or sludge from a waste treatment plant, water supply treatment plant, or air pollution control facility, or other discarded material, including solids, liquids, semi-solids, or contained gaseous material resulting from industrial, commercial, mining, and agricultural operations and from community activities. In addition, solid waste does not include radioactive source, special nuclear, and/or their byproduct material.

Sorption Removal of an ion or molecule from solution by adsorption and/or absorption; term often used when the exact nature of the mechanism of removal is not known.

Springing Springing is separation of acid oils, either phenolic or naphthenic, by neutralization of spent caustic solutions. The acid oils are known as "sprung acids."

Storm Drain A pipeline or channel system that carries surface water and/or run-off to public waters, but does not feed into the sewer system.

Stormwater Storm water is rain water discharged from a catchment area as a result of a storm.

Stormwater Run-Off (SRO) The pulse of surface water following a rain storm. The water carries sediment, gas, oil, animal feces, glass, and other waste from the watershed to receiving waters creating a difficult urban/suburban wastewater problem.

Sublaterals Sublaterals are sewer branches (min. 150 mm (6 inch) diameter nominal size) collecting effluents from catch basins and conveying them to the laterals.

Subsurface Flow Wetland (SF) A type of constructed wetland in which primarily treated waste flows through deep gravel or other porous substrate planted with wetland vegetation. The water is not exposed to the air, avoiding problems with odor and direct contact.

Surface Loading One of the guidelines for the design of settling tanks and clarifiers in treatment plants. Used by operators to determine if tanks and clarifiers are hydraulically (flow) over- or under-loaded. Also called overflow rate.

Surface Run-Off The precipitation that cannot be absorbed by the soil and flows across the surface by gravity. The water that reaches a stream by traveling over the soil surface or falls directly into the stream channels, including not only the large permanent streams but also the tiny rills and rivulets. Water that remains after infiltration, interception, and surface storage has been deducted from total precipitation.

Surface Water Surface water is natural rain water from the ground surface, paved areas, and roofs plus occasional courtyard and car washing wastewaters and incidental fire fighting water.

Surfactant   Abbreviation for surface-active agent. The active agent in detergents that possesses a high cleaning ability.

Susceptibility to Separation (STS) Number   STS number is the oil contents in parts per million, of the suspended water after the specified settling period.

Suspended Solids   Solids physically suspended in sewage that can be removed by proper laboratory filtering.

Suspended Solids (SS) in Sewage   Solid in suspension in sewage liquors as measured by filtration either through a glass fiber filter paper followed by washing and drying at 105 °C, or by centrifuging flow by washing and removal of the supernatant liquid.

Suspended Growth Processes   Wastewater treatment processes in which the microorganisms and bacteria treating the wastes are suspended in the wastewater being treated. The wastes flow around and through the suspended growths. The various modes of the activated sludge process make use of suspended growth reactors. These reactors can be used for BOD removal, nitrification, and denitrification.

Suspended Solids   1. Solids that either float on the surface or are suspended in water, wastewater, or other liquids, and which are largely removable by laboratory filtering. 2. The quantity of material removed from water in a laboratory test, as prescribed in Standard Methods for the Examination of Water and Wastewater, and referred to as Total Suspended Solids Dried at 103 to 105 °C.

Suspension   A solution having small particles dispersed throughout.

TMDL: Total Maximum Daily Load   The sum of the allowable loads of a single pollutant from all contributing point and non-point sources. It is a calculation of the maximum amount of a pollutant that a water body can receive on a daily basis and still meet water quality standards; and an allocation of that amount to the pollutant's sources.

Tank   An artificial container in which liquids are held or detained.

Tank, Dosing   Tank or compartment that provides storage of effluent and contains a device (pump or siphon) and associated appurtenances used to convey effluent to another pretreatment process or a final treatment and dispersal component.

Tank, Flow Equalization   Dosing tank that provides storage of effluent and uses timed dosing to allow for uniform delivery to a subsequent component over time, usually a day or more; *also known as* a surge tank.

Tank, Processing   Term applied to a septic tank when it is configured to receive a combination of raw sewage and recirculated effluent in order to enhance nitrogen removal.

Tank, Pump   Dosing tank that provides storage of effluent and houses a pump and associated appurtenances used to convey effluent to another pre-treatment process or a final treatment and dispersal component.

Tank, Recirculation   Dosing tank that mixes effluent from two or more components within the treatment train and allows a portion of partially treated effluent to pass through one or more treatment components again.

Tank, Septic   Water-tight, covered receptacle for treatment of sewage; receives the discharge of sewage from a building, separates settleable and floating solids from the liquid, digests organic matter by anaerobic bacterial action, stores digested solids through a period of detention, allows clarified liquids to discharge for additional treatment and final dispersal, and attenuates flows.

Tank, Siphon Dosing tank or compartment that provides storage of effluent, and contains a siphon to convey effluent from the tank to another pre-treatment process or to a final treatment and dispersal component.

Temperature Controller Device that responds directly or indirectly to deviation from a desired temperature by actuating a control or initiating a control sequence.

Temperature Sensor A device that opens and closes a switch in response to changes in the temperature. This device might be a metal contact, or a thermocouple that generates minute electric current proportional to the difference in heat, or a variable resistor whose value changes in response to changes in temperature.

Tertiary Treatment Any process of water renovation that upgrades treated wastewater to meet specific reuse requirements. May include general cleanup of water or removal of specific parts of wastes insufficiently removed by conventional treatment processes. Typical processes include chemical treatment and pressure filtration. Also called "advanced waste treatment."

Tertiary Treatments (Effluent Polishing) A further stage of treating sewage by removing suspended solids. Consequential removal of residual BOD may occur. The use of filtration to remove microscopic particles from wastewater that has already been treated to a secondary level.

Tertiary Waste Treatment Following secondary treatment, the clarified effluent may require additional aeration and/or other chemical treatment to destroy bacteria remaining from the secondary treating stage, and to increase the content of dissolved oxygen needed for oxidation of the residual BOD. Tertiary treatment can also be used to remove nitrogen and phosphorous.

Tertiary Wastewater Treatment (Advanced) Biological or chemical polishing of wastewater to remove organics, solids, and nutrients. Tertiary wastewater effluent limits are generally 10 mg/L BOD_5 and 10 mg/L TSS.

Thickening Treatment to remove water from the sludge mass to reduce the volume that must be handled.

Toe Walls Toe walls are raised curbs that control spillage and drainage of storm, process, and fire water.

Top Water Level (TWL) The maximum water level in a settlement tank, an aeration tank, or a sludge storage tank.

Total Dissolved Solids Total Dissolved Solids (TDS) is the combined total of all dissolved solids in wastewater, both organic and inorganic and very fine, such as colloidal minerals. Generally particles must be smaller than two micrometers to be considered a dissolved solid. For example, salt dissolved in water is a dissolved solid. Therefore TDS will "survive" screening or other coarse filtration.

Total Organic Carbon (TOC) TOC is a measure of the amount of carbon in a sample originating from organic matter only. The test is run by burning the sample and measuring the CO_2 produced.

Total Solids The total amount of solids in solution and suspension.

Toxic Pollutant Pollutants or combinations of pollutants, including disease-causing agents, which after discharge and upon exposure, ingestion, inhalation, or assimilation into any organism, either directly from the environment or indirectly by ingestion through food chains, will, on the basis of information available to the Administrator of EPA, cause death, disease, behavioural abnormalities, cancer, genetic mutations, physiological malfunctions, (including malfunctions in reproduction), or physical deformations, in such organisms or their offspring. Toxic pollutants also include those pollutants listed by the Administrator under CWA Section 307(a)(1) or any pollutant listed under Section 405(d) that relates to sludge management.

Toxicity Test A procedure to determine the toxicity of a chemical or an effluent using living organisms. A toxicity test measures the degree of effect on exposed test organisms of a specific chemical or effluent.

Treatability How treatable a water sample is with a given substance.

Treatment Method, technique, or process designed to remove solids and/or pollutants from wastewater.

Treatment, Aerobic Digestion of organic matter in an environment containing molecular (or dissolved) oxygen (O_2).

Treatment, Advanced Secondary Level of treatment that achieves 95% reduction in BOD and TSS, generally to levels below 10 mg/L.

Treatment, Anaerobic Digestion of organic matter in an environment without molecular (or dissolved) oxygen (O_2).

Treatment, Biological Process involving the metabolic activities of bacteria and other microorganisms in the breakdown of complex organic materials into simpler, more stable substances.

Treatment, Chemical Process involving the addition of chemicals to obtain a desired result, such as precipitation, coagulation, flocculation, pH adjustment, disinfection, or sludge conditioning.

Treatment, Physical Treatment that involves only the physical means of solid–liquid separation, such as filtration, flotation, and sedimentation; chemical and biological reactions do not play an important role in physical treatment.

Treatment, Primary Physical treatment processes involving removal of particles, typically by settling and flotation with or without the use of coagulants (e.g., a grease interceptor or a septic tank provides primary treatment).

Treatment, Secondary Biological and chemical treatment processes designed to remove organic matter; a typical standard for secondary effluent is BOD and TSS less than or equal to 20 mg/L each on a 30-day average basis.

Treatment, Tertiary Advanced treatment of wastewater for enhanced organic matter removal, pathogen reduction, and nutrient removal; typical standards for tertiary effluent vary according to regulatory requirements.

Trickling-Filter An aerobic biological wastewater treatment process used as secondary treatment of sewage. Effluent from the primary clarifier is distributed over a bed of rocks. As the liquid trickles over the rocks, a biological growth on the rocks breaks down the organic matter in the sewage. The effluent is then taken to a clarifier to remove biological matter coming from the filter.

TSS Total suspended solids in wastewater. As the name implies, the total solid particles that are suspended (as opposed to dissolved) in the wastewater. TSS must be filtered out, flocculated, digested, and so on for removal in the treatment of wastewater. Though not necessarily pollutants, TSS is considered to be a measure of pollutants in water by the EPA in the USA.

Turbidity The cloudy appearance of water caused by the presence of suspended and colloidal matter. In the Handbook on Wastewater Management waterworks field, a turbidity measurement is used to indicate the clarity of water. Technically, turbidity is an optical property of the water based on the amount of light reflected by suspended particles. Turbidity cannot be directly equated to suspended solids because white particles reflect more light than dark-colored particles and many small particles will reflect more light than an equivalent large particle. Any finely divided, insoluble impurities that mar the clarity of the water. A measure of the clarity of water. Typically turbidity is measured by determining light transmission through the water.

Turbidity Meter An instrument for measuring and comparing the turbidity of liquids by passing light through them and determining how much light is reflected by the particles in the liquid. The normal measuring range is 0 to 100 and is expressed as Nephelometric Turbidity Units (NTUs).

Turbidity Units (TU) Turbidity units are a measure of the cloudiness of water. If measured by a nephelometric (deflected light) instrumental procedure, turbidity units are expressed in nephelometric turbidity units (NTU) or simply TU. Those turbidity units obtained by visual methods are expressed in Jackson Turbidity Units (JTU) which is a measure of the cloudiness of water; they are used to indicate the clarity of water. There is no real connection between NTUs and JTUs. The Jackson turbidimeter is a visual method and the nephelometer is an instrumental method based on deflected light.

Ultra-violet Disinfection (UV) A disinfection method in which final wastewater effluent is exposed to ultra-violet light to kill pathogens and microorganisms.

Ultra Filtration A membrane filter process used for the removal of some organic compounds in an aqueous (watery) solution.

Unit or Units Unit or units refer to one or all process, off-site, and/or utility units and facilities as applicable to form a complete operable refinery and/or complex.

Unit Operations, Physical Treatment methods in which the application of physical force predominates as a means for removal of wastewater constituents; includes flocculation, sedimentation, flotation, filtration, screening, mixing, and gas transfer.

Unit Processes, Biological Treatment methods in which the removal or conversion of constituents is brought about by biological activity; primarily used to remove the biodegradable organic constituents through conversion to cell tissue or gases; also used to remove nutrients (nitrogen and phosphorous).

Unit Processes, Chemical Treatment methods in which the removal or conversion of constituents is brought about through the addition of chemicals or by other chemical reactions; includes precipitation, adsorption, and disinfection.

Upset An exceptional incident in which there is unintentional and temporary non-compliance with the permit limit because of factors beyond the reasonable control of the permittee. An upset does not include non-compliance to the extent caused by operational error, improperly designed treatment facilities, inadequate treatment facilities, lack of preventive maintenance, or careless or improper operation.

Volatile A volatile substance is one that is capable of being evaporated or changed to a vapor at relatively low temperatures. Volatile substances can also be partially removed by air stripping. In terms of solids analysis, volatile refers to materials lost (including most organic matter) upon ignition in a muffle furnace for 60 minutes at 550 °C. Natural volatile materials are chemical substances usually of animal or plant origin. Manufactured or synthetic volatile materials such as ether, acetone, and carbon tetrachloride are highly volatile and not of plant or animal origin.

Volatile Acids   Fatty acids produced during digestion that are soluble in water and that can be steam-distilled at atmospheric pressure. Also called "organic acids." Volatile acids are commonly reported as equivalent to acetic acid.

Volatile Solids   Those solids in water, wastewater, or other liquids that are lost on ignition of the dry solids at 550 °C for 60 minutes.

Volume, Dose   1. Amount of effluent delivered to the distribution system during a dosing event including the drainback volume, pipe fill volume, and the delivered dose volume; 2. Amount of effluent delivered as determined by the pump on and pump off levels in a demand dosed system.

Vulnerability Assessment (Water)   An evaluation of drinking water source quality and its vulnerability to contamination by pathogens and toxic chemicals.

Waste Activated Sludge (WAS), mg/L   The excess growth of microorganisms which must be removed from the process to keep the biological system in balance. That portion of sludge from the secondary clarifier in the activated sludge process that is wasted to avoid a buildup of solids in the system.

Wastewater   A community's used water and water carried solids (including used water from industrial processes) that flow to a treatment plant. Stormwater, surface water, and groundwater infiltration may also be included in the wastewater that enters a wastewater treatment plant. The term "sewage" usually refers to household wastes, but this word is being replaced by the term "wastewater."

A community's used water and water carried solids (including used water from industrial processes) that flow to a treatment plant. Stormwater, surface water, and groundwater infiltration may also be included in the wastewater that enters a wastewater treatment plant. The term "sewage" usually refers to household wastes, but this word is being replaced by the term "wastewater."

The liquid-borne waste products of domestic, industrial, agricultural, and manufacturing activities. In a community, an average of 50 to 100 gallons of wastewater is generated per person per day.

Wastewater Facilities   The pipes, conduits, structures, equipment, and processes required to collect, convey, and treat domestic and industrial wastes, and dispose of the effluent and sludge.

Wastewater Ordinance   The basic document granting authority to administer a pretreatment inspection program. This ordinance must contain certain basic elements to provide a legal framework for effective enforcement.

Wastewater Stabilization Pond   Constructed basin lined with either soil with very low permeability or a synthetic material, surrounded by berms and which contains at least three feet of wastewater which utilizes sunlight, wind, or mechanical aeration, and natural bacteria to break down waste via physical, chemical, and biological processes to stabilize wastewater; typically consists of two or more basins with operational controls allowing or facilitating flow through the basins.

Wastewater Treatment Plant   An arrangement of pipes, equipment, devices, tanks, and structures for treating wastewater and industrial wastes. A water pollution control plant.

Wastewater, Industrial   Water or liquid-carried waste from an industrial process resulting from industry, manufacture, trade, automotive repair, vehicle wash, business, or medical activity; this wastewater may contain toxic or hazardous constituents.

Wastewater, Residential Strength Effluent from a septic tank or other treatment device with a BOD_5 less than or equal to 170 mg/L; TSS less than or equal to 60 mg/L; and fats, oils, and grease less than or equal to 25 mg/L.

Wastewater, Raw Any wastewater leaving a source.

Wastewater Reclamation Treatment or processing of wastewater to produce water of a quality appropriate for another use, including recycling or reuse.

Wastewater Recycling: Reclamation process of collection and treatment of wastewater on-site for return and use back into the same site; for example, collection and reclamation of gray water from an establishment for subsequent toilet flushing in that same establishment.

Wastewater Reuse Reclamation process of collection and treatment of wastewater for the deliberate application of that treated wastewater for a beneficial purpose such as turf irrigation.

Wastewater Treatment System Assembly of components for collection, treatment, and dispersal of sewage or effluent.

Wastewater Treatment System, Cluster Wastewater treatment systems designed to serve two or more sewage-generating dwellings or facilities with multiple owners; typically includes a comprehensive, sequential land-use planning component and private ownership.

Wastewater Treatment System, Collector Wastewater treatment system that conveys sewage or effluent from multiple sources to a location where treatment and dispersal occurs.

Wastewater Treatment System, Community Publicly owned wastewater treatment system for the collection, treatment, and dispersal of wastewater from two or more lots, or two or more equivalent dwelling units.

Wastewater Treatment System, Decentralized Wastewater treatment system for collection, treatment, and dispersal/reuse of wastewater from individual homes, clusters of homes, isolated communities, industries, or institutional facilities, at or near the point of waste generation.

Wastewater Treatment System, Individual Wastewater treatment system designed to serve one sewage-generating dwelling or facility.

Wastewater Treatment System, On-site (OWTS) Wastewater treatment system relying on natural processes and/or mechanical components to collect and treat sewage from one or more dwellings, buildings, or structures and disperse the resulting effluent on property owned by the individual or entity.

Water Pollution A general term signifying the introduction into water of microorganisms, chemicals, wastes, or sewage which renders the water unfit for its intended use.

Water Quality Standard (WQS) A law or regulation that consists of the beneficial use or uses of a waterbody, the numeric and narrative water quality criteria that are necessary to protect the use or uses of that particular waterbody, and an antidegradation statement.

Water Quality-Based Effluent Limit (WQBEL) A value determined by selecting the most stringent of the effluent limits calculated using all applicable water quality criteria (e.g., aquatic life, human health, and wildlife) for a specific point source to a specific receiving water for a given pollutant.

Water Softening Reduction in the number of and/or removal of polyvalent cations that are the principal cause of hardness in water.

Water Table Upper surface of groundwater or that level in the ground where the water is at atmospheric pressure.

Water Treatment Discharge Byproduct from a water treatment device, such as regeneration water from an ion-exchange unit, reject water from a reverse-osmosis unit, or the backwash from an iron filter.

Weir Device designed to measure or control flow; consists of a wall or obstruction of known geometric shape placed perpendicular to the direction of flow.

Well Hole bored or drilled into the ground.

Well, Monitoring Well constructed for the purpose of determining groundwater level or constituents.

Well, Water Well constructed for the purpose of extracting potable water.

Well-sorted Material of uniform size with maximum void space.

Whole Effluent Toxicity (WET) The total toxic effect of an effluent measured directly with a toxicity test.

Zone Portion of a component that is separately managed as a single unit.

Zone of Dispersal Layers of soil or rock material surrounding the zone of treatment through which the effluent moves away from the final treatment and dispersal component.

Zone of Saturation Layer in the ground in which interstitial voids (cracks, crevices, holes, etc.) are filled with water; the level at the top of this zone is the water table.

Zone of Treatment Soil or fill material that removes pollutants from pre-treated effluent by processes that include physical filtration of bacteria and other constituents, adsorption of viruses and bacteria by clay and organic matter, biological destruction of pathogens by soil microorganisms, sorption or precipitation of phosphorus, biochemical transformations of nitrogen compounds, and biological assimilation of phosphorus and nitrogen.

Bibliography and Further Reading

Abdel-Gawad, S., Abdel-Shafy, M. 2002 Pollution control of industrial wastewater from soap and oil industries: A case study. *Water Science and Technology* 46 (4–5), pp. 77–82.

Abdrakhimov, Yu. R. 1989 Treating wastewater from catalyst manufacturing to remove metal ions. *Chemistry and Technology of Fuels and Oils* 24, pp. 458–60.

Abilov, F.A., Orudjev, A.G., Lange, R. 1999 Optimization of oil-containing wastewater treatment processes. *Desalination* 124, pp. 225–29.

Adams, M., Campbell, I., Robertson, P.K.J. 2008 Novel photocatalytic reactor development for removal of hydrocarbons from water. *International Journal of Photoenergy* 1–7 (Article ID 674537).

Allen, E.W. 2008a Process water treatment in Canada's oil sands industry: I. Target Pollutants and treatment objectives. *Journal of Environmental Science* 138, pp. 123–38.

Allen, E.W. 2008b Process water treatment in Canada's oil sands industry: II. A review of emerging technologies. *Journal of Environmental and Engineering Science* 7, pp. 499–524.

Alta, L., Büyükgüngör, H. 2008 Sulfide removal in petroleum refinery wastewater by chemical precipitation. *Journal of Hazardous Materials* 153, pp. 462–69.

AltelaRain™ 2007 System ARS–4000: New Patented Technology for Cleaning Produced Water On-Site. Altela Information 26 January.

Al-Tell, N., Lueders, R. 1994 Texas refiner starts up new waste water treatment plant. *Oil and Gas Journal* 92, pp. 72–78.

Amosov, V.V., Zilberman, A.G., Kucheryavykh, E.I. 1976 Experience in local treatment of wastewater from petrochemical production. *Chemistry and Technology of Fuels and Oils* 12, pp. 850–52.

API Publ. 420 1990a Monographs on Refinery Environmental Control, Management of Water Discharges, The Chemistry and Chemicals of Coagulation and Flocculation, 1st Edn.

API Publ. 421 1990b Monographs on Refinery Environmental Control, Management of Water Discharges, Design and Operation of Water Separators, 1st Edn.

API 1988 Std. 650 Welded Steel Tanks for Oil Storage, 8th Edn.

API Publ. 4296 1978 Analysis of Refinery Waste Waters for the EPA Priority Pollutants, 1st Edn.

API Publ. 4346 1981 Refinery Waste Water Priority Pollutant Study-Sample Analysis and Evaluation of Data, 1st Edn.

Waste Management in the Chemical and Petroleum Industries, First Edition. Alireza Bahadori.
© 2014 John Wiley & Sons, Ltd. Published 2014 by John Wiley & Sons, Ltd.

API Publ. 4388 1988 Land Treatment–Safe and Efficient Disposal of Petroleum Waste, 1st Edn.

API Vol. 1, 1969 Manual on Disposal of Refinery Wastes, Liquid Wastes 1st Edn.

API 1980 API Manual on Waste Disposal of Refinery.

Arensdorf, J.J., Miner, K., Ertmoed, R., Clay, B., Stadnicki, P., Voordouw, G. 2009 Mitigation of reservoir souring by nitrate in a produced water re-injection system in Alberta. *Proceedings – SPE International Symposium on Oilfield Chemistry* 2, pp. 999–1006.

Asia, I.O., Enweani, I.B., Eguavoen, I.O. 2006 Characterization and treatment of sludge from the petroleum industry African. *Journal of Biotechnology* 5, pp. 461–66.

ASTM D 1888 1978 Test Methods for Particulate and Dissolved Matter, Solids, or Residue in Water. 2nd Edn.

ASTM ASTM D 270, ASTM D 4687 General Planning of Waste Sampling, ASTM D 4844 Air Monitoring at Waste Management Facilities for Workers, ASTM E 300.

AWWA 1998 *Water Treatment Plant Design*. 3rd Edn. McGraw-Hill; pp. 221–80.

Azetsu-Scott, K., Yeats, P., Wohlgeschaffen, G., et al. 2007 Precipitation of heavy metals in produced water: influence on contaminant transport and toxicity. *Marine Environmental Research*, 63, pp. 146–67.

Bahadori, A., Zahedi, G., Zendehboudi, S., Bahadori, M. 2013 Estimation of air concentration in dissolved air flotation (DAF) systems using a simple predictive tool. *Chemical Engineering Research and Design* 91, pp. 184–90.

Bahadori, A., Clark, M.W., Boyd, B. 2013 *Essentials of Water Systems Design in the Oil, Gas and Chemical Processing Industries*. Springer.

Bahadori, A., Vuthaluru, H.B., Mokhatab, S. 2009 Simple correlation accurately predicts aqueous solubility of light alkanes, *Energy Sources, Part A: Recovery, Utilization and Environmental Effects* 31, pp. 761–66.

Bahadori, A., Vuthaluru, H.B., Tadé, M.O., Mokhatab, S. 2008 Predicting water–hydrocarbon systems mutual solubility. *Chemical Engineering and Technology* 31, pp. 1743–47.

Baker, J.R., Milke, M.W., Mihelcic, J.R., 1999 Relationship between chemical and theoretical oxygen demand for specific classes of organic chemicals. *Water Research* 33, pp. 327–34.

Bakhshian, S., Dashtian, H., Abuzar Mirzai, P., Al Anazi, B.D. 2009 A review on impacts of drilling mud disposal on environment and underground water resources in south of Iran, Proceedings of the SPE/IADC Middle East Drilling Technology Conference and Exhibition, pp. 447–54.

Ball, H.L. 1994 Nitrogen reduction in an on-site trickling filter/upflow filters wastewater treatment system. *Proceedings of the 7th International Symposium on Individual and Small Community Sewage Systems*. American Society of Agricultural Engineers.

Bates, R.L., McDaniel, R., and Luckianow, B. 1993 "Chemical Oxidation by H_2O_2 Addition to Satisfy Wastewater Oxygen Demands During Production Field Startup Operations," *Proc., International Coalbed Methane Symposium*, Vol. I, Birmingham, Alabama pp. 365–74.

Bazua CD, Wilke CR. 1977 Ethanol effects on the kinetics of a continuous fermentation with Saccharomyces cerevisiae. *Biotechnol Bioeng Symp.* 7, pp. 105–18.

Belkin, S., Brenner, A., Abeliovich, A.1992 Effect of inorganic constituents on chemical oxygen demand –I. Bromides are unneutralized by mercuric sulfate complexation. *Water Research*, 26, pp. 1577–81.

Berkowitz, N. 1979 *An Introduction to Coal Technology*, pp. 30–32. Academic Press.

Berkowitz, J., Dickey, H., Prince, F., Stalzer, R. 1987 Performance of treatment technologies for refinery hazardous wastes. *Preprints Symposia* 32, p. 751.

Bessa, E., Sant'Anna, G.L., Dezotti, M. 2001 Photocatalytic/H_2O_2 treatment of oil field produced waters. *Applied Catalysis B: Environmental* 29, pp. 125–34.

Bilstad, T., Espedal, E, 1996 Membrane separation of produced water. *Water Science Technology* 34, pp. 239–46.

Boysen, J.E., Harju, J.A., Shaw, B., et al. 1999 SPE/EPA Exploration and Production Environmental Conference. Austin, TX: The current status of commercial deployment of the freeze thaw evaporation treatment of produced water; pp. 1–3. SPE 52700.

Boysen, J. 2007 TX: The freeze–thaw/evaporation (FTE) process for produced water treatment, disposal and beneficial uses. 14th Annual International Petroleum Environmental Conference. Houston, pp. 5–9.

British Standard, BS 262.375 1983 Design and Installation of Small Sewage Treatment Works and Cesspools.

British Standard BS 59.91666667 1971(1983) Media for Biological Percolating Filters.

Buller, A.T., Johnsen, S., Frost, K. 2003 Offshore produced water management-knowledge, tools and procedures for assessing environmental risk and selecting remedial measures. Memoir 3. Statoil Research and Technology Offshore.

Çakmakce, M., Kayaalp, N., Koyuncu, I. 2008 Desalination of produced water from oil production fields by membrane processes. *Desalination*, 222, pp. 176–86.

Campos, J.C., Borges, R.M.H., Oliveira Filho, A.M., Nobrega, R., Sant'Anna Jr., G.L. 2002 Oilfield wastewater treatment by combined microfiltration and biological processes. *Water Research* 36, pp. 95–104.

Canadian Association of Petroleum Producers (CAPP) 2008 *Environmental Challenges and Progress in Canada's Oil Sands*. CAPP.

Cassidy, A.L. 1993 Advances in flotation unit design for produced water treatment. Production Operations Symposium. Oklahoma.

Chan, L-H., Starinsky, A., Katz, A. 2002 The behavior of lithium and its isotopes in oilfield brines: evidence from the Heletz-Kokhav field, Israel. *Geochimica et Cosmochimica Acta* 66, pp. 615–23.

Chapman, D., Grammas, J. 1989 Management and analysis of data from petroleum refinery wastewater treatment plants. *Water Pollution Research Journal of Canada* 24, pp. 391–410.

Chaubal, S. 2006 The wastewater challenge. *Hydrocarbon Engineering* 11, pp. 87–88.

Chen, C., Guo, J., He, X., Zeng, Q. 2011 Analyzing and treatment of waste water from petroleum industry, 5th International Conference on Bioinformatics and Biomedical Engineering, iCBBE 2011, art. no. 5781163.

Chiang, H.-L., Choa, C.-G., Chen, S.-Y., Tsai, M.-C. 2003 The reuse of biosludge as an adsorbent from a petrochemical wastewater treatment plant. *Journal of the Air and Waste Management Association* 53, pp. 1042–51.

Chin, K.K. 1994 Evaluation of treatment efficiency of processes for petroleum refinery wastewater. *Water Science and Technology* 29, pp. 47–50.

Cline, J.T. 1998 Treatment and discharge of produced water for deep offshore disposal. API Produced Water Management Technical Forum and Exhibition.

Coelho, A., Castro, A.V., Dezotti, M., Sant'Anna Jr., G.L. 2006 Treatment of petroleum refinery sourwater by advanced oxidation processes. *Journal of Hazardous Materials* 137, pp. 178–84.

Colorado School of Mines. 2009 Technical Assessment of produced water treatment technologies. pp. 8–128. An Integrated Framework for Treatment and Management of Produced Water. RPSEA Project 07122–12, Colorado.

Crawford, D.W., Bonnevie, N.L., Wenning, R.J. 1995 Sources of pollution and sediment contamination in Newark Bay, New Jersey. *Ecotoxicology and Environmental Safety* 30, pp. 85–100.

Daniels, R., Davies, J., Gravell, A., Rowland, P. 2000 Movement of petroleum hydrocarbons in sandy coastal soils. *Journal of Environmental Monitoring* 2, pp. 645–50.

Darwish, M.A., Al Asfour, F., Al-Najem, N. 2003 Energy consumption in equivalent work by different desalting methods: case study for Kuwait. *Desalination* 152, pp. 83–92.

DeJohn, P.B., Adams, A.D. 1975 Activated carbon improves wastewater treatment. *Hydrocarbon Processing* 54, pp. 104–7.

Demirci, S., Erdogan, B., Ozcimder, R. 1998 Wastewater treatment at the petroleum refinery, Kirikkale, Turkey using some coagulants and Turkish clays as coagulant aids. *Water Research* 32, pp. 3495–99.

DIN 997 1970 Tracing Dimensions for Bars and Rolled Steel Sections.

Diya'Uddeen, B.H., Daud, W.M.A.W., Abdul Aziz, A.R. 2011 Treatment technologies for petroleum refinery effluents: A review. *Process Safety and Environmental Protection* 89, pp. 95–105.

Dold, P.L. 1989 Current practice for treatment of petroleum refinery wastewater and toxics removal. *Water Pollution Research Journal of Canada* 24, pp. 363–90.

Dold, P. 1992 Activated sludge treatment of petroleum refinery wastewater: Part 1 – Experimental behaviour. *Water Science and Technology* 26, pp. 333–43.

Doran, G., Leong, L.Y.C. 2000 Developing a cost effective solution for produced water and creating a 'new' water resource. United States Department of Energy; DOE/MT/95008-4.

Doran, G.F., Williams, K.L., Drago, J.A., et al. 1998 Pilot-study results to convert oilfield produced water to drinking-water or reuse quality. Proceedings of the SPE Annual Technical Conference. pp. 403–17. New Orleans, LA Production Operations and Engineering/General.

Eckenfelder, W.W., Bowers, A.R., Roth, J.A., et al. 1997 *Chemical Oxidation: Technologies for the Nineties*. Vol. 6. pp. 1–12. Technomic Publishing Co. Inc.

El-Naas, M.H., Al-Zuhair, S., Al-Lobaney, A., Makhlouf, S. 2009 Assessment of electrocoagulation for the treatment of petroleum refinery wastewater. *Journal of Environmental Management* 91, pp. 180–85.

Ely, J.W., Horn, A., Cathey, R., Fraim, M., Jakhete, S. 2011 Game changing technology for treating and recycling frac water. *Proceedings – SPE Annual Technical Conference and Exhibition* 1, pp. 506–17.

EPA. 1980 *Onsite Wastewater Treatment and Disposal Systems Design Manual*. EPA.

Ettouney, H.M., El-Dessouky, H.T., Gowin, P.J., et al. 2002 Evaluating the economics of desalination. *Chemical Engineering Progress* 98, pp. 32–9.

Faibish, R.S., Cohen, Y. 2001a Fouling-resistant ceramic-supported polymer membranes for ultrafiltration of oil-in-water microemulsions. *Journal of Membrane Science* 185, pp. 129–43.

Faibish, R.S., Cohen, Y. 2001b Fouling and rejection behavior of ceramic and polymer-modified ceramic membranes for ultrafiltration of oil-in-water emulsions and microemulsions. *Colloids and Surfaces A: Physicochemical and Engineering Aspects* 191, pp. 27–40.

Fakhru'l-Razi, A., Pendashteh, A., Abdullah, L.C., et al. 2009 Review of technologies for oil and gas produced water treatment. *Journal of Hazardous Material* 170, pp. 530–51.

Fang, H.-Y., Chou, M.-S., Huang, C.-W. 1993 Nitrification of ammonia–nitrogen in refinery wastewater. *Water Research* 27, pp. 1761–65.

Fujishima, A., Honda, K. 1972 Electrochemical photolysis of water at a semiconductor electrode. *Nature* 238, pp. 37–8.

Galil, N., Rebhun, M. 1992 Multiple technological barriers combined with recycling of water and oil in wastewater treatment of a coastal petrochemical complex. *Water Science and Technology* 25, pp. 277–82.

Galil, N., Rebhun, M. 1993 Combined aerated ponds and chemical clarification in the treatment of petrochemical wastewater. *Water Science and Technology* 27, pp. 79–88.

Gallenkemper, B., Voss, U. 1991 Reduction of oil in sludges from light petroleum separators. *Waste Management and Research* 9, pp. 293–304.

Gandurina, L.V., Gervits, E.I. 1987 Treatment of oil–containing wastewater using activated silicic acid as a flocculant. *Chemistry and Technology of Fuels and Oils* (English translation of Khimiya i Tekhnologiya Topliv i Masel) 23, pp. 449–50.

Gao, W., Smith, D.W., Habib, M. 2008 Petroleum refinery secondary effluent polishing using freezing processes – Toxicity and organic contaminant removal. *Water Environment Research* 80, pp. 517–23.

Gaya, U.I., Abdullah, A.H. 2008 Heterogeneous photocatalytic degradation of organic contaminants over titanium dioxide: a review of fundamentals, progress and problems. *Journal of Photochemistry and Photobiology C: Photochemistry Review* 9, pp. 1–12.

Gerber, Ya., V., Lukyanov, V.I., Popova, I.A. 1979 Biochemical treatment of refinery wastewater. *Chemistry and Technology of Fuels and Oils* 15, pp. 254–59.

Gerber, Ya., Gorobets, V., Ioakimis, E.G. 1979 Effect of ammonium nitrogen on biochemical treatment of refinery wastewater. *Chemistry and Technology of Fuels and Oils* 15, pp. 259–62.

Gheorghe, C.G., Pantea, O., Matei, V., Bombos, D., Borcea, A.-F. 2011 The efficiency of flocculants in biological treatment with activated sludge. *Revista de Chimie* 62, pp. 1023–26.

Ghose, T.K., Tyagi, R.D., 1979 Rapid ethanol fermentation of cellulose hydrolysate: II. Product and substrate inhibition and optimization of fermentor design. *Biotechnology and Bioengineering* 21, pp. 1401–20.

Gloyna, E.F., Tischler, L.F. 1979 Design of waste stabilization pond systems. *Progress in Water Technology* 11, pp. 47–70.

Grant, A., Briggs, A.D. 2002 Toxicity of sediments from around a North Sea oil platform: are metals or hydrocarbons responsible for ecological impacts? *Marine Environmental Resources* 53, pp. 95–116.

Gray, N.F. 1990 *Activated Sludge – Theory and Practice*. Oxford University Press.

Grini, P.G., Hjelsvold, M., Johnsen, S. 2002, HSE Conference. Kuala Lumpur, Malaysia; Choosing produced water treatment technologies based on environmental impact reduction. 20–22 March, SPE paper 74002.

Grutsch, J.F., Quanstrom, W.R. 1985 A short history of refinery process wastewater treatment in the USA. *Industry and Environment* 8, pp. 16–20.

Gu, Q., Peng, C., Li, F., et al. 2005 Treatment of oil–contaminated wastewater through bed coalescence using a new filter medium. *Environmental Engineering Science* 22, pp. 472–78.

Guan, W.-S., Lei, Z.-X., Zhu, J.-H. 2000 Petrochemical wastewater treatment by biological process. *Journal of Environmental Sciences* 12, pp. 220–24.

Guo, H., Li, X., Wen, K., Lin, B. 2000 Discussion on design of solid waste disposal site in petroleum refining enterprise. *Petroleum Refinery Engineering* 30, pp. 52–55.

Gutierrez, G., Lobo, A., Allende, D., et al. 2008 Influence of coagulant salt addition on the treatment of oil-in-water emulsions by centrifugation, ultrafiltration, and vacuum evaporation. *Separation Science and Technology* 43, pp. 1884–95.

Han, K. and Levenspiel, O. 1988 Extended monod kinetics for substrate, product, and cell Inhibition. *Biotechnology and Bioengineering* 32, pp. 430–43

Han, R., Zhang, S., Xing, D., et al. 2010 Desalination of dye utilizing copoly(phthalazinone biphenyl ether sulfone) ultrafiltration membrane with low molecular weight cut-off. *Journal of Membrane Science* 358, pp. 1–6.

Han, S.-L., Zaotang, L., Yandun, G. 1999 Application of desulfurized activated carbon for desulfurizing ammonia in refinery plant, Linchan Huaxue Yu Gongye. *Chemistry and Industry of Forest Products* 19, pp. 55–60.

Hanor, J.S. 1984 Variation in the chemical composition of oilfield brines with depth in Northern Louisiana and Southern Arkansas: Implications for mechanisms and rates of mass transport and diagenetic reaction. *Transactions of the Gulf Coast Association of Geological Societies*, 34, pp. 55–61.

Hansen, B.R., Davies, S.R.H. 1994 Review of potential technologies for the removal of dissolved components from produced water. *Chemical Engineering Research and Design* 72, pp. 176–88.

Hart, J.A. 1973 Waste water recycled for use in refinery cooling towers. *Oil and Gas Journal* 71, pp. 92–96.

Hayes, T., Gowelly, S., Moon, P., et al. 2006 International Petroleum Environmental Conference. San Antonio, TX; electrodialysis treatment of coal bed methane produced water: application issues and projections of costs. 17–20 October.

He, Y., Jiang, Z.W. 2008 Technology review: treating oilfield wastewater. *Filter and Separation*, 45, pp. 14–16.

Heidarzadeh, N., Gitipour, S., Abdoli, M.A. 2010 Characterization of oily sludge from a Tehran oil refinery. *Waste Management and Research* 28, pp. 921–27.

Heins, W.F., McNeill, R. 2007 Vertical–tube evaporator system provides SAGD–quality feed water. *World Oil Magazine*, p. 228.

Henshaw, P.F., Bewtra, J.K., Biswas, N., Franklin, M. 1992 Biological removal of hydrogen sulfide from refinery wastewater and conversion to elemental sulfur. *Water Science and Technology* 25, pp. 265–67.

Hobson, T. 1996 Skimming oily wastewater. *Pollution Engineering* 28, pp. 52–54.

Holubar, P., Grudke, T., Moser, A., Strenn, B., Braun, R. 2000 Effects of bacterivorous ciliated protozoans on degradation efficiency of a petrochemical activated sludge process. *Water Research*, 34, pp. 2051–60.

Hou, F.-S. 2005 Optimizing refining processing for the development of China's petroleum refining industry, Shiyou Xuebao. *Shiyou Jiagong/Acta Petrolei Sinica (Petroleum Processing Section)* 21, pp. 7–16.

Hu, X.-J., Yang, Y.-N., Liu, H., et al. 2007 Study on the treatment of oilfield produced wastewater with different salinities using halophile enhanced biological activated carbon. *Huanjing Kexue/Environmental Science* 28, pp. 2213–18.

Hu, Z., and Grasso, D. 2006 Chemical oxygen demand. In *Encyclopedia of Analytical Science Analysis*, pp. 325–30. Elsevier Ltd.

Huchler, L.A. 2004 Understanding refinery wastewater treatment basics – Part 2. *Hydrocarbon Processing* 83 (2 Section 1), pp. 108.

Hudgins, C.M. 1994 Petrotech Consultants Inc. Chemical use in North Sea oil and gas E&P. *Journal of Petroleum Technology*, 46, pp. 67–74.

Hwang, S., Moore, I. 2011 Water network synthesis in refinery. *Korean Journal of Chemical Engineering* 28, pp. 1975–85.

Igunnu, E.T, Chen, G.Z. 2012 Produced water treatment technologies. *International Journal of Low-Carbon Technologies* 0 p. 1–21.

Ishak, S., Malakahmad, A., Isa, M.H. 2012 Refinery wastewater biological treatment: A short review. *Journal of Scientific and Industrial Research* 71, pp. 251–56.

Jensen, K. 1983 Sources and routes of removal of petroleum hydrocarbons in a Danish Marine inlet. *Oil and Petrochemical Pollution* 1, pp. 207–16.

Jerez Vegueria, S.F., Godoy, J.M., Miekeley, N. et al. 2002 Environmental impact studies of barium and radium discharges by produced waters from the 'Bacia de Campos' oil-field offshore platforms. *Brazilian Journal of Environmental Radioactivity* 62, 29–38.

Johnson, K.D., Martin, C.D., Davis, T.G. 1999 Treatment of wastewater effluent from a natural gas compressor station. *Water Science and Technology* 40, pp. 51–56.

Judd, S., Jefferson, B. 2003 *Membranes for Industrial Wastewater Recovery and Re–use*. pp. 14–169. Elsevier Ltd.

Kaiser, W.R. 1992 Geologic evaluation of critical production parameters for coalbed methane resources. *Quarterly Review of Methane from Coal Seams Technology* 9, pp. 25–31.

Kalbfus, W. 1986 Analyze the hydrocarbons in liquid refinery wastes. *Hydrocarbon Processing* 65, pp. 77–78.

Karelin, Ya., A., Zhukov, D.D., Saidaminov, I.A. 1979 Biochemical treatment of oil–containing wastewater in aeration tanks. *Chemistry and Technology of Fuels and Oils* 15, pp. 540–44.

Kaur, G., Mandal, A.K., Nihlani, M.C., et al. 2009 Control of sulfidogenic bacteria in produced water from the Kathloni oilfield in northeast India. *International Biodeterioration and Biodegradration* 63, pp. 151–55.

Kawaguchi, H., Li, Z., Masuda, Y., Sato, K. 2012 Dissolved organic compounds in reused process water for steam-assisted gravity drainage oil sands extraction. *Water Research* 46, pp. 5566–74.

Kfir, O., Tal, A., Gross, A., Adar, E. 2012 The effect of reservoir operational features on recycled wastewater quality. *Resources, Conservation and Recycling* 68, pp. 76–87.

Khawaji, A.D., Kutubkhanah, I.K., Wie, J. 2008 Advances in seawater desalination technologies. *Desalination* 221, pp. 47–69.

Khemakhem, S., Larbot, A., Ben Amar, R. 2009 New ceramic microfiltration membranes from Tunisian natural materials: application for the cuttlefish effluents treatment. *Ceramics International* 35, pp. 55–61.

Khosravi, J., Alamdari, A. 2009 Copper removal from oil–field brine by coprecipitation. *Journal of Hazardous Material* 166, pp. 695–700.

Knight, R.L., Kadlec, R.H., Ohlendorf, H.M. 1999 The use of treatment wetlands for petroleum industry effluents. *Environmental Science and Technology* 33, pp. 973–80.

Kochetkova, R.P., Nikitishina, V.A., Kolisnyk, G.P. 1981 Effect of petroleum sulfones on biochemical treatment of wastewater from petrochemical production. *Chemistry and Technology of Fuels and Oils* 17, pp. 235–37.

Konieczny, K., Bodzek, M., Rajca, M. 2006 A coagulation–MF system for water treatment using ceramic membranes. *Desalination* 198, pp. 92–101.

Kozawa, T., Wueki, T., Kobayashi, H., Matsui, S. 1992 Management of toxics for the Fukashiba industrial wastewater treatment plant of the Kashima petrochemical complex. *Water Science and Technology* 25, pp. 247–54.

Kriipsalu, M., Marques, M., Maastik, A. 2008 Characterization of oily sludge from a wastewater treatment plant flocculation-flotation unit in a petroleum refinery and its treatment implications. *Journal of Material Cycles and Waste Management* 10, pp. 79–86.

Kubsad, V., Gupta, S.K., Chaudhari, S. 2005 Treatment of petrochemical wastewater by rotating biological contactor. *Environmental Technology* 26, pp. 1317–26.

Kujawski, D., Wong, A. 2011 Optimizing biological water treatment in petroleum refining. *Chemical Engineering* 118, pp. 60–63.

Kulik, N., Trapido, M., Veressinina, Y., Munter, R. 2007 Treatment of surfactant stabilized oil–in–water emulsions by means of chemical oxidation and coagulation. *Environmental Technology* 28, pp. 1345–55.

Kwok, Dong N., Lo, S. W., Wong, K. 2005 An optical biosensor for multi-sample determination of biochemical oxygen demand (BOD). *Sensors and Actuators B* 110, pp. 289–98.

Laffly, G. 1989 Developments in legislation related to treatment of petroleum refinery effluents. A U.S. overview. *Water Pollution Research Journal of Canada* 24, pp. 355–62.

Lawrence, A.W., Miller, J.A., Miller, D.L. 1995 A regional assessment of produced water treatment and disposal practices and research needs. *SPE/EPA Exploration and Production Environmental Conference*. pp. 373–92. Houston, TX.

Lawrence, A.W. 1993 Coalbed methane produced-water treatment and disposal options. *Quarterly Review of Methane from Coal Seams Technology* 11, pp. 6–17.

Lefebvre, O., Moletta, R. 2006 Treatment of organic pollution in industrial saline wastewater: A literature review. *Water Research* 40, pp. 3671–82.

Legovic, B., Nikolic, O., 1986 Systems and solutions for wastewater treatment in the biggest Yugoslav oil refinery. *Water Science and Technology* 18, pp. 85–90.

Leng, W.H., Zhu, W.C., Ni, J. et al. 2006 Photoelectrocatalytic destruction of organics using TiO2 as photoanode with simultaneous production of H2O2 at the cathode. *Applied Catalysis A: General* 300, pp. 24–35.

Letterman, R.D., Clifford, D.A. 1999 Ion exchange and inorganic adsorption. In: Letterman, R.D. (ed.) *Water Quality and Treatment*. McGraw-Hill.

Li, G., An, T., Chen, J., et al. 2006 Photoelectrocatalytic decontamination of oilfield produced wastewater containing refractory organic pollutants in the presence of high concentration of chloride ions. *Journal of Hazardous Material* 138, pp. 392–400.

Li, G., An, T., Nie, X., et al. 2007 Mutagenicity assessment of produced water during photoelectrocatalytic degradation. *Environmental Toxicology and Chemistry* 26, pp. 416–23.

Li, Z., Lai, G. 1999 Application of vortex separator in wastewater treatment of petroleum refining unit. *Petroleum Refinery Engineering* 29, pp. 45.

Li, Y., Xie, Q., Zhang, X., et al. 2011 Study on new method of microorganism acclimation for biological treatment of high-salinity oilfield wastewater. *Proceedings – International Conference on Computer Distributed Control and Intelligent Environmental Monitoring*, art. no. 5748158, pp. 1760–63. CDCIEM.

Li, Y., Yao, P.-J. 2003 Wastewater minimization in petroleum refineries. *Shiyou Xuebao, Shiyou Jiagong/Acta Petrolei Sinica (Petroleum Processing Section)* 19, pp. 57–63.

Li, R., He, C., Chen, L., et al. 2011 Flocculation treatment of wastewater from low-level penetrative oil fields in Northern Shaanxi. *Proceedings of 2011 International Symposium on Water Resource and Environmental Protection* 2, art. no. 5893232, pp. 1207–10. ISWREP.

Lin, K.-H., Hsu, H.-T., Ko, Y.-W., et al. 2009 Pyrolytic product characteristics of biosludge from the wastewater treatment plant of a petrochemical industry. *Journal of Hazardous Materials* 171, pp. 208–14.

Lindquist, S.E., Fell, C. 2009 Fuels – Hydrogen Production. *Encyclopedia of Electrochemical Power Sources*. Vol. 3, pp. 369–83. Elsevier.

Lindsay, J.T., Prather, B.V. 1977 Solving petroleum refinery wastewater problems. *Journal of the Water Pollution Control Federation* 49, pp. 1779–85.

Lobo, A., Cambiella, Á., Benito, J.M. et al. 2006 Ultrafiltration of oil-in-water emulsions with ceramic membranes: influence of pH and crossflow velocity. *Journal of Membrane Science* 278, pp. 328–34.

Lu, M., Zhang, Z., Yu, W., Zhu, W. 2009 Biological treatment of oilfield–produced water: A field pilot study. *International Biodeterioration and Biodegradation* 63, pp. 316–21.

Luckianow, B.J., Hall, W.L. 1991 Water storage key factor in coalbed methane production. *Oil & Gas Journal* 89, pp. 79–84.

Ludzack, F.J., Noran, D.K. 1965 *Tolerance of High Salinities by Conventional Wastewater Treatment Processes*. Water Environment Federation.

Ma, H., Wang, B. 2006 Electrochemical pilot–scale plant for oil field produced wastewater by M/C/Fe electrodes for injection. *Journal of Hazardous Material* 132, pp. 237–43.

Ma, Y., Zhang, J.-Y., Wong, M.-H., Wu, W.-Z. 2001 Optimization of control parameters for petroleum waste composting. *Journal of Environmental Sciences* 13, pp. 385–90.

Madaeni, S.S. 1999 The application of membrane technology for water disinfection. *Water Resources* 33, pp. 301–8.

Madaeni, S.S., Eslamifard, M.R. 2010 Recycle unit wastewater treatment in petrochemical complex using reverse osmosis process. *Journal of Hazardous Materials* 174, pp. 404–9.

Majid, A., Sparks, B.D. 1999 Potential applications of oil sands industry wastes. *Journal of Canadian Petroleum Technology* 38, pp. 29–33.

Malato, S., Blanco, J., Richter, C. et al. 1998 Enhancement of the rate of solar photocatalytic mineralization of organic pollutants by inorganic oxidizing species. *Applied Catalysis B: Environment* 17, pp. 347–56.

Mamma, D., Gerontas, S., Philippopoulos, C.J., Christakopoulos, P., Macris, B.J., Kekos, D. 2004 Combined photo-assisted and biological treatment of industrial oily wastewater. *Journal of Environmental Science and Health – Part A Toxic/Hazardous Substances and Environmental Engineering* 39, pp. 729–40.

Mansuy-Huault, L., Regier, A., Faure, P. 2009 Analyzing hydrocarbons in sewer to help in PAH source apportionment in sewage sludges, *Chemosphere* 75, pp. 995–1002.

Maqueda, M.A.M., Martinez, D., S.A., Narváez, D., Rodriguez, R., M.G., Aguilar, L., R., Herrero, V., V.M. 2006 Dynamical modelling of an activated sludge system of a petrochemical plant operating at high temperatures. *Water Science and Technology* 53, pp. 135–42.

Mark, W. 2007 *The Guidebook to Membrane Desalination Technology: Reverse Osmosis, Nanofiltration and Hybrid Systems Process, Design, Applications and Economics.* 1st Edn, pp. 160–80. L'Aquila Desalination Publications.

Mattson, J.S. 1973 Preparation of granular activated carbons from petroleum residues. *Industrial & Engineering Chemistry Product Research and Development* 12, pp. 312–17.

McKinney, R.E. 1967 Biological treatment systems for refinery wastes. *Journal of the Water Pollution Control Federation* 39, pp. 346–59.

Meijer, D.T., Madin, C. 2010 Removal of dissolved and dispersed hydrocarbons from oil and gas produced water with mppe technology to reduce toxicity and allow water reuse. *APPEA Journal*, 1–11.

Meijer, D.T., Kuijvenhoven Cor, A.T., Karup, H. 2004 NEL Produced Water Workshop. Aberdeen, UK; Results from the latest MPPE field trials at NAM and total installations; p. 21–22.

Merlo, R., Gerhardt, M.B., Burlingham, F., et al. 2011 Petroleum refinery stripped sour water treatment using the activated sludge process. *Water Environment Research* 83, pp. 2067–78.

Metcalf and Eddy (2003). *Wastewater Engineering: Treatment Disposal Reuse.* 4th Edn. McGraw-Hill.

Mijaylova Nacheva, P., Ramírez Camperos, E., Sandoval Yoval, L., 2008 Treatment of petroleum production wastewater for reuse in secondary oil recovery. *Water Science and Technology* 57, pp. 875–82.

Miskovic, D., Dalmacija, B., Zivanov, Z. 1986 An investigation of the treatment and recycling of oil refinery wastewater. *Water Science and Technology* 18, pp. 105–14.

Mondal, S., Wickramasinghe, S.R. 2008 Produced water treatment by nanofiltration and reverse osmosis membranes. *Journal of Membrane Science* 322, pp. 162–70.

Monod, J. 1949 The growth of bacterial cultures. *Annual Review of Microbiology* 0.125, 371–94.

Mount, D.R., O'Neil, P.E., and Evans, J.M. 1993 discharge of coalbed produced water to surface waters – assessing, predicting, and preventing ecological effects. *Quarterly Review of Methane from Coal Seams Technology* 11, pp. 18–25.

Murray-Gulde, C., Heatley, J.E., Karanfil, T., et al. 2003 Performance of a hybrid reverse osmosis-constructed wetland treatment system for brackish oil field produced water. *Water Research* 37, pp. 705–13.

Myasnikov, N., Gandurina, L.V., Butseva, L.N. 1985 Russian use of flotation for wastewater treatment. *Effluent & Water Treatment Journal* 25, pp. 202–6.

Nadav, N. Boron removal from seawater reverse osmosis permeate utilizing selective ion exchange resin. *Desalination* 124, 131–35.

National Energy Board (NEB) 2006 Canada's Oil Sands – Opportunities and Challenges to 83.95833333 An Update. June 2006.

O'Neil, P.E., Harris, S.C., and Mettee, M.F. 1989 Stream monitoring of coalbed methane produced water from the cedar cove degasification field, Alabama. Proceedings of the Coalbed Methane Symposium, Tuscaloosa, Alabama, pp. 355–61.

Ontario Ministry of the Environment 1992 Background Document on the Development of the Draft Petroleum Refining Sector Effluent Limits Regulation, Water Resources Branch, Ontario Ministry of the Environment, August 1992, Canada.

Ortiz-Gallarza, S.M., Ramírez-López, J.A. 2003 Water quality of the Tula River related to the petroleum refining industry: Accumulation factors and treatments. *Progress in Water Resources* 9, pp. 67–77.

OSHA n.d Interim Final Rule for Hazardous Waste Operations and Emergency Response – Recommended Practice D 1605 ASTM, Annual Book of ASTM Standard Volume 11.03, NIOSH Manual of Analytical Methods, – OSHA Analytical Method Manual.

Otadi, N., Hassani, A.H., Javid, A.H., Khiabani, F.F. 2010 Oily compounds removal in wastewater treatment system of pars oil refinery to improve its efficiency in a lab scale pilot. *Journal of Water Chemistry and Technology* 32, pp. 370–77.

Parker, W., Farquhar, G.J. 1989 Treatment of a petrochemical wastewater in an anaerobic packed bed reactor. *Water Pollution Research Journal of Canada* 24, pp. 195–205.

Pars, H.M., Meijer, D.T. 1998 Removal of dissolved hydrocarbons from production water by macro porous polymer extraction (MPPE). SPE International Conference on Health, Safety and Environment in Oil and Gas Exploration and Production. Caracas, Venezuela; 1998 June, SPE paper no. 46577.

Parsons, W.A. 1978 Polishing biological plant effluents. *Hydrocarbon Processing* 57 (10 Section 1), pp. 125–26.

Pashin, J.C., Ward, W.E., Winston, R.B. et al. 1990 Geologic Evaluation of Critical Production Parameters for Coalbed Methane Resources, Annual Report, Part II – Black Warrior Basin. Gas Research Institute.

Pavlova, A., Ivanova, R. 2003 Determination of petroleum hydrocarbons and polycyclic aromatic hydrocarbons in sludge from wastewater treatment basins. *Journal of Environmental Monitoring* 5, pp. 319–23.

Pelizzetti, E., Pramauro, E., Minero, C. et al. 1990 Sunlight photocatalytic degradation of organic pollutants in aquatic systems. *Waste Management* 10, 65–71.

Petzet, G.A. (ed.) 1990 Devon pressing fruitland coal seam program. *Oil & Gas Journal* 88, pp. 28–30.

Pinzón Pardo, A.L., Brdjanovic, D., Moussa, M.S., et al. 2007 Modelling of an oil refinery wastewater treatment plant. *Environmental Technology* 28, pp. 1273–84.

Piubeli, F., Grossman, M.J., Fantinatti-Garboggini, F., Durrant, L.R. 2012 Enhanced reduction of COD and aromatics in petroleum-produced water using indigenous microorganisms and nutrient addition. *International Biodeterioration and Biodegradation* 68, pp. 78–84.

Potumarthi, R., Mugeraya, G., Jetty, A. 2008 Biological treatment of toxic petroleum spent caustic in fluidized bed bioreactor using immobilized cells of thiobacillus RAI01. *Applied Biochemistry and Biotechnology* 151, pp. 532–46.

Rajkumar, K., Muthukumar, M., Sivakumar, R. 2010 Novel approach for the treatment and recycle of wastewater from soya edible oil refinery industry – an economic perspective. *Resources, Conservation and Recycling* 54, pp. 752–58.

Ramalho, R.S. 1978 Principles of activated sludge treatment. *Hydrocarbon Processing* 57 (10 Section 1), pp. 112–18.

Rana, S. 2009 Environmental risks – Oil & gas operations reducing compliance cost using smarter technologies Society of Petroleum Engineers – SPE/IATMI Asia Pacific Health Safety, Security and Environment Conference and Exhibition 2009, *APHSSEC* 9, pp. 30–40.

Ray, J.P., Rainer Engelhardt. F. 1992 Produced water: technological/environmental issues and solutions. *Environmental Science and Pollution Research* 46, pp. 1–5.

Ray, J.P., Engelhardt, F.R., Stephenson, M.T. 1992 A survey of produced water studies. In: Ray, J.P., Engelhardt, F.R. (eds) *Produced Water: Technological/Environmental Issues and Solutions*. pp. 1–12. Plenum Publishing Corp.

Reed, B.E., Carriere, P., Lin, W., et al. 1998 Oily wastewater treatment by ultrafiltration: pilot–scale results and full–scale design. *Practice Periodical of Hazardous, Toxic, and Radioactive Waste Management* 2, pp. 100–7.

Reis, J.L.R., Dezotti, M., Sant'Anna Jr., G.L. 2007 Toxicity evaluation of the process effluent streams of a petrochemical industry. *Environmental Technology* 28, pp. 147–55.

Restrepo, R., Forero, J.E., Cardenosa, M., Gonzalez, J. 2009 Application of combined technologies for the treatment of phenols in wastewater oil productions. *SPE Latin American and Caribbean Petroleum Engineering Conference Proceedings* 2, pp. 977–81.

Reynolds, R.R. 2003 *Produced Water and Associated Issues: A Manual for the Independent Operator*. Vol. 6 Oklahoma Geological Survey Open-file Report, p. 1–56.

Roach, R.W., Carr, R.S., Howard, C.L. et al. 1993 An assessment of produced water impacts at two sites in the Galveston Bay system. United States Fish and Wildlife Service, Clear Lake Field Office unpublished report. Houston, Texas.

Saeedi, M., Khalvati–Fahlyani, A. 2011 Treatment of oily wastewater of a gas refinery by electrocoagulation using aluminum electrodes. *Water Environment Research* 83, pp. 256–64.

Saien, J., Nejati, H. 2007 Enhanced photocatalytic degradation of pollutants in petroleum refinery wastewater under mild conditions. *Journal of Hazardous Materials* 148, pp. 491–95.

Salahi, A., Mohammadi, T. 2011 Oily wastewater treatment by ultrafiltration using Taguchi experimental design. *Water Science and Technology* 63, pp. 1476–84.

Santo, C.E., Vilar, V.J.P., Botelho, C.M.S. et al. 2012 Optimization of coagulation-flocculation and flotation parameters for the treatment of a petroleum refinery effluent from a Portuguese plant. *Chemical Engineering Journal* 183, pp. 117–23.

Santos, F.V., Azevedo, E.B., Sant'Anna Jr., G.L., Dezotti, M. 2006 Photocatalysis as a tertiary treatment for petroleum refinery wastewaters. *Brazilian Journal of Chemical Engineering* 23, pp. 451–60.

Sarathy, B.P., Hoy, P.M., Duff, S.J.B. 2002 Removal of oxygen demand and acute toxicity during batch biological treatment of a petroleum refinery effluent. *Water Quality Research Journal of Canada* 37, pp. 399–411.

Sarathy, B.P., Hoy, P.M., Duff, S.J.B. 2002 Removal of oxygen demand and acute toxicity during batch biological treatment of a petroleum refinery effluent. *Water Quality Research Journal of Canada* 37, pp. 399–411.

Scheren, P.A., Ibe, A.C., Janssen, F.J., Lemmens, A.M. 2002 Environmental pollution in the Gulf of Guinea – A regional approach. *Marine Pollution Bulletin* 44, pp. 633–41.

Schultz, T.E. 2005, Cover story: Biological wastewater treatment. *Chemical Engineering* 112, pp. 44–51.

Scott, A.R. and Kaiser, W.R.: Relation between Basin Hydrology and Fruitland Gas Composition, San Juan Basin, Colorado and New Mexico. *Quarterly Review of Methane from Coal Seams Technology* 9, pp. 10–18.

Sedlukho, Y.P. 1991 Application of new coalescence method for treatment of emulsified petroleum products wastewater. *Water Science and Technology* 24, pp. 261–68.

Seidle, J.P. 1991 Long-Term Gas Deliverability of a Dewatered Coalbed. Paper SPE 21488 presented at the 1991 SPE Gas Technology Symposium, Houston, Texas, January 1991.

Sengupta, P., Saikia, N., Borthakur, P.C. 2002 Bricks from petroleum effluent treatment plant sludge: Properties and environmental characteristics. *Journal of Environmental Engineering* 128, pp. 1090–94.

Shut'ko, A.P., Sorochenko, V.F.1988 Reagent treatment of wastewater in petroleum refineries. *Chemistry and Technology of Fuels and Oils* 24, pp. 184–86.

Siedlecka, E.M., Stepnowski, P., Jastorff, B. 2002 Effect of H2O2 on characteristics and biological treatment of petroleum refinery wastewater. *Fresenius Environmental Bulletin* 11, pp. 223–26.

Siedlecka, E.M., Behrend, P., Jastorff, B. 2002 Enhanced photo-degradation of contaminants in petroleum refinery wastewater. *Water Research* 36, pp. 2167–72.

Siedlecka, E.M., Stepnowski, P. 2006 Treatment of oily port wastewater effluents using the ultraviolet/hydrogen peroxide photodecomposition system. *Water Environment Research* 78, pp. 852–56.

Sirivedhin, T., McCue, J., Dallbauman, L. 2004 Reclaiming produced water for beneficial use: salt removal by electrodialysis. *Journal of Membrane Science* 243, pp. 335–43.

Sokolov, V.P., Chikunova, L.A., Kudrin, V.B., Gustov, V.A. 1979 Treatment and utilization of oil sludge from flotation treatment of refinery wastewater. *Chemistry and Technology of Fuels and Oils* 15, pp. 295–98.

Sokolov, V.P., Chikunova, L.A., Gustov, V.A. 1988 Treating catalyst plant water. *Soviet Journal of Water Chemistry and Technology* (English Translation of Khimiya i Tekhnologiya Vody) 10, pp. 81–84.

Sokolov, V.P., Chikunova, L.A., Gustov, V.A. 1991 Purification of drainage waters of catalyst production. *Khimiya i Tekhnologiya Topliv i Masel* 2, pp. 34–35.

Sorkin, Ya., G. 1975 Crude oil pretreatment and environmental protection. *Chemistry and Technology of Fuels and Oils* 11, pp. 913–15.

Sorkin, Ya., G. 1968 Methods for improving the utilization of water in a petroleum refinery. *Chemistry and Technology of Fuels and Oils* 4, pp. 77–81.

Sorochenko, V.F., Shutko, A.P., Skalozub, F.I. 1984 Treatment of wastewater from No. I. sewer system. *Chemistry and Technology of Fuels and Oils* 20, pp. 614–16.

Spellman, F.R. 2003 *Handbook of Water and Wastewater Treatment Plant Operations.* pp. 3–630. CRC Press.

Spiegler, K.S., Kedem, O. 1966 Thermodynamics of hyperfiltration (reverse osmosis): criteria for efficient membranes. *Desalination* 1, pp. 311–26.

Stamper, D.M., Montgomery, M.T. 2008 Biological treatment and toxicity of low concentrations of oily wastewater (bilgewater). *Canadian Journal of Microbiology* 54, pp. 687–93.

Steiner, J.L., Bennett, G.F., Mohler, E.F., Clere, L.T. 1978 Air flotation treatment of refinery waste water. *Chemical Engineering Progress* 74, pp. 39–45.

Stephen, G. 1989 Optimization and control petroleum refinery wastewater treatment systems – status, trends and needs. *Water Pollution Research Journal of Canada* 24, pp. 463–77.

Stephenson, J.P. 1989 Online instrumentation of petroleum refinery wastewater treatment plants. *Water Pollution Research Journal of Canada* 24, pp. 435–50.

Stergar, V., Zagorc-Konan, J., Zgajnar-Gotvanj, A. 2003 Laboratory scale and pilot plant study on treatment of toxic wastewater from the petrochemical industry by UASB reactors. *Water Science and Technology* 48, pp. 97–102.

Strømgren, T., Sørstrøm, S.E., Schou, L. et al. 1995 Acute toxic effects of produced water in relation to chemical composition and dispersion. *Marine Environmental Research* 40, pp. 147–69.

Su, D., Wang, J., Liu, K. et al. 2007 Kinetic performance of oil-field produced water treatment by biological aerated filter. *Chinese Journal of Chemical Engineering* 15: 591–4.

Svarovsky, L. 1992 *Hydrocyclones: Analysis and Applications.* p. 1–3. Kluwer Academic Publishers.

Tien, H.T., Chen, JW. 1992 Photoelectrolysis of water in semiconductor septum electrochemical photovoltaic cells. *Solar Energy* 48, pp. 199–204.

Tony, M.A., Purcell, P.J., Zhao, Y. 2012 Oil refinery wastewater treatment using physicochemical, Fenton and Photo–Fenton oxidation processes. *Journal of Environmental Science and Health – Part A Toxic/Hazardous Substances and Environmental Engineering* 47, pp. 435–40.

Triet, L.M., Viet, N.T., Thinh, T.V., et al. 1991 Application of three step–biological pond with the use of aquatic plant for post treatment of petroleum wastewater in Vietnam. *Water Science and Technology* 23, pp. 1503–07.

United State Resource Conservation and Recovery Act (RCRA) 1974 and Hazardous Solid Waste 1984 (HSWA).

Urban, D., Frisbie, S., Croce, S. 1997 Compliance strategy for cyanides in petroleum refinery wastewater: Part 1 – Source characterization and treatment. *Environmental Progress* 16, pp. 171–78.

Vagapov, R.R., Khokhlov, N.G., Shagitov, Z.M. 2006 Field treatment of waste waters on Askinskaya group of petroleum deposits. *Neftyanoe Khozyaistvo – Oil Industry* 3, pp. 126–28.

Veil, J.A., Puder, M.G., Elcock, D. et al. 2004 A White Paper Describing Produced Water from Production of Crude oil, Natural Gas, and Coal Bed Methane. US Department of Energy, Argonne National Laboratory.

Velmurugan, V, Srithar, K. 2008 Prospects and scopes of solar pond: a detailed review. *Renewable Sustainable Energy Review* 12, pp. 2253–63.

Velz, C.J. 1948 A Basic Law for the Performance of Biological Filter. *Sewage Works Journal* 20, pp. 607–17.

Ventresque, C., Turner, G., Bablon, G. 1997 Nanofiltration: from prototype to full scale. *Journal of the American Water Works Association* 89, pp. 65–76.

Vlasopoulos, N., Memon, F.A., Butler, D., Murphy, R., 2006 Life cycle assessment of wastewater treatment technologies treating petroleum process waters. *Science of the Total Environment* 367, pp. 58–70.

Wang, S., Dai, X., Sun, Y. 2011 Study on the treatment of wastewater of oil shale retorting by photocatalytic degradation with TiO 2/ *International Conference on Remote Sensing, Environment and Transportation Engineering, RSETE 2011 – Proceedings*, art. no. 5966025, pp. 7190–93.

Wang, L., Barrington, S., Kim, J.-W. 2007 Biodegradation of pentyl amine and aniline from petrochemical wastewater. *Journal of Environmental Management* 83, pp. 191–97.

Welz, M.L.S., Baloyi, N., Deglon, D.A. 2007 Oil removal from industrial wastewater using flotation in a mechanically agitated flotation cell, *Water* SA33–4, pp. 453–58.

Whittaker, M., Pollard, S.J.T., Fallick, T.E. 1995 Characterisation of refractory wastes at heavy oil-contaminated sites: A review of conventional and novel analytical methods. *Environmental Technology* 16, pp. 1009–33.

Wicks, Z.W. 2007 *Organic Coatings: Science and Technology*. John Wiley & Sons, Ltd.

Wimberley, W.F. 1989 To dispose of waste wisely. *Hydrocarbon Processing* 68, pp. 45–49.

Wong, J., Maroney, P., Diepolder, P., et al. 1992 Petroleum effluent toxicity reduction – From pilot to full–scale plant. *Water Science and Technology* 25, pp. 221–28.

Wong, J.M. 1999 Petrochemicals. *Water Environment Research* 71, pp. 828–33.

Xianling, L., Jianping, W., Qing, Y., Xueming, Z. 2005 The pilot study for oil refinery wastewater treatment using a gas–liquid–solid three-phase flow airlift loop bioreactor. *Biochemical Engineering Journal* 27, pp. 40–44.

Xie, W., Chen, J., Zhong, L., Zhong, H. 2008 Treatment of slightly polluted wastewater in a petroleum refinery by combined process of circulating biological aerated filter and filtration. *Huagong Xuebao/Journal of Chemical Industry and Engineering (China)* 59, pp. 1251–56.

Xie, W., Chen, J., Zhong, H., Li, D. 2011 Treatment of alkaline wastewater from oil refinery using circulating biological aerated filter. *Proceedings – International Conference on Computer Distributed Control and Intelligent Environmental Monitoring (CDCIEM)*, art. no. 5748298, pp. 2360–65.

Xu, P., Drewes, J.E. 2006 Viability of nanofiltration and ultra-low pressure reverse osmosis membranes for multi–beneficial use of methane produced water. *Separation and Purificiation Technology* 52, pp. 67–76.

Xu, P., Cath, T., Wang, G, et al. 2009 *Critical Assessment of Implementing Desalination Technology*. Water Research Foundation.

Yang, Y., Zhang, G., Yu, S. et al. 2010 Efficient removal of organic contaminants by a visible light driven photocatalyst Sr6Bi2O9. *Chemical Engineering Journal* 162, pp. 171–7.

Yao, D., Zhang, Y., Yang, Q., Li, J. 2004 Commercial application of refinery sewage deep treating technology for wastewater reuse. *Petroleum Processing and Petrochemicals* 35, pp. 76–78.

Yusupov, E.A. 1988, Application of screen filters in water treatment in petroleum refineries. *Chemistry and Technology of Fuels and Oils* 24, pp. 92–94.

Zhou, W., Li, Y. Min, M., et al. 2011 Local bioprospecting for high–lipid producing microalgal strains to be grown on concentrated municipal wastewater for biofuel production. *Bioresource Technology* 102, pp. 6909–19.

Zhou, Q., Shen, B. 2010 Biodegradation potential and influencing factors of a special microorganism to treat petrochemical wastewater. *Petroleum Science and Technology* 28, pp. 135–45.

Zhou, Q., Sun, F., Liu, R. 2005 Joint chemical flushing of soils contaminated with petroleum hydrocarbons. *Environment International* 31, pp. 835–39.

Index

absorption, 3, 18, 174
activated carbon, 82, 85, 87, 101, 102
activated sludge, 105, 112, 113, 114, 116, 118, 120, 123, 207, 220, 221, 226, 228, 231, 232, 236
adsorption, 10, 14, 15, 72, 82, 84, 85, 86, 89, 101, 114
aerobic, 107, 108, 111, 112, 113, 120, 121, 125, 128
aerobic digester, 231, 236
aggregation, 15, 26
air oxidation, 170
alum, 87, 93
aluminium chloride, 173
ammonia, 1, 11, 14, 16, 86, 90, 96, 101, 107, 115, 116
anaerobic, 105, 107, 108, 112, 118, 124, 177
anaerobic digester, 231
anaerobic sludge digestion, 233, 235
anthracite, 73, 76
API separator, 29, 30, 31, 33, 34, 57, 73
aromatic, 16, 66, 119

backwash, 47, 72, 75, 76, 77, 78, 79
bacteria, 1, 20
baffle, 39, 42, 43, 45, 46, 49, 54, 55, 58, 60, 62, 64, 66, 77
basin, 53, 54, 55, 59, 60, 63
belt press filter, 12, 238
biochemical oxygen demand, 1, 4, 22, 107

biodegradable organics, 1, 2, 9, 14, 15
biokinetic, 102, 104
biological filter, 203, 204, 205
biological oxidation, 114, 120, 121
biological sludge, 234, 237, 253, 255
biological treatment, 99, 101, 102, 103, 104, 105
BOD, 1, 2, 22, 26, 29, 48, 72, 73, 86, 87, 89, 90, 91, 97, 105, 106, 107, 112, 113, 114, 115, 117, 120, 122, 123, 125, 181, 192

carbon dioxide, 1, 106, 107, 113, 124
carcinogenicity, 5, 6
caustic scrubs, 167
centrifugation, 238, 261
chemical oxidation, 140, 142
chemical precipitation, 81, 85, 87, 95, 97
chemical sewer, 164, 168, 173
chemical stabilization, 84
chlorination, 6, 14, 82, 85, 95, 96, 97, 117
clarifier, 90, 91, 92, 93, 94
coagulant, 82, 88, 90, 93
coagulation, 10, 14, 15, 16, 17, 49, 52, 53, 54, 58, 81, 82, 88, 89
coal bed water, 131, 132, 133, 135, 136
coal seam gas, 129, 130, 135
coke, 217, 252, 257
colloidal, 4, 15, 55, 87, 88, 90
comminution, 26, 28
compatibility, 21

composting, 220, 236, 244
cyanide, 8, 11, 13, 82, 96, 115, 117, 172
cycle time, 40, 61, 73, 74, 76

detention time, 53, 54, 55, 59
dewatering, 7, 12, 224, 237
dipper, 269, 271
disinfection, 12, 17, 24, 81, 85, 94, 95, 96, 97, 224, 237
disposal pond, 169
dissolved air flotation, 49, 56, 57, 66
drying bed, 202

effluent weir, 43, 58
evaporation pond, 136

ferric chloride, 173
filtration, 10, 14, 15, 16, 17, 18, 26, 47, 53, 54, 71, 72, 73, 74, 75, 76, 77, 78, 141, 144, 145
flocculation, 26, 48, 49, 54, 55, 58, 60, 62, 65, 81, 85, 87, 88, 89, 90, 92, 93, 116
flotation, 9, 12, 14, 15, 17, 26, 47, 49, 56, 57, 58, 60, 61, 63, 64, 65, 66, 67, 93, 116
flow equalization, 26, 46, 47
fluoride, 13
fly ash, 218, 242, 247, 265
Freeze–Thaw Evaporation, 140

garnet, 73, 76
granular media filter, 71, 72, 73, 76
grease, 8, 9, 33, 38, 86, 96, 97
grid system, 269
grit removal, 9, 20, 28, 29, 51
groundwater, 2, 9, 10, 16, 34

hazardous waste reduction, 258
hazardous wastes, 248, 250, 258
heat drying, 224, 237, 242, 244
heavy metal, 1, 2, 4, 8, 10, 14, 16
hydrogen sulfide, 90, 91

incineration, 170, 171, 176, 177, 247
infiltration, 16, 136

inorganic solids, 1, 4, 14
ion exchange, 8, 10, 82, 83, 137, 138, 141, 142, 153

lagoon, 9, 12, 14, 17, 19, 224, 241, 247
land farming, 255, 266
landfill, 10, 12, 247, 249, 255
land treatment, 257, 262, 264, 265
Langmuir isotherm, 85, 86
lime, 14, 86, 87, 97
liquefied gas spill, 178

mechanical aerator, 121
mineral filter, 206
mixing, 26, 48, 49, 54, 88, 93

neutralization, 60, 62, 65, 81, 88, 170, 172
nitrogen, 1, 2, 10, 14, 24, 75
non-hazardous waste, 251
nutrient, 1, 2, 9, 10, 14, 195

odour, 15, 20, 21, 195, 197, 209
oil sand, 153, 155, 157, 160
oil trap, 29, 45, 46
oil–water separator, 29, 30, 31, 32, 33, 34, 35, 43, 44, 47
oxidizing agent, 13, 22, 82

parallel-plate separators, 30, 43, 44, 45
particle, 25, 54, 75
particle size, 54, 75, 81, 82, 88, 89
pathogen, 1, 2, 4, 14, 95, 96, 195
phenol, 90, 91, 96, 166, 168, 182, 189
phosphoric acid, 173
phosphorus, 1, 2, 14, 85, 87
pollutant, 4, 6, 10, 11, 16, 22, 82, 85
polyelectrolyte, 173
preliminary wastewater treatment, 9
priority pollutant, 1, 3, 6, 16
pyrolysis, 242, 262

reaction jet inlet, 39
refractory organic, 1, 2, 14
reverse osmosis, 10, 14, 139, 137, 138, 146, 154, 156

rotary drum, 37, 38, 43
rotating biological, 105, 126, 127

SAGD, 155, 157
sand filters, 74, 75, 77
sanitary sewer, 165
scale, 8, 16, 37, 53, 59
screening, 7, 9, 14, 20, 25, 26, 27, 31
scum, 217, 226, 227, 231
secondary treatment, 1, 7, 9, 10, 15, 16, 17, 20
sedimentation, 9, 10, 14, 15, 16, 18, 20, 26, 45, 47, 49, 51, 52, 53, 54, 55, 56, 72, 82, 85, 87, 88, 89, 95, 97
settlement tank, 200, 201, 203, 204
settling, 32, 50, 52, 53, 54, 55, 56
settling velocity, 32, 49, 50, 51, 53
sewage, 1, 4, 6
sewage oxidation pond, 126, 128
sewer, 6, 7, 13, 161
shale gas, 129, 130, 135
skimmer, 30, 35, 36, 37, 38, 39, 43, 45, 46, 57, 58, 60, 61, 64, 66
sludge, 2, 4, 6, 9, 10, 11, 12, 14, 16, 17, 19, 20, 21, 26, 28, 34, 38, 40, 41, 42, 44, 49, 52, 53, 54, 56, 58, 60, 61, 65, 65, 76, 223, 245, 250, 260
sludge blending, 229
sludge degritting, 229
sludge digester, 202
sludge drying bed, 224, 240, 241
sludge grinding, 223, 229
sludge storage, 230
solid waste, 215

stabilization, 224, 233
stokes, 32, 33, 50, 51, 59
stormwater, 162, 164, 166, 174, 178, 189
stripping, 6, 10, 14, 17, 26
surface water, 6
surfactant, 2, 6
suspended solid, 1, 2, 9, 10, 14, 26, 27, 29, 52, 54, 58, 60, 64, 71, 73, 85, 87, 90, 91

tertiary treatment, 7, 10, 17, 94, 207
thermal reduction, 224, 242
thickening, 7, 12, 26, 47, 52, 56, 224, 230, 242
thief sampler, 269, 271
toxic organic, 4, 5
trickling filter, 105, 107, 110, 114, 123, 124, 125, 126
trier sampler, 270, 272

ultimate disposal, 224, 237, 242, 254, 255, 259, 260
unconventional oil and gas, 129

vacuum filtration, 238
volatile organic compound, 6, 7, 16, 26, 33

waste degradation, 263, 264
waste minimization, 255, 256
waste sampling, 268, 275
wax tailing, 218, 252
weighted bottle, 269, 270
wetland, 141, 154, 155, 156
wetting rate, 125, 126